W9-CRB-428

A C S S Y M P O S I U M S E R I E S **574**

Anthracycline Antibiotics

New Analogues, Methods of Delivery, and Mechanisms of Action

Waldemar Priebe, EDITOR
The University of Texas
M. D. Anderson Cancer Center

Developed from a symposium sponsored
by the Division of Carbohydrate Chemistry
at the 205th National Meeting
of the American Chemical Society,
Denver, Colorado,
March 28–April 2, 1993

American Chemical Society, Washington, DC 1995

Library of Congress Cataloging-in-Publication Data

Anthracycline antibiotics: new analogues, methods of delivery, and mechanisms of action / Waldemar Priebe, editor.

p.   cm.—(ACS symposium series, ISSN 0097–6156; 574)

"Developed from a symposium sponsored by the Division of Carbohydrate Chemistry at the 205th National Meeting of the American Chemical Society, Denver, Colorado, March 28–April 2, 1993."

Includes bibliographical references and index.

ISBN 0–8412–3040–4

1. Anthracyclines—Congresses.

I. Priebe, Waldemar, 1948–   . II. American Chemical Society. Division of Carbohydrate Chemistry. III. Series.

RC271.A63A5677   1995
616.99′4061—dc20   94–38929
                                                                     CIP

The paper used in this publication meets the minimum requirements of American National Standard for Information Sciences—Permanence of Paper for Printed Library Materials, ANSI Z39.48–1984. ∞

Foreword

THE ACS SYMPOSIUM SERIES was first published in 1974 to provide a mechanism for publishing symposia quickly in book form. The purpose of this series is to publish comprehensive books developed from symposia, which are usually "snapshots in time" of the current research being done on a topic, plus some review material on the topic. For this reason, it is necessary that the papers be published as quickly as possible.

Before a symposium-based book is put under contract, the proposed table of contents is reviewed for appropriateness to the topic and for comprehensiveness of the collection. Some papers are excluded at this point, and others are added to round out the scope of the volume. In addition, a draft of each paper is peer-reviewed prior to final acceptance or rejection. This anonymous review process is supervised by the organizer(s) of the symposium, who become the editor(s) of the book. The authors then revise their papers according to the recommendations of both the reviewers and the editors, prepare camera-ready copy, and submit the final papers to the editors, who check that all necessary revisions have been made.

As a rule, only original research papers and original review papers are included in the volumes. Verbatim reproductions of previously published papers are not accepted.

M. Joan Comstock
Series Editor

Contents

Preface

DESPITE MORE THAN 20 YEARS OF EFFORT by the pharmaceutical industry and academia to develop new anticancer agents and to find a "better doxorubicin," doxorubicin remains a frontline chemotherapeutic agent in the treatment of human cancers, especially solid tumors such as carcinomas and sarcomas. The failure to find a clearly better doxorubicin has led in turn to the decline of interest in anthracyclines and to a reduced emphasis on such projects throughout the pharmaceutical industry. However, the clinical use of two other anthracycline antibiotics, epirubicin and idarubicin, has increased significantly in recent years, especially in Europe. Also, difficulties in developing biologicals as important chemotherapeutic drugs and the promising results of clinical studies of drugs such as Taxol (paclitaxel) have revived interest in anthracycline antibiotics and in more traditional approaches to drug development.

The development of new anthracyclines was hampered by several factors, especially a lack of knowledge of the exact mechanism of action. Early studies focused on the structure–activity relationship, and trials to establish the exact mechanism of action on a molecular level were not successful. Attempts to identify a mechanism of action for doxorubicin were complicated by the fact that anthracycline antibiotics exert their antitumor properties not through a single mechanism but more probably through an array of unrelated biochemical processes involving numerous cellular targets. Even such basic questions as why daunorubicin and doxorubicin differ in antitumor activity and how this relates to a single chemical change, i.e., the replacement of hydrogen at C-14 with hydroxyl, still remain unanswered. For years these questions have had enormous clinical and commercial significance because daunorubicin's use is limited to leukemia, whereas doxorubicin shows both antileukemic and solid tumor activity.

Development of new analogues was also influenced by self-imposed limitations based on initial structure–activity relationship studies by leading groups and the belief that water solubility was necessary for pharmaceutical formulation and development. Thus, in practice, water-insoluble, lipophilic compounds were neglected and insufficiently evaluated. This neglect was unfortunate because, despite the obvious problems of formulation, such analogues can offer desirable properties, such as a different spectrum of activity and reduced toxicity. Some of these properties are due to different intracellular, tissue, and organ distributions.

In recent years, problems associated with the development of new anthracyclines and methods of their delivery have been addressed directly by many groups. New developments in other fields have led to the discovery of an important mechanism of action of anthracyclines by topoisomerases, and a more detailed picture of interaction of anthracyclines with DNA is emerging. Other new mechanisms of action have been proposed and are discussed in this book. Development of resistance is now perceived as one of the major obstacles to effective chemotherapy, and as such, it already significantly influences the design of new drugs and methods for evaluating them.

The main goal of the symposium on which this book is based was to bring together researchers who are involved in the direct design and synthesis of new drugs with researchers who are investigating biochemical processes and mechanisms of action. We hoped that such a meeting, by reviewing recent developments in related disciplines, would help to point out future directions in research and to facilitate close interdisciplinary efforts. The meeting was attended by chemists, medicinal chemists, biochemists, pharmacologists, and clinicians. We hope that this book will help to achieve the symposium's goals on a global scale and to stimulate further research and new discoveries in this area.

This book covers research relevant to the development of novel anthracyclines, such as the synthesis of promising new analogues, studies of mechanisms of action, and new approaches to improving properties for this class of compounds by using different drug-delivery and tumor-targeting systems. Although it occurred too late to be included in this volume, Bristol-Myers Squibb recently began clinical studies on a doxorubicin covalently bound to a chimeric (mouse–human) monoclonal antibody, BR96. The results of preclinical in vivo evaluation are promising (Trail, P. A. et al. *Science (Washington, D.C.)* **1993**, *261*, 212), and the success or failure of this study could influence this approach in years to come.

It is becoming apparent that to achieve success at the clinical stage, the development of new analogues should parallel and be closely connected to the development of methods to increase tumor targeting, to avoid organ-targeted toxicities, and to limit other undesired side effects. Because anthracyclines are highly flexible with regard to structural modifications, the physicochemical properties and potency of new analogues can be altered without sacrificing efficacy. In consequence, anthracyclines can be designed to be compatible with a specific delivery or targeting system, thus allowing one to take full advantage of a particular system.

The studies presented in this book clearly indicate that the recent progress in studies of the mechanism of action, the development of new anthracyclines, and the use of drug delivery to increase tumor targeting will soon result in new, more selective chemotherapeutics.

Acknowledgments

I thank the contributors to this book for their efforts, Federico Arcamone for valuable suggestions, and Jude Richard for his great help in preparing this book.

WALDEMAR PRIEBE
The University of Texas
M. D. Anderson Cancer Center
Houston, TX 77030

August 16, 1994

Chapter 1

Unresolved Structure–Activity Relationships in Anthracycline Analogue Development

Edward M. Acton

Consultant, 281 Arlington Way, Menlo Park, CA 94025

Continuing interest in quinone-altered anthracyclines, for example, can be explored by screening against a panel of 60 human tumor cell lines in culture. 5-Iminodaunorubicin gave a pattern of cytotoxic potencies that was distinct from the pattern produced by doxorubicin, daunorubicin, and other anthracyclines. This was consistent with accumulated evidence for an altered mechanism of action with the modified quinone. Previously, 5-iminodaunorubicin was active but not superior to the parent quinone when screened against leukemia P388 in the mouse. As a model quinone isostere synthesized in 7 steps, 1-hydroxy-7, 8-bis-(morpholinomethyl)phenazine-5,10-dioxide was cytotoxic across the 60-line panel. Without N-oxidation to the quinone isostere level, the phenazine was noncytotoxic.

It is ironic that the volume of anthracycline research has declined in recent years. The outlook for progress toward specific goals is better that ever, yet major objectives of anthracycline analogue development have yet to be attained. This is despite the voluminous clinical and experimental work done since the introduction of daunorubicin and doxorubicin in the 1960s. These molecules continue to offer an excellent point of departure for studies to develop better cancer drugs and explore antitumor mechanisms.

Needed Improvements

More than a thousand analogues have been obtained and tested, but anthracyclines with better activity and less toxicity are still needed (1, 2). It is precisely because of the attention given this series that the various properties needed in an improved analogue have been defined (1-3). Doxorubicin has been called the most active single agent against cancer because of its broad antitumor spectrum, but there are some important tumor types (e.g., colon, lung) that do not respond; an even broader spectrum is needed. Among responding tumors, the rate of response is often low--for doxorubicin as a single agent, sometimes 30%; even a doubling of this rate could give an enormous benefit. In many cases, initially responding tumors become resistant to anthracyclines over time, and this resistance often extends to chemically unrelated

0097–6156/95/0574–0001$08.00/0
© 1995 American Chemical Society

drugs (multidrug resistance); analogues effective in resistant tumors are widely sought. Clinicians have learned how to manage the toxic side effects--acute myelosuppression and chronic cardiotoxicity (*1, 2, 4*)--that were recognized early in the clinical application of daunorubicin and doxorubicin, but they still impose treatment limitations (on dose level and on the number of drug courses); less toxic analogues would provide significant therapeutic benefit.

There are basically two needs in an approach to these problems through analogue development: identification of significant alterations in the parent structure, and comparative testing of new compounds. Design of the structural changes can be based on observed structure-activity relationships together with considerations of synthetic accessibility. Equally productive may be to consider unresolved questions of structure-activity relationships in the results accumulated to date. Also, a survey of the analogues obtained so far will show what types of structure changes are underrepresented. For example, few changes have been made in the carbon skeleton of the aglycone. Some will be presented (*5*) at this symposium, but they are rare. This is surprising in view of the creative studies of aglycone total synthesis that have come from various groups of leading synthetic chemists (*6, 7*). Furthermore, there have been few changes of any kind at the quinone function. This is ironic because the quinone is an important site of mechanistic action in the anthracycline molecule. It could be useful to explore the effect on activity of various changes at the quinone.

Screening Methods

Once certain structure changes have been selected, the question is what standardized screen to use for analogue comparison. Cancer screening has been the subject of recent reviews (*8-11*). For many years, mouse leukemia P388 provided a good standard test in the mouse. Potency was measured in terms of the required drug dose, and efficacy in terms of host survival time. Quantitative data were generated on many compounds, and if test parameters like tumor inoculum and dosing schedule were kept the same (as in the National Cancer Institute [NCI] screen), the compounds could be quantitatively compared. However, predictive value for clinical activity was at best only qualitative. Efficacy and potency in the mouse did not correlate directly with specific aspects of clinical activity like spectrum or rate of response. Development of the immune-suppressed nude mouse permitted testing against human tumor lines, but this has been an expensive animal model, and its predictiveness is still unsettled. For one thing, it is not clear how many human lines should be used to evaluate a given compound. And there are the unavoidable differences in pharmacology and pharmacokinetics between any animal model and man. Perhaps for these reasons (and others), high-volume antitumor screening has recently tended toward in vitro testing.

That may seem surprising, but methodology has been developed for the use of panels of human tumors in culture, which should give representative average values for cytotoxic potencies and may also give an indication of antitumor spectrum. The human tumor clonogenic assay (also called the Salmon stem-cell assay) tests for the inhibition of colony formation using fresh tumor explants (*11*). An encountered shortcoming was that many patient tumors could not be grown for the assay. The NCI has recently organized a screen using a panel of 60 established human tumor cell lines (*9*). It is called a "disease-oriented" screen because it is comprised of eight subpanels representing important tumor types found in human disease. The assay is based on the measurement of cell growth. Of course, the 60 lines in any test will vary in their responses. A working hypothesis of this new assay at NCI was that compounds of interest would show specificity (in terms of increased potency) for a particular subpanel. Another aspect of the assay is that the pattern of the 60 responses (apart from any subpanel specificity) may be characteristic of the compound, or the class of compound. Statistical methods have been developed for comparing the patterns and quantifying their degree of similarity (*8*). This in vitro assay, then, gives a type of result that is quite different from that obtained in the mouse. It may be of interest to compare results obtained from such different screens as mouse leukemia and the 60-cell-line panel on one drug candidate. A useful example should be 5-iminodaunorubicin .

X = H, daunorubicin
X = OH, doxorubicin

X = H, 5-iminodaunorubicin
X = OH, 5-iminodoxorubicin

Iminoquinones

The iminoquinones are among the relatively few quinone-altered anthracyclines. Iminoquinones have been prepared (*12, 13*) from both daunorubicin and doxorubicin, but most studies (*14-16*) have been carried out with 5-iminodaunorubicin because it is obtainable in a single step. 5-Iminodoxorubicin, on the other hand, requires five semisynthetic steps. 5-Iminodaunorubicin is a good example of a mechanistically altered and active analogue that was not selected for development because it failed to stand out in a mouse screen. Amination of the quinone causes suppression of the usual redox cycling and generation of oxygen radicals. Probably because of this, 5-iminodaunorubicin was found to be noncardiotoxic in rats and dogs, confirming a number of predictive tests. Less attention has been paid to the evident deletion of the reductive deglycosidation that is the normal metabolic process for anthracycline deactivation. No evidence for occurrence of this step could be identified when metabolism was studied in the rat (*17, 18*). The altered pharmacokinetics thereby predicted for 5-iminodaunorubicin were not considered when (at an earlier time) the screening regimens for this compound were chosen. Beginning with leukemia P388, 5-iminodaunorubicin was tested against various tumors in the mouse. It was active in 16 different regimens but was not superior to daunorubicin or doxorubicin. But if mouse activity parameters do not correlate directly with clinical activity factors, results from a quite different type of screen may be of equal interest.

Figure 1 shows the 60 dose-response curves for 5-iminodaunorubicin (NSC 254681) in a single test from NCI's new disease-oriented screen. Cell growth (in percent relative to untreated control cells) is plotted against drug concentration (in molarity on a logarithmic scale). Activity of the drug across the panel can be compared at any selected level of growth: e.g., +50% = 50% growth inhibition; 0% = cytostasis; and -50% = 50% cell kill, or LC_{50}. The pattern of responses will be

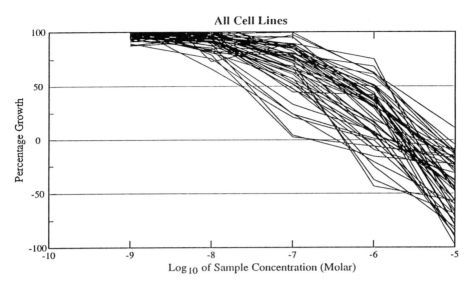

Figure 1. Dose-response curves for 5-iminodaunorubicin (NSC 254681; Exp ID 90009NS60) against 60 human tumor cell lines in disease-oriented NCI screen.

somewhat different at each level of growth. For development of cytotoxic agents, interest is primarily at the level of cell kill. In Figure 1, the highest concentration was 10^{-5} M, and many of the cell lines underwent 50% cell kill only above that level. In a test where the highest concentration was raised to 10^{-4} M, all but two of the lines (colon subpanel) underwent 50% cell kill and gave measured LC_{50} values. Results from this test are shown in Figure 2, in bar-graph form, with each log LC_{50} represented by a bar extending from the mid value (-4.74) either to the right (lower values, more sensitive lines) or left (higher values, more resistant lines). The cell lines in Figure 2 are grouped by subpanel but are not named. The purpose here is to show the distinctive overall pattern produced by 5-iminodaunorubicin in comparison with daunorubicin (NSC 82151; averaged from several dozen tests). Most anthracyclines--including doxorubicin--give a pattern resembling that shown for daunorubicin in Figure 2 (19). One of the few to give a pattern that is different is 5-iminodaunorubicin. The difference that is visually evident in Figure 2 was also observed in a preliminary use of the COMPARE pattern recognition program (8) being developed by NCI. Existence of the difference seems consistent with various results (14) on the alteration in biochemical mechanisms with the imino quinone. Perhaps screening results from this panel of cell lines can be used together with results from the mouse screen in choosing distinctive members of an analogue series for further study. It has always been easier to find active analogues than to choose among them.

Quinone Isosteres

Some years ago, we proposed to synthesize another type of quinone-altered analogue. These were isosteric structures (Figure 3) in which the quinone carbonyls were to be replaced by N-oxide functions (20).
This approach was partly justified by modest antitumor activity observed for some simple phenazine-5,10-dioxides. Work toward these targets was initiated but not completed because of certain synthetic difficulties and delays. Intermediates that were synthesized are summarized in Scheme I. The linear tetracycles with two

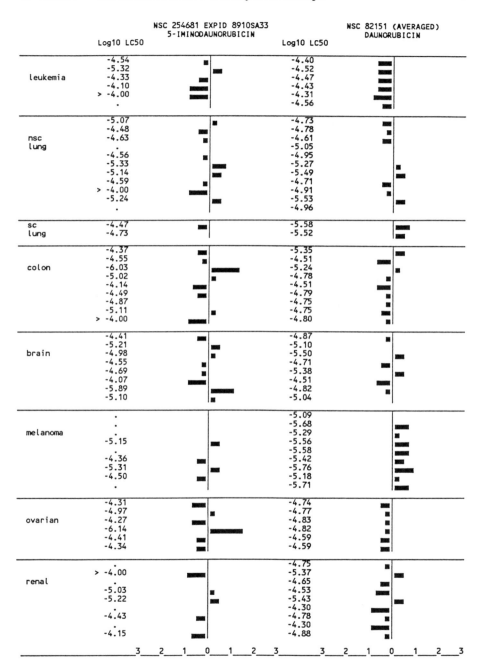

Figure 2. Comparison of 5-iminodaunorubicin and daunorubicin in the NCI screen. Cell lines grouped by subpanel. Responses expressed as bargraphs giving a pattern for each compound.

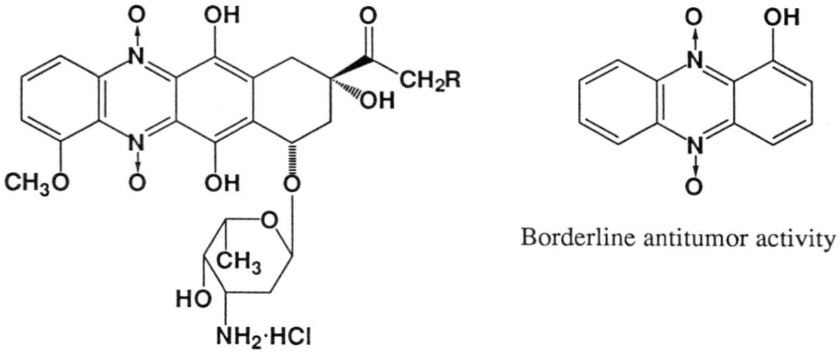

Figure 3. Proposed quinone isosteres and a known phenazine dioxide (*20*)

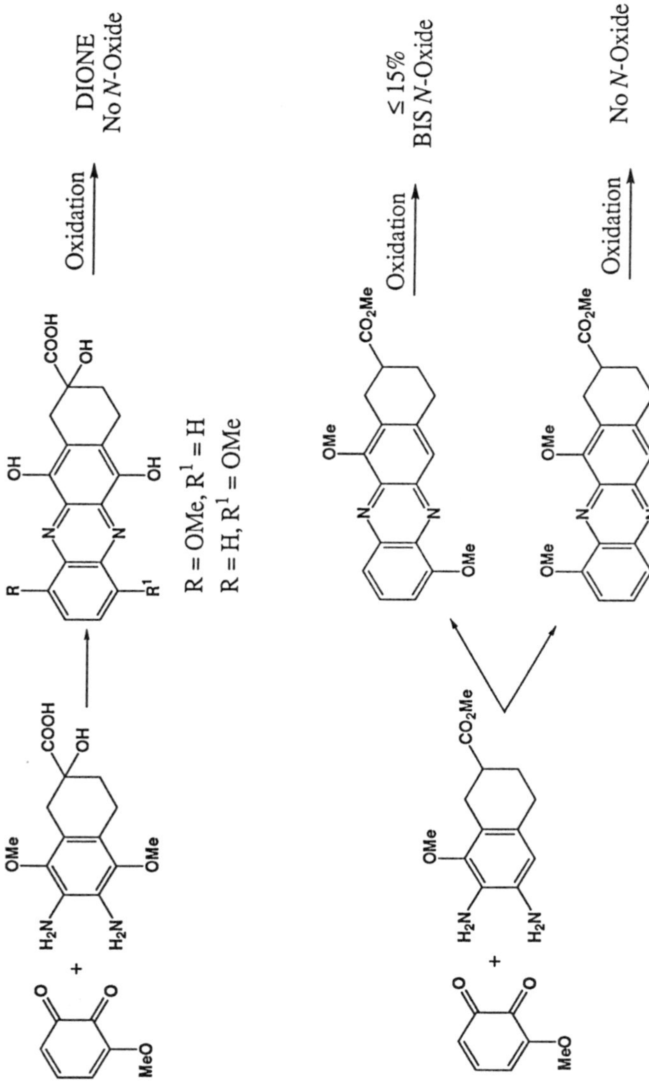

SCHEME I
Intermediates Synthesized (20, 21)

nitrogen atoms in the anthracycline ring C are tetrahydrobenzophenazines, and phenazines are readily synthesized by condensing *o*-quinones with *o*-phenylenediamines.

The products in Scheme I were constructed with side chains at position 9 that could be elaborated. Where difficulties were encountered was in the *N*-oxidations. First (top line), it was found (*20*) that attempts to oxidize the Ns in ring C adjacent to a *p*-hydroquinone in ring B invariably gave oxidation to the B-ring *p*-quinone, regardless of the absence or presence of OH blocking (*O*-methyl or *O*-acetyl). There was formed "no *N*-oxide" at all. Redesign of the target to delete the 6-OH (bottom line) gave some improvement (*21*). A 4,11-di-OMe tetracycle gave "<15% bis-*N*-oxide" but only after retreatment of a mono-*N*-oxide mixture and isolation by chromatography. It was clear we could not complete a synthesis through this step. Oxidation of ring B had been successfully avoided, but MeO substituent effects appeared to hinder the *N*-oxidation. This explanation was suggested by the fact that the 1,11-di-OMe isomer gave the mono 5-*N*-oxide (less hindered position) in a 64% yield, but yielded "no bis-*N*-oxide" at all. It appeared that target redesign to delete all but one peri-OH might be required (this should be compatible with retention of activity)(*21*).

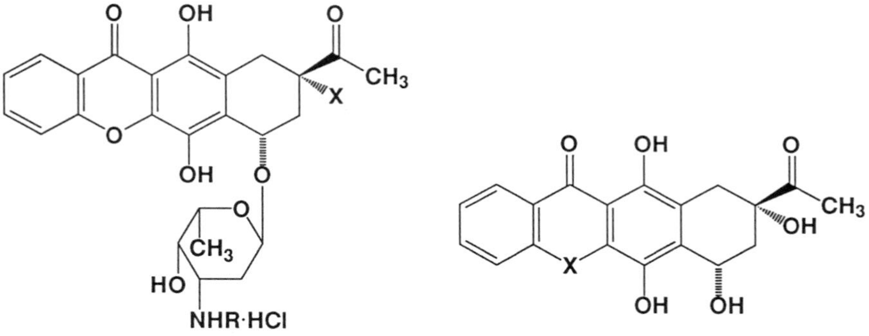

R = H, X = H
R = COCF$_3$, X = OH

X = O
X = S

Figure 4. Approaches to other quinone-altered analogues (*22, 23*)

Work on other changes in the carbon skeleton at the quinone was being described in the meantime (Figure 4). Xanthone- and thioxanthone-derived aglycones were reported (*22*), and a glycosylated xanthone (*23*) which was an anomeric mixture retaining an *N*-trifluoroactyl protecting group. Biological data were very limited but these compounds appeared to be cytotoxic. Absence of any stimulation of O$_2$ consumption in a microsomal test system (*23*) was consistent with the fact that these were analogues having the quinone deleted.

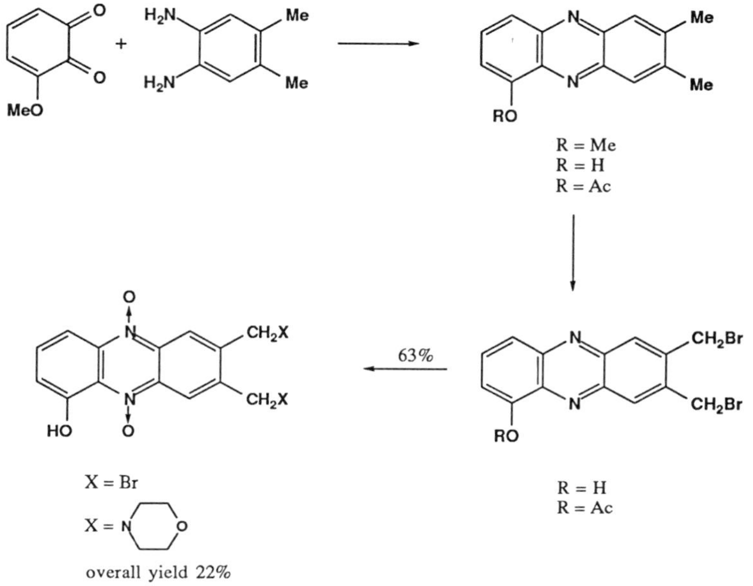

SCHEME II
Synthesis of Model Phenazine Dioxides

More recently, a model phenazine dioxide with a single peri-OH was synthesized and tested (Scheme II). We chose this as a simpler structure with which to explore a successful bis-*N*-oxidation, and to test for biological activity with a simple amino side chain instead of an amino sugar. 3-Methoxy-*o*-quinone underwent condensation with 4,5-dimethyl-*o*-phenylene-diamine to give 1-methoxy-7,8-dimethylphenazine in a 62% yield. Completion of the synthesis was found to be quite specific for the sequence shown. This MeO-phenazine gave a di-*N*-oxide in good yield, but it was unstable and not a useful intermediate. Hence, the oxidation needed to be done on a free OH compound late in the sequence. Bromination of the methyls with *N*-bromosuccinimide could not be done in the presence of a free OH, and the protecting group had to be converted from OMe to OAc, for easy cleavage with 95% trifluoroacetic acid after the bromines were introduced. The di-*N*-oxide was then formed in a 63% yield, using 10 equivalents of *m*-chloroperbenzoic acid at 60° for 12 h. The morpholine side chain was introduced last, using a minimal amount of dichloromethane to dissolve the phenazine, and then adding 100 equivalents of morpholine (higher dilution of the phenazine permitted accompanying quaternization of the dibromo side chains with a single morpholine). The overall yield was 22%, and the product was isolated as the di-HCl salt. Hence, a workable process involving phenazine di-*N*-oxidation was achieved. Diagnostic spectral evidence for the di-*N*-oxidation can be found in the proton NMR, shown in Figure 5. Protons peri to the incipient *N*-oxide functions undergo a noticeable downfield shift. Downfield shift and H-bonding of the peri-OH is truly striking and fulfills the analogy with peri-OH quinones.

Compound	X = H			X = Br		
	H-6; H-9	H-4	OH	H-6; H-9	H-4	OH
A	8.00; 7.96	7.74 -7.71 (with H-3)	8.18	8.30; 8.28	7.82-7.78 (with H-3)	8.13
B	8.41; 8.38	8.06	14.63	8.67; 8.66	8.05	14.28
B-A	0.41; 0.42	0.32	6.45	0.37; 0.38	0.23	6.15

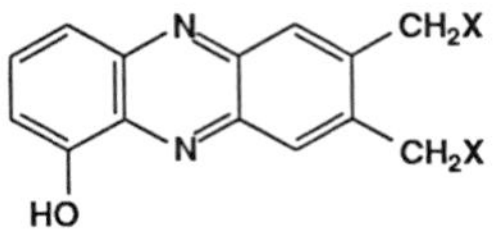

A 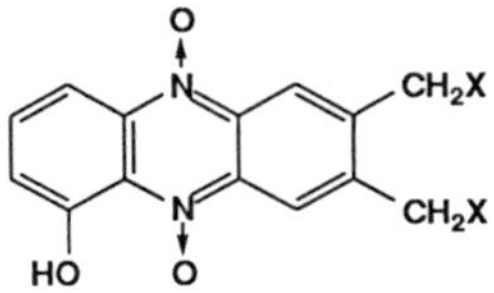   B

Figure 5. Proton NMR downfield shifts diagnostic for *N*-oxidation

The target product **NSC 614049** and its unoxidized counterpart **NSC 647138** were compared in the NCI cell-line screen. Dose-response curves are shown in Figure 6. The importance of having the phenazine at the di-*N*-oxide level that is isosteric with quinone is very clear. Almost all the cell lines undergo 50% cell kill (LC$_{50}$) with the di-*N*-oxide, but only one with the unoxidized phenazine. The bar-graph presentation in Figure 7 shows even more clearly the lack of response with the unoxidized phenazine (**NSC 647138**) and the pattern of good responses (average log LC$_{50}$ = -4.68) with the di-*N*-oxide (**NSC 614049**). No further testing has been completed at this time, but the activity observed with **NSC 614049** suggests such phenazine di-*N*-oxides may be acceptable models for studying quinone isosteres in the anthracyline series.

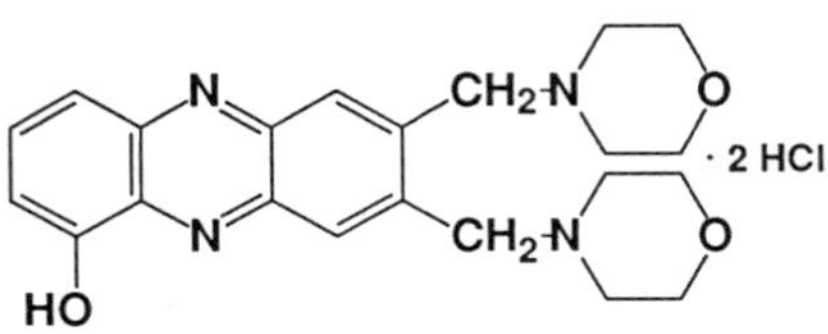

NSC 647138 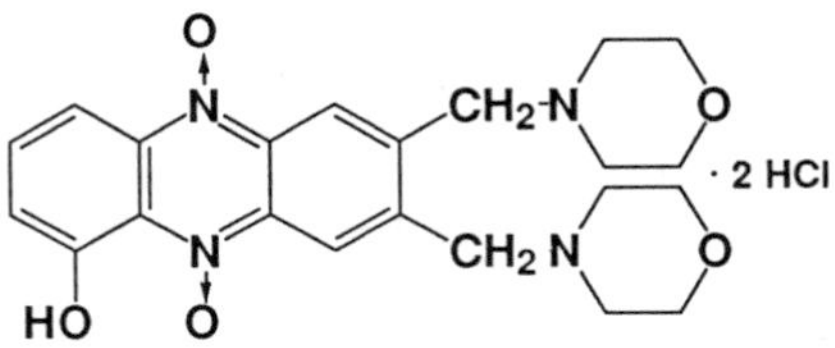   NSC 614049

The case for continuing anthracycline analogue development has been made at various times. The need remains. Such efforts can be better informed now than ever before. Interesting new chemistry can be planned at various sites in these complex molecules, designed to address unsettled questions from the structure-activity relationships accumulated to date. A more varied array of screening methods is at hand for evaluating and comparing new compounds. One difficulty is to design

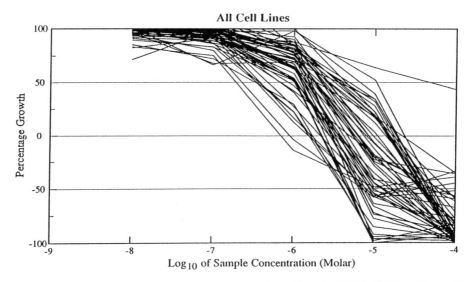

Figure 6A. Dose-response curves for phenazine dioxide NSC 614049 (Exp. ID 9109MD97) against 60 human tumor cell lines in NCI screen.

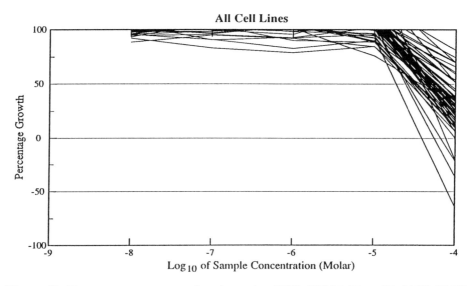

Figure 6B. Dose-response curves for phenazine NSC 647138 (Exp. ID 9112MD22) against 60 human tumor cell lines in NCI screen.

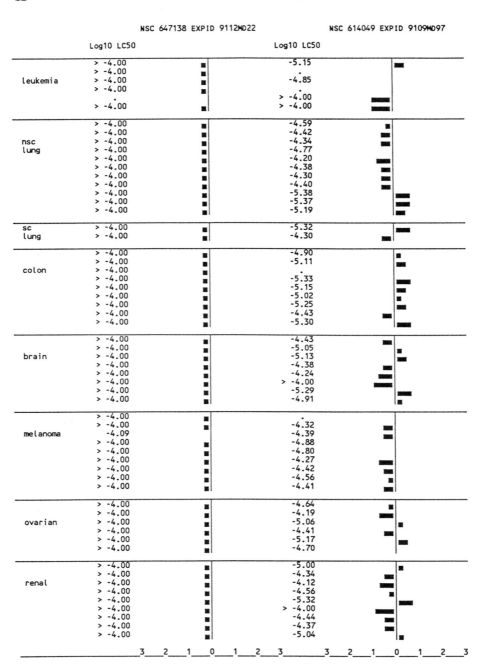

Figure 7. Comparison of phenazine NSC 647138 and phenazine dioxide NSC 614049 in NCI screen. Responses expressed as bargraphs.

projects on a multidisciplinary basis, when investigators increasingly are obliged to specialize within a discipline. The scientific advances and therapeutic benefits still to come from anthracycline studies, however, can be very great.

Acknowledgments. The previously unpublished synthesis in Scheme II was carried out by Drs. Francois Lefoulon and Alina Sen while at The University of Texas M. D. Anderson Cancer Center and by Dr. Michael Tracy of SRI International. Synthetic work was funded by NIH Grants CA24168 and CA43328. Compounds were screened by the NCI, Division of Cancer Treatment, Developmental Therapeutics Program.

Literature Cited

1.   Weiss, R.B. *Semin. Oncol.* **1993**, *19*, 670-686.
2.   Mross, K. *Eur. J. Cancer* **1991**, *27*, 1542-1544.
3.   Carter, S. K. *Cancer Chemother. Pharmacol.* **1980**, *4*, 5-10.
4.   Basser, R.L.; Green, M.D. *Cancer Trt. Rev.* **1993**, *17*, 57-77.
5.   Attardo, G.; Courchesne, M.; Xu, Y. C.; Rej, R.; Lavalee, J-F.; Lebeau, E.; Kraus, J-L.; Lamothe, S.; Nguyen, D.; Wang, W,; St-Denis, Y.; Belleau, B. (deceased). Discovery of novel heteroanthracycline oncolytics. Vide infra.
6.   Krohn, K. *Angew. Chem. Int. Ed. Engl.* **1986**, *25*, 790-807.
7.   *Recent aspects of anthracycline chemistry.*; Kelly, T. R., Ed. Tetrahedron (Tetrahedron Symposia-in-Print No. 17) **1984**, *40* (no. 22), 4537-4793.
8.   Grever, M. R.; Schepartz, S. A.; Chabner, B. A. *Semin. Oncol.* **1993**, *19*, 622-638, and leading references.
9.   Boyd, M. R. *Princip. Pract. Oncol.* **1989**, *3*, 1-12.
10.   Boven, E.; Winograd, B.; Berger, D. P.; Dumont, M. P.; Braakhuis, B. J. M.; Fodstad, O.; Langdon, S.; Fiebig, H. H. *Cancer Res.* **1992**, *52*, 5940-5947.
11.   Shoemaker, R. H.; Wolpert-Defilippes, M. K.; Kern, D. H.; Lieber, M. M.; Makuch, R. W.; Melnick, N. R.; Miller, W. T.; Salmon, S. E.; Simon, R. M.; Venditti, J. M.; VonHoff, D. D. *Cancer Res.* **1985**, *45*, 2145-2153.
12.   Tong, G. L.; Henry, D. W.; Acton, E. M. *J. Med. Chem.* **1979**, *22*, 36-39.
13.   Acton, E. M.; Tong, G. L. *J. Med. Chem.* **1981**, *24*, 669-673.
14.   Acton, E. M. *Drugs Exptl. Clin. Res.* **1985**, *11*, 1-8.
15.   Acton, E. M. *Cancer Bull.* **1985**, *37*, 173-179.
16.   Myers, C. E.; Muindi, J. R. F.; Zweier, J.; Sinha, B. K. *J. Biol. Chem.* **1987**, *262*, 11571-11577.
17.   Peters, J. H.; Gordon, G. R.; Kashiwase, D.; Acton, E. M. *Cancer Res.* **1984**, *44*, 1453-1459.
18.   Reductive deglycosidation of 5-iminodaunorubicin has been observed in a chemical reduction with dithionite and upon electrochemical reduction. Bird, D. M.; Boldt, M.; Koch, T. H. *J. Amer. Chem. Soc.* **1987**, *109*, 4046-4053.
19.   The pattern for daunorubicin (and doxorubicin) suggests specificity for the melanoma subpanel, although clinical melanoma is not described as responsive to anthracycline treatment: this apparent anomaly is not mentioned in publications on the NCI screen.
20.   Acton, E. M.; Tong, G. L. *J. Het. Chem.* **1981**, *18*, 1141-1147.
21.   Tracy, M.; Acton, E. M. *J. Org. Chem.* **1984**, *49*, 5116-5124.
22.   Wong, C-M.; Mi, A-Q.; Ren, J.; Haque, W.; Lam, H-Y.; Marat, K. *Can. J. Chem.* **1984**, *62*, 1600-1607.
23.   Lown, J. W.; Sondhi, S. M. *J. Org. Chem.* **1985**, *50*, 1413-1418.

RECEIVED July 13, 1994

Chapter 2

Non-Cross-Resistant Anthracyclines with Reduced Basicity and Increased Stability of the Glycosidic Bond

Waldemar Priebe, Piotr Skibicki, Oscar Varela[1], Nouri Neamati,
Marcos Sznaidman[2], Krzysztof Dziewiszek, Grzegorz Grynkiewicz[3],
Derek Horton[4], Yiyu Zou, Yi-He Ling, and Roman Perez-Soler

M. D. Anderson Cancer Center, The University of Texas, 1515 Holcombe Boulevard, Box 80, Houston, TX 77030

Anthracycline antibiotics with reduced basicity and increased stability of the glycosidic bond were synthesized, and their structure-activity relationship (SAR) was studied. Increased stability of the glycosidic bond was achieved by introduction of halogen at C-2'. Using selected model compounds we investigated the role of the basic amino group and the effects of changes in basicity on the drugs' activity against multidrug resistant (MDR) tumors and correlated this with the drugs' affinity to P-glycoprotein, the transport protein responsible for efflux of drugs out of MDR cells. We summarize here our synthetic efforts in this area and present results regarding the SAR of substituents affecting glycosidic bond stability and basicity of the amino function. We also present an analysis of in vitro cytotoxicity, cellular uptake, and efflux against sensitive and MDR cells, both murine and human, for selected congeners.

Anthracyclines like doxorubicin (DOX, **1**) and daunorubicin (**2**) are the most effective anticancer agents against leukemias, lymphomas, breast carcinoma, and sarcomas (*1*). However, development of resistance (acquired resistance) after initially effective systemic chemotherapy often limits the effectiveness of chemotherapy. Also, the existence of tumors like adenocarcinomas of the

[1]Current address: Department of Chemistry, University of Buenos Aires, 1428 Buenos Aires, Argentina
[2]Current address: Burroughs Wellcome Company, Research Triangle Park, NC 27709
[3]Current address: Pharmaceutical Research Institute, 01–793 Warsaw, Poland
[4]Current address: Department of Chemistry, The American University, Washington, DC 20016

0097–6156/95/0574–0014$10.34/0
© 1995 American Chemical Society

gastrointestinal tract or lung, which are de novo resistant to chemotherapy, indicates the necessity of focusing on the design and synthesis of drugs active against resistant tumors. The following phenomena are associated with the development of resistance: (a) reduction of intracellular drug accumulation, (b) changes in drug metabolism, (c) alteration of intracellular drug targets (for example Topo II), and (d) increased DNA repair potential [for reviews see (2-5)]. The most studied and understood mechanism of resistance is the mechanism associated with decreased intracellular drug accumulation, presence of a membrane-bound energy-dependent efflux system mediated by overproduction of P-glycoprotein (P-gp), and overexpression of the *mdr1* gene. This mechanism is often referred to as typical or classic MDR (2, 6, 7). Structurally diverse drugs such as DOX, daunorubicin, mitoxantrone, vincristine, vinblastine, vindesine, etoposide (VP-16), teniposide (VM-26), dactinomycin (actinomycin D), gramicidin D, plicamycin (mithramycin), mitomycin C, trimetrexate, and taxol are all involved in MDR.

1 - R=OH -- Doxorubicin
2 - R=H   -- Daunorubicin

Our efforts to overcome MDR focus on identification of chemical modifications that minimize drug efflux by reducing drug affinity to P-gp. This approach led us to the formation of hypothesis that the amino group in the sugar portion of DOX is an important functionality for substrate recognition by the P-gp multidrug transporter (8-12). We assumed that studies to confirm such an hypothesis would also be very useful in elucidating the MDR mechanism, thus possibly leading to the identification of a drug's structural requirement for "binding" to P-gp. These efforts were recently reviewed (13) and will not be discussed here in detail.

In this presentation we will review the design and synthesis of non-basic anthracyclines and structure-activity relationship (SAR) of carbohydrate- and aglycon-modified antibiotics with special emphasis on the 2'-substituted and 3'-deaminated anthracyclines. Using selected compounds we will point out structural features responsible for non-cross-resistant properties of anthracycline congeners.

Importance of the Amino Group at C-3'

The 3'-amino group has been perceived as the key structural element of anthracycline antibiotics. It plays an important role in the interactions of anthracyclines with different macromolecules and is involved in (a) stabilization of anthracycline interaction with DNA, (b) increased affinity of anthracyclines for negatively charged phospholipids (formation of a complex between DOX and cardiolipin was shown to be strongly stabilized by interaction between the positively charged sugar amine and the negatively charged phosphate groups), (c) determining membrane transport and cytotoxicity, and (d) mutagenic activity of anthracyclines (derivatization of the amino group reduces mutagenicity without altering the antitumor properties) (14).

The above observations stress the importance of the 3'-amino group in the interactions of anthracyclines with different macromolecules. Therefore, drug-macromolecule binding should be significantly affected by the deamination or reduction of basicity at the C-3' position. Consequently, any alteration of basicity at C-3' should noticeably affect the antitumor activity and toxicity and possibly other biological properties of DOX. It was therefore our initial goal to study what role the amine group plays in the binding of the drug to P-gp and to what extent the reduction of basicity increases accumulation of the drug in MDR cells and, consequently, increases the cytotoxicity against MDR cells. At the same time it was important to examine to what extent the amine group is responsible for the antitumor properties of DOX. To achieve these goals we decided to select a series of analogues with reduced basicity for comparative studies with parent aminated anthracyclines.

Synthesis of Anthracyclines Hydroxylated at C-3'

First, 3'-deamino-3'-hydroxy anthracyclines were prepared by El Khadem et al. (ε-rhodomycinone glycosides) (15) and Fuchs et al. (daunomycinone glycosides) (16) and although they did not show impressive activity, results indicated that 3'-hydroxy derivatives merit further investigation. These studies were extended by synthesis of 2-deoxy-L-fucopyranosyl-ε-pyrromycinone and 2-deoxy-L-fucopyranosyl-carminomycinone (17, 18) and 3'-hydroxy sugar modified analogues of daunorubicin (19). However, 3'-hydroxydoxorubicin (8) remained a difficult synthetic target for several years. Initial synthesis of 3'-hydroxy-doxorubicin led to 3,4-di-O-acetyl derivatives (20), and not until a 14-O-silyl blocking group was introduced could 3'-deamino-3'-hydroxydoxorubicin 8 (hydroxyrubicin) be synthesized (21) (Scheme 1). A similar approach was used to synthesize 3'-deamino-3'-hydroxy-epidaunorubicin 13 and 3'-deamino-3'-hydroxy-epirubicin 14 (hydroxyepirubicin) (22) (Scheme 2).

Hydroxyrubicin was synthesized by Koenigs-Knorr coupling reaction of selectively blocked adriamycinone 6 with glycosyl chloride 4. Glycosyl chloride 4 was obtained by electrophilic addition of dry hydrogen chloride to 3,4-di-O-acetyl-L-fucal (3). Coupling reaction gave the α anomer 7 with good yield, and no β anomer was detected. The deacetylation with sodium methoxide gave a 3',4'-dihydroxy derivative, which then was desilylated with tetrabutylammonium fluoride in tetrahydrofuran solution to hydroxyrubicin (8) (Scheme 1).

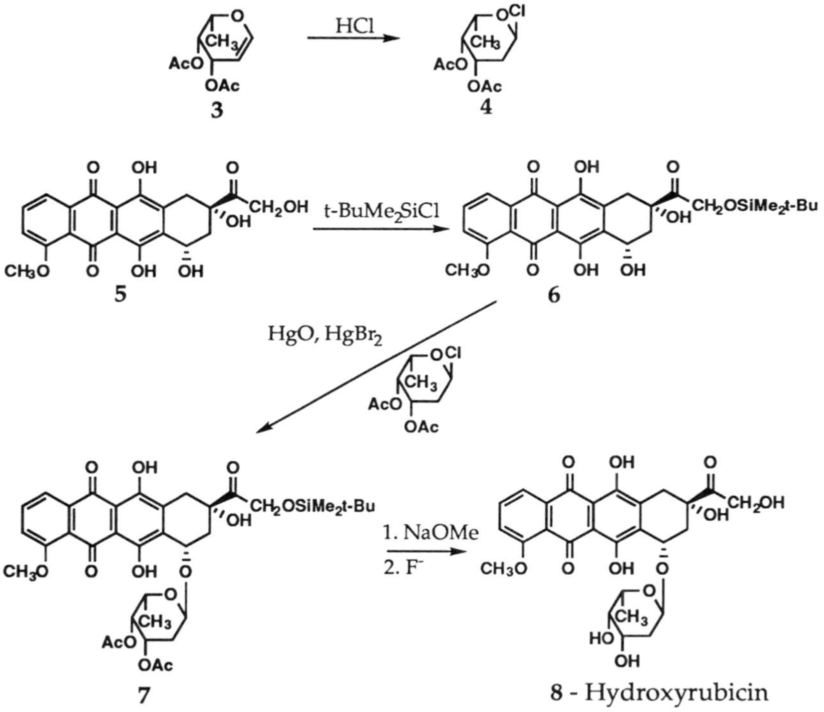

SCHEME 1

Glycosyl chloride 10 derived from 3,4-di-O-acetyl-L-rhamnal was coupled with 14-O-*tert*-butyldimethylsilyladriamycinone (6) or with daunomycinone (9) in conditions similar to that for hydroxyrubicin. However, in contrast to hydroxyrubicin, for reactions starting from 6 or 9 both α and β anomers for 11 and 12 were formed. Using conventional column chromatography it was impossible to separate them. To overcome that problem, (a) separation of anomers, although difficult, was achieved at the stage of deacetylated products, and (b) silver triflate was used as coupling reagent to increase stereoselectivity and α anomer 11 was isolated, though in relatively low yield. Hydroxyepirubicin (**14**) was prepared using the same deblocking procedure as for hydroxyrubicin (22) (Scheme 2).

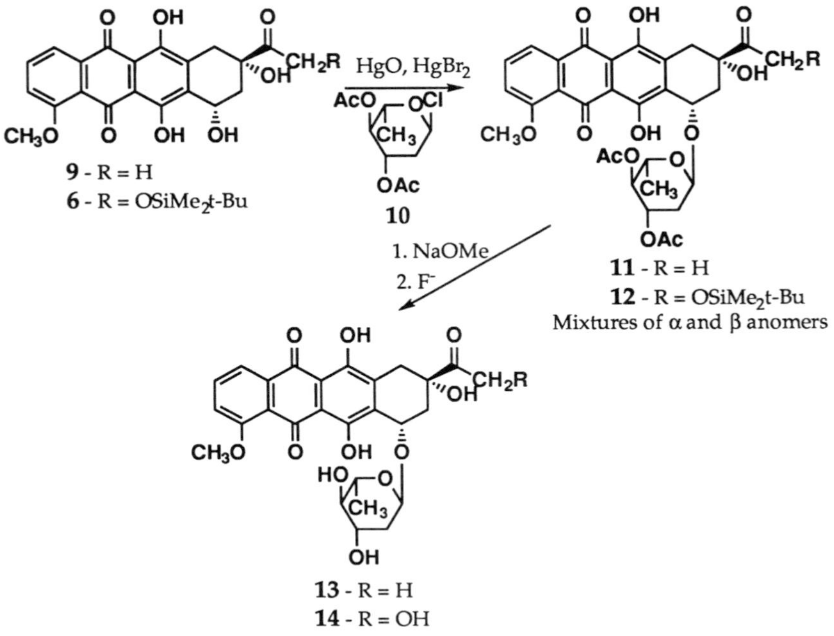

O
OH
O
CH₂R
OH
CH₃O
O
OH
OH
9 - R = H
6 - R = OSiMe₂t-Bu
HgO, HgBr₂
AcO
Cl
CH₃
OAc
10
11 - R = H
12 - R = OSiMe₂t-Bu
Mixtures of α and β anomers
1. NaOMe
2. F⁻
HO
13 - R = H
14 - R = OH

SCHEME 2

Both hydroxyrubicin (**8**) and hydroxyepirubicin (**14**) showed very good antitumor activity in the initial in vitro and in vivo evaluation, although the activity in vivo of hydroxyrubicin was consistently higher than that of hydroxyepirubicin. Hydroxyrubicin was also less toxic than hydroxyepirubicin (*21, 22*).

More complicated was synthesis of 3′-hydroxy analogue **26**, a 3′-deaminated analogue of esorubicin, the reason being that a possible sugar substrate, 4-deoxy-L-rhamnal or 4-deoxy-L-fucal **24**, was not known and the use of described procedures failed to produce the required glycal. The 3-*O*-acetyl-4-deoxy glycal **24** was prepared in five steps, with high yield from L-rhamnal (**15**) or L-fucal (**16**) using a combination of new blocking groups and free radical deoxygenation (Scheme 3) (23). Initially a radical deoxygenation reaction was attempted on thiocarbonyl imidazolide **18**; however, the only product isolated from this reaction was free hydroxyl substrate **17**. Change of blocking group from acetyl to *t*-butyldimethylsilyl led to derivatives **21** or **22** (depending on starting material), which were reduced with high yield to 4-deoxy-3-*O*-silylated glycal **23**. The overall yield of the three-step reaction from L-rhamnal **15** to 4-deoxy glycal **23** was 74%. A lower yield (56%) than that from L-rhamnal was obtained from L-fucal **16**, primarily because of the lower yields of selective silylation to **20** (71%) and the reductive deoxygenation of **22** to **23** (78%). 3-*O*-Acetyl-4-deoxy glycal **24** was prepared from **23** by desilylation and subsequent acetylation with 89% yield as a distillable liquid, which decomposes upon contact with silica gel (Scheme 3).

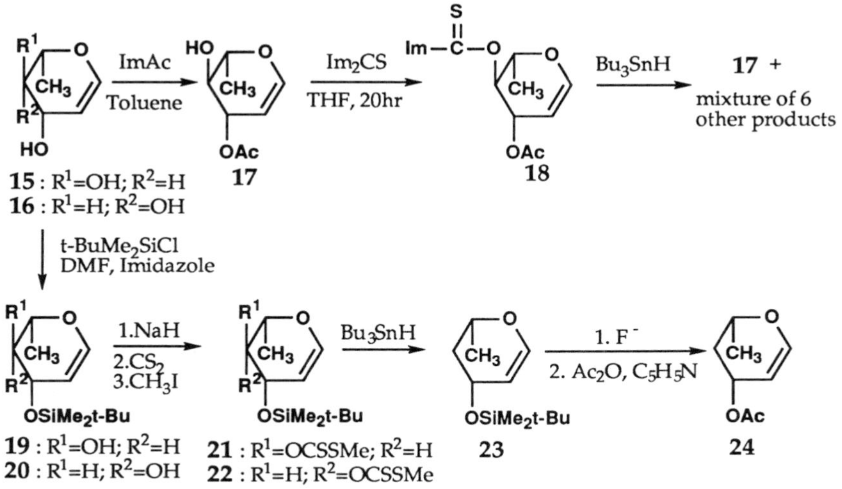
ImAc
Toluene
Im_2CS
THF, 20hr
Bu_3SnH
17 +
mixture of 6
other products
15 : R^1=OH; R^2=H
16 : R^1=H; R^2=OH
17
18
t-$BuMe_2SiCl$
DMF, Imidazole
1.NaH
2.CS_2
3.CH_3I
Bu_3SnH
1. F^-
2. Ac_2O, C_5H_5N
19 : R^1=OH; R^2=H
20 : R^1=H; R^2=OH
21 : R^1=OCSSMe; R^2=H
22 : R^1=H; R^2=OCSSMe
23
24

SCHEME 3

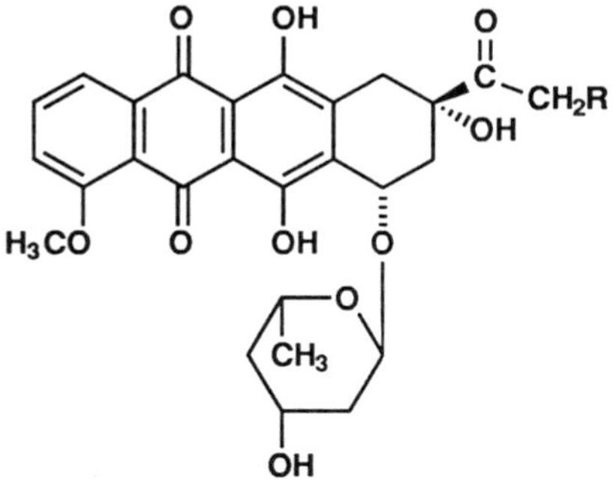

25 - R = H
26 - R = OH

Addition of hydrogen chloride to **24** gave glycosyl chloride, which was used immediately for coupling with daunomycinone (**9**) or 14-*O*-*tert*-butyldimethylsilyladriamycinone (**6**). Coupling was carried out in dichloromethane solution in the presence of mercuric bromide, yellow mercuric oxide, and molecular sieves 3 Å. Typical deblocking procedures gave 4'-deoxy-3'-hydroxydaunorubicin (25) and 3'-hydroxyesorubicin (26) (24).

Daunorubicin analogue 25 appeared to be significantly less active than 3'-hydroxyesorubicin 26, which showed cytotoxic potential in vitro similar to that of DOX and significantly better activity against P388 leukemia in vivo than DOX (T/C is a ratio of the test median survival time to median survival time of the control x 100%; T/C for 26 in two experiments was >600, and 100% and 90% of long-term survivals were noted; for DOX T/C at a maximum tolerated dose of 10 mg/kg was 172 and 300, and 0% and 20% long-term survivals, respectively, were noted) (24).

27 (WP474)

An interesting lipophilic analogue 27 (WP474) of hydroxyrubicin was obtained by selective acylation of the 14-hydroxyl of 8 with palmitoyl chloride (25). 7-O-(2,6-Dideoxy-α-L-lyxo-hexopyranosyl)14-O-palmitoyladriamycinone was obtained in 32% yield. Its antitumor activity was assessed in vivo in liposome form. Particularly interesting was its good activity in vivo against M-5076 reticulosarcoma when the drug was administered i.v.; DOX was basically inactive at the maximum tolerated dose, whereas WP474 showed a T/C of 175 at 20 mg/kg. Ester WP474 was also highly active against L1210 lymphoid leukemia (T/C > 600 at 60 mg/kg vs T/C 337 at 10 mg/kg for DOX) (25).

Biological Effects of Deamination and Introduction of Hydroxyl at C-3'

To evaluate the role of amino group and to study the effects of basicity of the sugar moiety on the toxicity and antitumor activity of anthracycline antibiotics, as well as to assess the potentials of deaminated anthracycline analogues as new anticancer drugs, we have selected for comparative studies hydroxyrubicin (8), an analogue whose amino group was replaced by hydroxyl and all of whose other structural features and configurations were identical with those of DOX.

In a series of studies of hydroxyrubicin and DOX (8, 9, 11, 12) we have shown that hydroxyrubicin has in vitro cytotoxicity similar or superior to that of DOX against P388, L1210, and M-5076 cells, as determined by an MTT assay, and against 8226 and CEM cells, as determined by a growth inhibition assay. Hydroxyrubicin was significantly more effective than doxorubicin in inhibiting the growth of multidrug-resistant CEM-VBL and 8226R cells (5 and 13 times, respectively). Hydroxyrubicin was equally cytotoxic against KB 3-1 and multidrug-resistant KB-V-1 cells. The resistance index (RI) for hydroxyrubicin was equal to 1.1, whereas the RI for DOX was >50 (Table I).

Table I. Cytotoxicity of Hydroxyrubicin Against Sensitive and MDR Human Carcinoma (KB-3-1, KB-V-1), Myeloma (8226, 8226R) and Leukemia (CEM, CEM-VBL) Cells[a]

Drug	ID_{50} ($\mu g/ml$)			ID_{50} (ng/ml)			ID_{50} (ng/ml)		
	KB-3-1	KB-V-1	RI	8226	8226R	RI	CEM	CEM-VBL	RI
Hydroxy-rubicin (8)	18.9	21.3	**1.1**	4.2	53	**13**	5.0	300	**60**
DOX (1)	4.0	> 50.0	**>25**	2.6	700	**269**	4.5	1,600	**356**

[a] Resistance index (RI) = ID_{50} for resistant cells/ID_{50} for sensitive cells. MTT assay for KB cells; growth inhibition assay for 8226 and CEM cells.

Flow cytometry study of cellular uptake and retention of hydroxyrubicin and DOX showed that in sensitive 8226 cells, 2-h uptake and retention of DOX were similar or higher than those of hydroxyrubicin (Figure 1). In 8226R cells, uptake and retention of hydroxyrubicin were about threefold higher than those of DOX (Figure 2). This indicated that the lack of the basic amino group is responsible for the increased cellular uptake and retention of hydroxyrubicin in multidrug-resistant cells, possibly because of the reduced interaction of hydroxyrubicin with P-gp, a multidrug transporter. Increased cellular uptake and retention of hydroxyrubicin correlated with a partial or total lack of cross-resistance of this analogue with the parent compound, DOX. The hydroxyrubicin activity against MDR tumors was confirmed in vivo against P388 leukemia resistant to DOX. Hydroxyrubicin at the optimal dose (37.5 mg/kg, i.p. on day 1) had significant activity, whereas DOX (10 mg/kg, i.p. on day 1) was inactive (%T/C 163-200 for 8 vs. 118-120 for DOX; data from two experiments) (12).

Hydroxyrubicin was also less toxic in mice, its LD_{50} was about threefold higher than that of DOX (79.1 mg/kg versus 25.7 mg/kg), and at equitoxic doses it was less cardiotoxic, as assessed by the Bertazzoli test. The lack of basic amine function was also responsible for the decreased affinity of hydroxyrubicin to negatively charged cardiolipin, which in part might be a reason for the lower cardiotoxicity of hydroxyrubicin (12).

These studies clearly indicate that the amino group at position 3' is not essential for DOX to exert its biological activity and, that as a matter of fact, its removal might have positive effects, i.e., increased activity against MDR tumors and reduced general toxicity and cardiotoxicity.

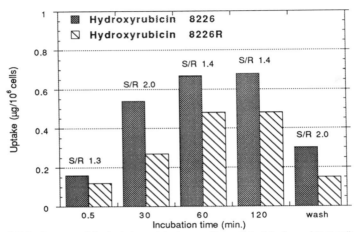

S/R (resistance uptake factor) = uptake by sensitive cells/uptake by resistant cells.

Figure 1. Uptake and retention of hydroxyrubicin by 8226 and 8226R cells.

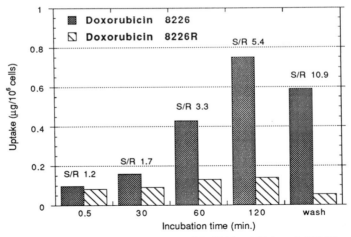

Figure 2. Uptake and retention of DOX by 8226 and 8226R cells.

Doxorubicin was shown to induce protein-associated DNA breaks by inhibition of DNA topoisomerase II and stabilization of cleavable complexes at specific sequences. Comparison of cleavage induced by the topoisomerase II alone or in the presence of DOX, hydroxyrubicin, the epipodophyllotoxin etoposide (VP-16), and the acridine *m*-AMSA showed that hydroxyrubicin induced a specific pattern of cleavage in the human c-*myc* origin similar to that of DOX (26). Studies of the DNA single-strand breaks (SSBs) induced by 1 h exposure to DOX or hydroxyrubicin in KB-3-1 and KB-V-1 cells revealed that hydroxyrubicin-induced SSBs (at 10 μM) in sensitive and resistant cells were similar, whereas DNA SSBs induced by DOX were twofold fewer in resistant cells (26).

Hydroxyrubicin was also used as a model compound to assess the electrostatic contribution to free energy of DOX binding to DNA. The DNA free energy binding measured for DOX was -8.8 kcal mol^{-1}, whereas for hydroxyrubicin under the same conditions the value was -7.2 kcal mol^{-1} (27). These comparative studies allowed us not only to evaluate the electrostatic contributions to the DNA binding free energy of anthracycline antibiotics but also to test experimentally current polyelectrolyte theory (28) as applied to ligand-DNA interactions. The measured value of $\delta \log K / \delta \log [M^+]$ changed from -0.95 for DOX to -0.18 for hydroxyrubicin. This is in excellent agreement with the Friedman-Manning theory (28), which predicts for the interaction of an uncharged intercalating ligand a value of -0.24.

Effects of Basicity Reduction on Non-Cross-Resistant Properties of Lipophilic Anthracyclines

Comparative studies of DOX and hydroxyrubicin showed that the replacement of the basic amino group with a hydroxyl resulted in a compound with distinctively different biological properties. Maybe the most interesting lead is the increased activity against MDR cells, which can be associated with only one structural change in the DOX molecule . Therefore, it was interesting to check if the same holds true for analogues having totally different physicochemical and biological properties. Previously reported highly lipophilic derivatives of DOX, that is, 3'-*N*-benzyl-14-*O*-valeroyl-doxorubicin (**AD198**) and its metabolite 3'-*N*-benzyl-doxorubicin (**AD288**), showed good activity against tumors resistant to DOX. Lothstein et al. developed cell lines resistant to **AD198** (29, 30). This created a good base to further test our hypothesis that the reduction of basicity correlates with increased drug activity against MDR cells. At the same time, the possibility existed that by checking in practice our rationale, we could obtain analogues with potentially good antitumor activity against MDR tumors. Because of the extremely high lipophilicity of model compounds, the rate and pattern of uptake are different from those of DOX and hydroxyrubicin; thus we believed that synthesis and evaluation of 3'-oxy analogues of **AD198** and **AD288** would also clarify whether the underlying generality of reduction of basicity is associated with factors like direct interaction with transport proteins or with altered lipophilicity of deaminated compounds.

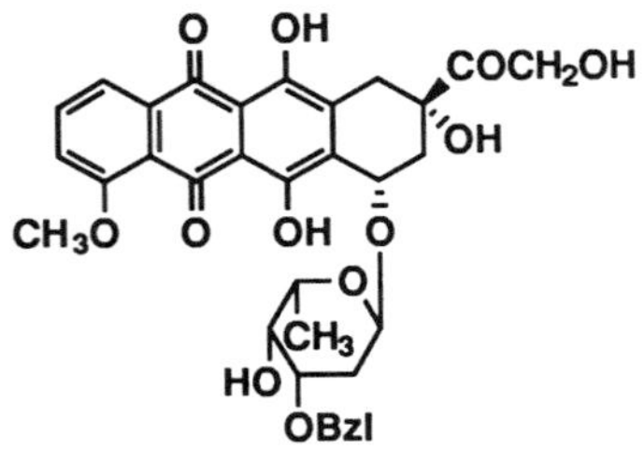

WP546

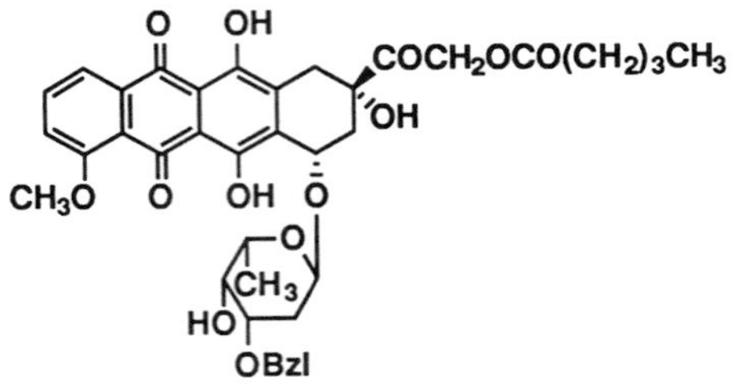

WP549

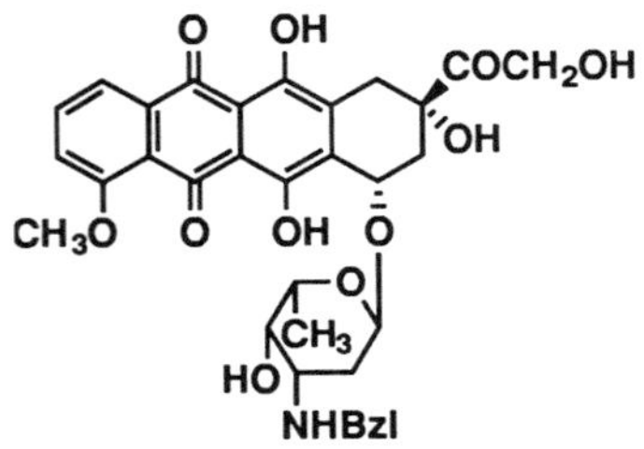

AD288

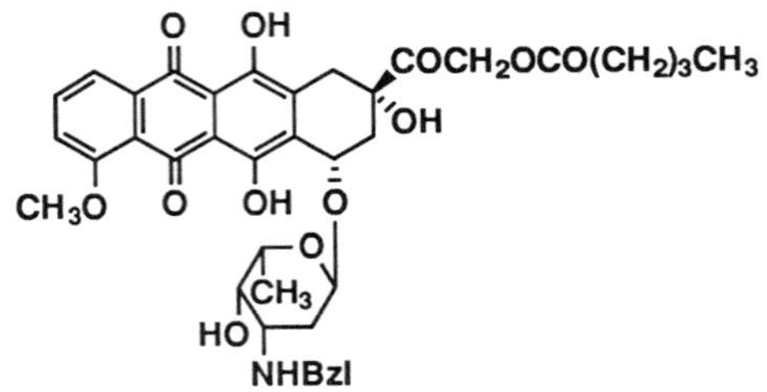

AD198

The 3'-*O*-benzyl analogues **WP546** and **WP549** were synthesized from 3'-*O*-benzyl-L-fucal and 14-*O*-substituted adriamycinone as a substrate (*31*). **WP546** and **WP549** analogues were tested against resistant cells selected by continuous exposure to **AD198** or vinblastine. Cells selected with **AD198** (J774.2/A300, J774.2/A750, P388/AD198) overexpressed the *mdr1b* (P-gp) gene and *mdr* mRNA and overproduced P-gp. These cells showed resistance to DOX, vinblastine, and **AD198** . While resistance to DOX and vinblastine was associated with reduced intracellular drug accumulation, the **AD198** net intracellular accumulation was unchanged compared with that of parental cells, thus indicating that resistance to **AD198** is not conferred through a P-gp mediated efflux, but through other cytoplasmic mechanisms (*32*).

Cytotoxicity assay (Table II) indicates that for highly lipophilic compounds also, the replacement of 3'-nitrogen with oxygen leads to non-cross-resistant or partially non-cross-resistant drugs. The 3'-oxy analogues are apparently not affected by the presence of P-gp or by the other mechanism of resistance specific to **AD198**. In some cases resistant cells were hypersensitive, thus making the RI less than 1. These results very clearly showed that the mechanism of resistance selected by lipophilic compounds having a basic center can be overcome by their congeners having a nitrogen atom replaced with oxygen. More detailed studies are in progress.

Table II. Resistance Indexes of 3'-*O*-Substituted Analogues (WP546, WP549) versus 3'-*N*-Anthracyclines (DOX, AD198, AD288) in J7/A300, J7/A750, J7/40V, P388/AD198, and KB-V-1 Resistant Cell Lines[a]

Drug	J774.2/ A300	J774.2/ A750	J774.2/ 40V	P388/ AD198	KB-V-1
DOX	15.0	43.3	31.5	-	58.2
AD198	15.3	33.1	13.5	2.5	-
AD288	16.5	-	-	-	-
WP546	2.2	2.6	2.2	0.8	0.9
WP549	1.9	1.9	2.3	0.5	0.5

[a]Cells were selected by continuous exposure to AD198 (J774.2/A300, J774.2/A750, P388/AD198) or vinblastine (J774.2/40V, KB-V-1) (*31, 32*).

Synthesis of 3'-Mercaptodoxorubicin

An interesting result of studies of analogues hydroxylated at C-3' prompted synthesis of DOX analogues in which a basic nitrogen atom was replaced with a sulfur atom. Initial efforts resulted in synthesis of 3'-thiomethyl-3'-deamino-daunorubicin derivatives. All analogues showed lower activity in vivo than corresponding acetoxy or hydroxy analogues (*33*). Final judgment about the potential of 3'-thio anthracyclines had to be postponed until analogues containing a free mercapto group could be synthesized.

Synthesis of the desired 3'-mercapto analogues was hindered by the lack of adequate methods for the preparation of 3-mercapto sugars. Initial efforts to prepare such sugars using a Ferrier rearrangement reaction, as previously described by us for 3-thioalkyl sugars (*34*), failed. Reaction of 3,4-di-*O*-acetyl-L-fucal with thioacetic acid or potassium thioacetate in the presence of Lewis acid (BF3·Et2O) resulted in an anomeric mixture of 2,3-unsaturated sugars instead of the desired 3-*S*-acetyl glycals, whose presence could not be detected (*35*).

To overcome this difficulty an alternative approach based on conjugate addition reaction was developed (*35*) (Scheme 4). The 3,4-di-*O*-acetyl-L-fucal (**3**) was transformed to 4-*O*-acetyl-2,3,6-trideoxy-L-*threo*-hex-2-enopyranose (**28**) by heating in water and without further purification treated with thioacetic acid. Conjugate addition products **29** were subsequently acetylated to 1,4-di-*O*-acetyl-3-*S*-acetyl-3-thio-hexopyranoses **30**. Refluxing of **30** in xylene containing silica gel for 1 h gave a pair of the 3-*S*-acetyl-2,3,6-trideoxy-L-*lyxo*- and -L-*xylo*-hex-1-enitols (**31** and **32**, respectively).

The 3-thioacetyl-glycal **31** with the desired L-*lyxo* configuration was used towards synthesis of 3'-thio-3'-deamino doxorubicin by first transforming **31** to glycosyl chloride using electrophilic addition of dry HCl in benzene. Then, coupling of 14-*O*-tert-butyldimethylsilyladriamycinone (**6**) with the excess of glycosyl donor was carried out in dichloromethane in the presence of HgBr2, HgO, and molecular sieves 3 Å (600 mesh) and gave the 3'-*S*-acetyl-14-*O*-(*tert*-

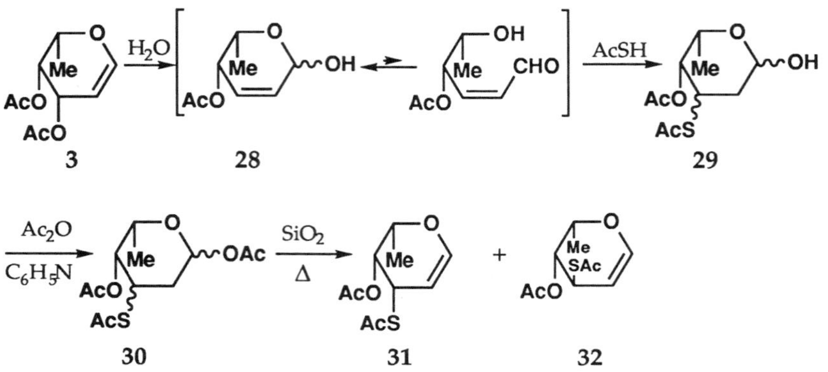

SCHEME 4

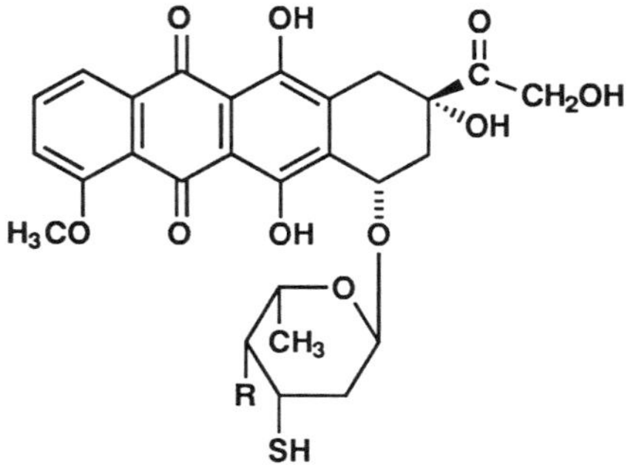

33 - R = OH
34 - R = OAc

butyldimethylsilyl)doxorubicin in 78% yield. Deblocking of the thiol group and desilylation gave 3′-mercapto-doxorubicin analogue **34**. The 4-O-acetyl group resisted conventional deblocking procedures, and alternative methods for the preparation of fully unblocked analogue **33** are now being used. In the meantime it was very interesting to observe that the 3′-mercaptodoxorubicin **34** was significantly more active than previously synthesized 3′-thiomethyl derivatives and showed cytotoxic potential similar to that of DOX (**35**). Identical conditions were used to prepare 3′-S-acetyldaunorubicin (**35**); however, its activity was significantly lower than that of DOX.

Anthracyclines Substituted with Halogen at C-2′

The concept of increasing the stability of the glycosidic bond in anthracyclines by the introduction of electron-withdrawing substituents at the C-2′ was for the first time presented by us in 1980 (*36*) and in more detail during the First ACS Anthracycline Symposium (*37*) and then during the International

Carbohydrate Symposium (*38*). Subsequently, a part of the initial results were described in 1982 in two chapters of the book based on the ACS Anthracycline Symposium (*33, 39*).

Inductive effects of halogens are significantly higher than those of other elements; therefore, in our approach we have focused on synthesis of 2'-halogeno anthracyclines (*33, 37, 38, 40-45*). Our analogues were the first 2'-halogeno-substituted anthracycline analogues and probably the first halogeno versions of 2-deoxy glycosides of biologically and clinically important drugs. The importance of this modification was later confirmed by others in synthesizing a variety of 2'-halo-anthracyclines (*46-51*).

In our studies of 2'-substituted anthracyclines we distinguished two different groups of 2'-halo analogues: (a) 2'-halo-3'-deamino and (b) 2'-halo-3'-amino analogues. Both groups of analogues will have the stability of the glycosidic bond increased due to inductive effects of halogen at C-2', and analogues of both groups will have reduced basicity at C-3'. However, analogues with a hydroxy or acetoxy group at C-3' cannot be protonated, whereas analogues with amine function at C-3' will form water-soluble ammonium salts. Halogens at C-2', by reducing basicity of the amino group (pKa ~6), will strongly affect equilibrium in physiological solution, and in contrast to daunorubicin or DOX (pKa ~8), less than 10% of drug will be in protonated form. It is known that anthracyclines enter into the cell as uncharged species by a passive diffusion process through the lipid bilayer; therefore, halogens at C-2' might significantly affect the rate of uptake of the drug as well as its subcellular distribution. In consequence this might strongly affect the biological properties of the drug.

3'-Hydroxy Anthracyclines Halogenated at the C-2' Position

2'-Iodo-3'-deamino Daunorubicins.   First compounds synthesized from

35

that series were 2'-iodo daunorubicins having α-L-*talo* and α-L-*manno* configurations (*33, 38, 41, 44*). In a one-step reaction, daunomycinone was coupled with 3,4-di-*O*-acetyl-L-fucal (**3**) in the presence of *N*-iodosuccinimide (NIS) to give 2'-iodo-3',4'-diacetoxy-α-L-*talo*-daunorubicin **35**.

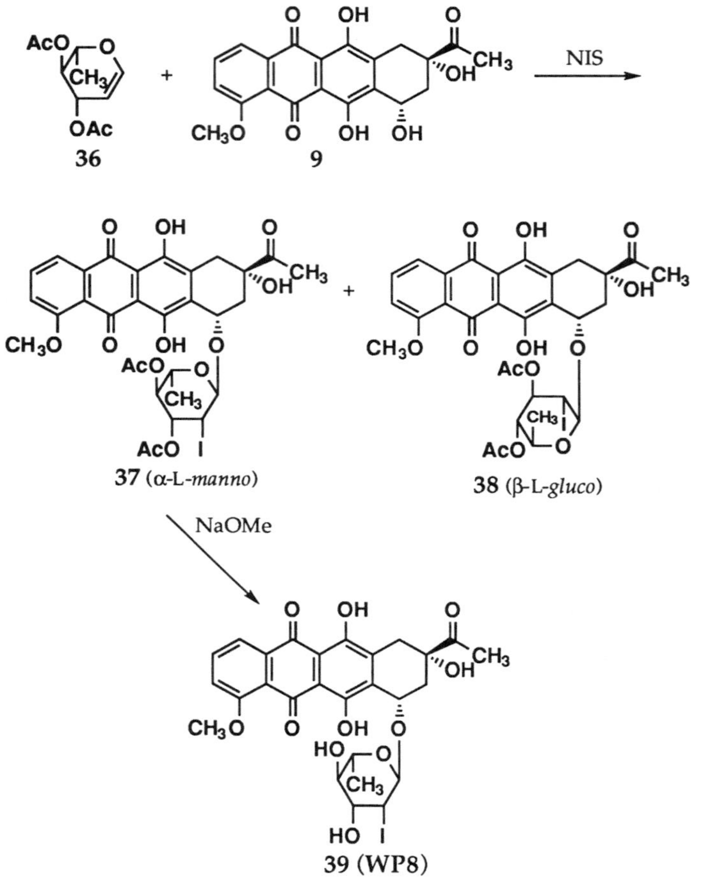

SCHEME 5

A similar reaction with 3,4-di-*O*-acetyl-L-rhamnal (**36**) (Scheme 5) gave two 2'-iodo *trans* addition products having α-L-*manno* (**37**) and β-L-*gluco* (**38**) configurations.

Daunorubicin analogue **37** in an α-L-*manno* configuration showed surprisingly good activity in vivo and was as active or more active than DOX. Testing showed that **37** is active against P388 leukemia (T/C 247 at 50 mg/kg and 208 at 25 mg/kg). It showed good activity against L1210 lymphoid leukemia (T/C 196 at 25 mg/kg) and Lewis lung carcinoma (T/C 187 at 25 mg/kg), and significant activity against B-16 melanoma (T/C 218 at 25 mg/kg). Compound **35** in a *talo* configuration was less efficacious than **37** and had activity comparable with daunorubicin (T/C 162), but was significantly less potent. The β-L-*gluco* analogue **38** did not display activity up to the highest tested dose of 50 mg/kg.

The most active analogue in this group, 2'-iodo daunorubicin **37**, was unblocked with sodium methoxide to 2'-iodo-2,6-dihydroxy-α-L-*manno*-daunorubicin **39** (**WP8**). In vivo evaluation against P338 indicated that unblocking increased the potency of the drug but not efficacy (T/C 222 at 12.5 mg/kg and 200 at 6.25 mg/kg). This compound also showed substantial activity in several tests at a dose of 3.12 mg/kg (T/C range 144-178).

2'-Bromo-3'-deamino Daunorubicins. Direct bromination of L-fucal **3** gave

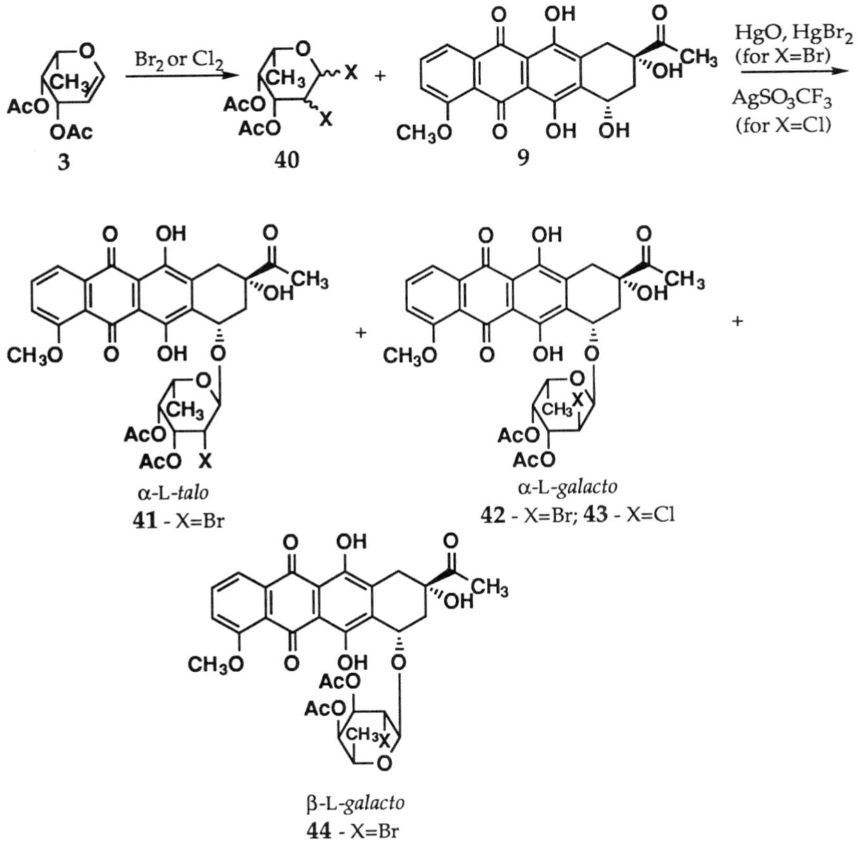

SCHEME 6

a mixture of glycosyl bromides epimeric at C-2 (40), which were coupled with daunomycinone under Koenigs-Knorr conditions (HgO, HgBr2) to give a mixture of 2′-bromo daunorubicins having α-L-*talo* (41), α-L-*galacto* (42), and β-L-*galacto* (44) configurations in the ratio 3:6:2, which was then separated by column chromatography (45) (Scheme 6).

Similarly 3,4-di-O-acetyl-L-rhamnal (36) was brominated in carbon tetrachloride to give α-L-*gluco* (45) and α-L-*manno* (46) glycosyl bromides in a 2:1 ratio. Subsequent coupling of the mixture of glycosyl bromides 45 and 46 with daunomycinone gave, after chromatography, α-*manno* analogue 47 in 13% yield, α-*gluco* isomer 49 in 28% yield, and β-*gluco* isomer 51 in 27% yield (Scheme 7).

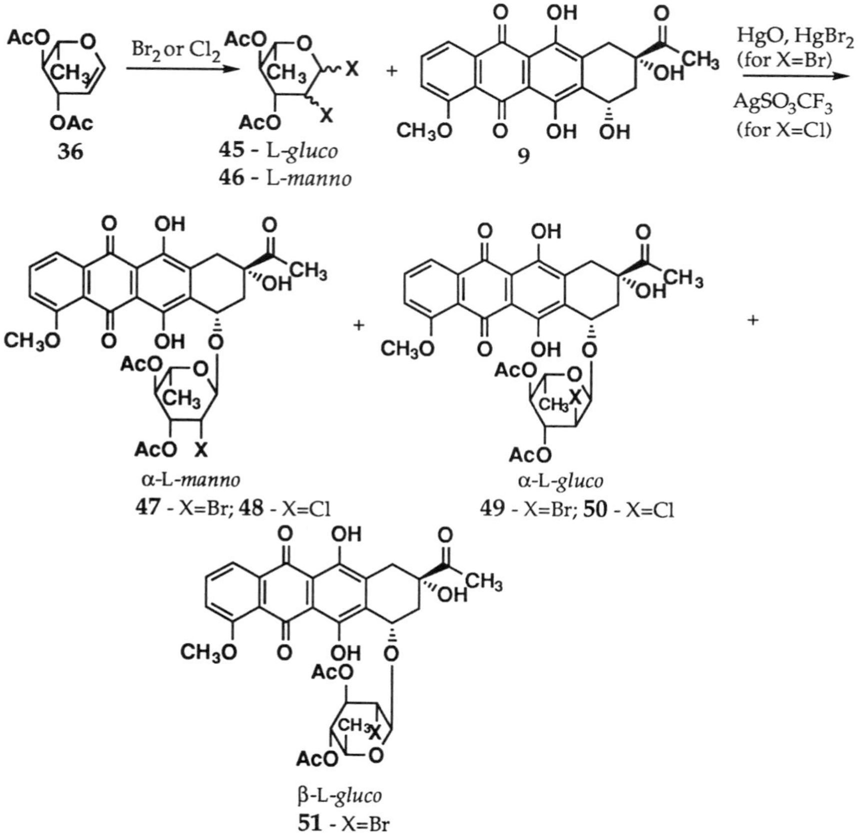

SCHEME 7

Biological evaluation in vivo against murine P388 leukemia showed for 2′-bromo-α-L-*talo*-daunorubicin 41 the T/C values of 145 at a dose of 50 mg/kg and 181 at 83 mg/kg; whereas the 2'-bromo-α-L-*manno*-daunorubicin 47 showed T/C 245 at 25 mg/kg, and in a separate experiment (injections on days 1, 5, and 9) showed T/C 278 at 25 mg/kg and T/C 229 at 12.5 mg/kg. The 2'-bromo analogues in L-*gluco* (49, 51) and L-*galacto* (42, 44) configurations did not display activity in vivo at doses up to 50 mg/kg in the P388 leukemia system.

2'-Chloro-3'-deamino Daunorubicins.   Chlorination of L-fucal 3 gave a mixture with the L-*galacto* isomer as the main component (~85%) and was followed by coupling with daunomycinone in the presence of silver triflate. After column chromatography, the α-L-*galacto* isomer 43 was isolated in 44% yield (Scheme 6).   Chlorination of 3,4-di-O-acetyl-L-rhamnal (36) gave a mixture of L-*gluco* and L-*manno* isomers (46) in a 4:1 ratio. Using the same coupling conditions as for 43, a mixture of dichlorides 46 gave only α glycosides having L-*manno* (48) and L-*gluco* (50) configurations (Scheme 7).

Again, the 2'-chlorodaunorubicin (48) in L-*manno* configuration showed good antitumor activity in vivo against P388 leukemia (T/C 248 at 25 mg/kg), whereas both L-*gluco* (50) and L-*galacto* (43) 2'-chloro analogues showed no activity against P388 leukemia at doses up to 50 mg/kg.

2'-Iodo-3'-deamino Daunorubicins Modified at C-5'.   Very promising results of biological evaluation of 2'-halo-3'-oxy-daunorubicins and a unique observed SAR prompted synthesis of analogues modified at C-5'.

52 - R = H (α-L-*lyxo*)
54 - R = CH$_2$OAc (α-L-*manno*)

53 - R = H (β-L-*xylo*)
55 - R = CH$_2$OAc (β-L-*gluco*)

Two modifications were considered: (a) a 6'-methyl group replaced with hydrogen and (b) a 6'-methyl group replaced with hydroxymethyl (CH2OH). Both type of analogues were synthesized by addition of daunomycinone to appropriate glycals in the presence of NIS. As a result, two *trans* addition products having α-L-*lyxo* (52) and β-*xylo* (53) configurations were obtained from 3,4-di-O-acetyl-L-xylal and two 2'-iodo glycosides, 54 and 55, from 3,4,6-tri-O-acetyl-L-glucal.

Analogue **52** (α-L-*lyxo*) showed T/C 143 and **54** (α-L-*manno*) displayed T/C 131 at the highest tested dose of 50 mg/kg, whereas β-*xylo* analogue **53** showed no activity up to 50 mg/kg. These results indicate that compounds having a methyl group at position C-5' are significantly more active than compounds with a hydrogen or CH_2OAc group at C-5'. Also, lack of activity of **53** further confirmed the deactivating effects of equatorial (2'*S*) halogen at C-2' in *trans* disposition to a β glycosidic bond.

2'-Halo-3'-hydroxy Analogues of Doxorubicin.   It is generally accepted and has been proven in clinical studies that DOX is a drug superior to daunorubicin, with a broader spectrum of activity resulting from the presence of a hydroxyl group at C-14. Therefore to fully evaluate the potentials of 2'-halo anthracyclines, we have prepared 2'-iodo- and 2'-chloro-3'-hydroxy-α-L-*manno*-doxorubicin (**56** and **57**, respectively).

56 - X = I
57 - X = Cl

2'-Iodo analogue **56** displayed in vivo against P388 leukemia T/C of 248 at 6.25 mg/kg and 186 at 3.12 mg/kg, whereas the 2'-chloro congener **57** displayed a T/C of 261 at 12.5 mg/kg and 197 at 6.25 mg/kg. Thus the overall activity was similar to that observed for daunorubicin analogues.

Structure-Activity Relationship of 2'-Halo Anthracyclines.   Analysis of in vivo data regarding a series of 2'-halo analogues is summarized in Figures 3 and 4. The largest set of compounds available for comparison were blocked daunorubicin analogues routinely tested when obtained (Figure 3). Analogues of daunorubicin shown in Figure 3 varied in configuration at C-1', C-2', and C-4' positions and in selection of halogen (*41, 44, 45*).

The 2'-iodo analogue **37**, 2'-bromo analogue **47**, and 2'-chloro analogue **48** showed similar potency and efficacy in vivo against P388 leukemia, indicating that the nature of halogen does not play a critical role. In contrast, orientation of the halogen was the most important factor influencing activity

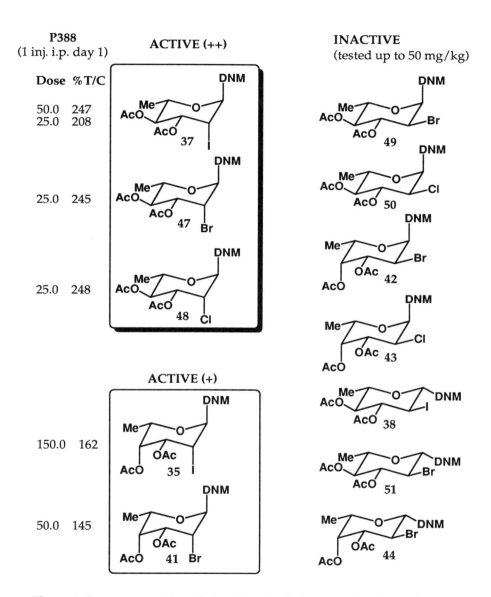

Figure 3. Structure-activity relationship of 2'-halogenated anthracyclines [DNM = daunomycinone (9)]

of anthracyclines. All analogues having an equatorially oriented halogen in the 2'S configuration (right column in Figure 3) were inactive up to 50 mg/kg, whereas all analogues having an axially oriented halogen (2'R configuration) were active. Our findings were also confirmed later by other investigators showing that in the case of 2'-fluoro substituted anthracyclines, equatorial orientation (2'S) of fluorine at C-2' led to inactive analogues (52).

Among the active analogues we distinguished two groups of compounds having α-L-*manno* (37, 47, 48) and α-L-*talo* (35, 41) configurations. Analogues 37, 47, and 48 having a substituent at C-4' *trans* disposed to 2'R-halogen were more potent and more efficacious than analogues (35, 41) epimeric at C-4'. The effect of orientation of substituent at C-4' was far greater for 2'-halo anthracyclines than for those in the natural 2'-deoxy series, where the differences between DOX and its 4'-epimer (epirubicin) are relatively smaller.

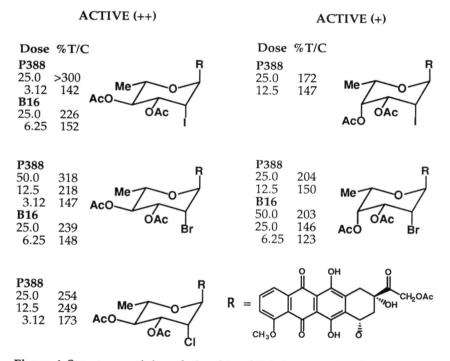

ACTIVE (++)   ACTIVE (+)

Dose	%T/C
P388	
25.0	>300
3.12	142
B16	
25.0	226
6.25	152

Dose	%T/C
P388	
25.0	172
12.5	147

Dose	%T/C
P388	
50.0	318
12.5	218
3.12	147
B16	
25.0	239
6.25	148

Dose	%T/C
P388	
25.0	204
12.5	150
B16	
50.0	203
25.0	146
6.25	123

Dose	%T/C
P388	
25.0	254
12.5	249
3.12	173

Figure 4. Structure-activity relationship of 2'-halogenated anthracyclines epimeric at C-4'.

The finding that orientation of substituent at both C-2' and C-4' is important for activity could influence our future design of new anthracyclines; therefore, a comparison was also made of DOX analogues (Figure 4). In DOX series, analogues having the α-L-*talo* configuration showed activity at doses lower than 50 mg/kg. However, they were less potent, and the T/C values for the α-*manno* series were consistently higher than for the α-*talo* series (Figure 4).

Water-Soluble Analogues of 2'-Halo Anthracyclines.   All deaminated analogues substituted with halogen at C-2' were highly insoluble in water. To overcome this problem we demonstrated that solubility can be achieved for this group of compounds by the introduction of esters containing water-solubilizing part of the 14-hydroxyl and that the ester bond at the 14 position does not affect efficacy (*40, 53*). Even in the extreme example, when sugar hydroxyl groups were still blocked, an acceptable water solubility was achieved.

58

A specific example is compound **58**, which was soluble as sodium salt and showed substantial antitumor activity and reduced toxicity . Tested against L1210 lymphoid leukemia it showed a T/C of 342 at 15 mg/kg and 166 against B16 melanocarcinoma. Reduced vesicant activity of **58** when compared with DOX was apparent after intradermal administration; 14 days after injection no animals injected with 0.2 mg of hemiadipate ester **58** showed lesions, whereas DOX at a dose of 0.1 mg caused lesions in all ten animals tested (*53*).

2'-Iodo-3'-deamino Anthracyclines Demethoxylated at the Aglycon at C-4. It was clear to us that any aglycon selected from aglycons previously shown to form active drug in combination with natural sugar should also give active drug when connected through a glycosidic bond with 2'-halo sugars in the α-*manno* or α-*talo* configuration. Interesting biological properties of idarubicin (4-demethoxydaunorubicin) and its increased potency led to the selection of 4-demethoxydaunomycinone to study synergistic effects of the aglycon and the 2'-halo sugar. To compare with the daunorubicin analogues **37** and **39** (**WP8**), we have prepared by an analogous method 2'-iodo-3'-deamino idarubicins **59** and **60**. In the initial in vivo evaluation, (P388) analogue **59**, which was very insoluble in water, showed T/C of 194 at 12.5 mg/kg and 167 at 6.25 mg/kg, whereas **60** displayed T/C of 232 at 6.25 mg/kg, 227 at 3.12 mg/kg, and 167 at 1.56 mg/kg. Comparison with the natural aglycon series indicates that demethoxylation significantly increases potency, although efficacy against P388 leukemia remains unchanged.

59 - R = Ac   61 - Annamycin
60 - R = H

A very interesting compound was produced when we used 4-demethoxyadriamycinone as the aglycon. Combination of this aglycone with 2'-iodo-α-L-*manno*-hexopyranose led to Annamycin (**61**), a compound non-cross-resistant or partially non-cross-resistant with DOX. Annamycin is currently in the last phase of preclinical development (*13, 41, 54-57*).

In Vitro Evaluation of Annamycin.   Free Annamycin (F-Ann) and liposomal Annamycin (L-Ann) were evaluated in vitro against a panel of sensitive and MDR cell lines (KB, KB-V-1, P388, P388/DOX, CEM, CEM/VBL, 8226, 8226/DOX). Annamycin and L-Ann displayed much lower RIs than DOX against all cell lines, indicating a good activity against MDR tumors. No difference in cytotoxicity was noticed between F-Ann and its liposomal formulation (*54*).

Uptake of Annamycin by sensitive P388 and resistant P388/DOX cells was higher than that of DOX, probably because of the higher lipophilicity of Annamycin. However, the efflux pattern of Annamycin, in contrast to that of DOX, was similar in both sensitive and resistant cells, suggesting that Annamycin's efflux was not mediated by P-gp (*57*). Annamycin uptake and efflux was also not affected by verapamil, thus further supporting the notion that efflux of Annamycin is not affected by the presence of P-gp. These results correspond well with in vitro evaluation showing that Annamycin is as cytotoxic as DOX against the P388 cell line but 50 to over 100 times more cytotoxic than DOX against the resistant P388/DOX line (*57*).

Annamycin was also a more potent inducer of single-strand DNA breaks, double-strand DNA breaks, and DNA-protein cross-links than DOX, both in sensitive and resistant cells. The level of DNA lesions in resistant cells caused by Annamycin was similar to or greater than those caused by DOX in sensitive cells (57). These results indicated that the higher activity against MDR cells of Annamycin might be caused by its increased accumulation and increased ability to induce DNA damage.

The search for other explanations of the higher activity of Annamycin than above that of DOX revealed Annamycin's ability to induce DNA degradation and programmed cell death (apoptosis) to a degree higher than DOX (56). DOX and Annamycin were effective in inducing DNA breakdown in P388 sensitive cells; however, in P388/DOX cells Annamycin caused a DNA cleavage effect, whereas DOX did not (56).

Annamycin Activity In Vivo.   Organ distribution and tumor uptake studies of F-Ann, L-Ann, and DOX in mice bearing advanced subcutaneous B16 melanoma tumors indicated that levels of L-Ann in lung and tumor were six and 10 times higher than those of DOX, respectively, whereas levels of F-Ann were five times higher than those of DOX. Levels of drugs in tumor were independent of tumor size. Interestingly, threefold higher levels of Annamycin were observed in plasma and brain (55).

F-Ann and L-Ann were evaluated in vivo in M5076 reticulosarcoma, Lewis lung carcinoma (LLC), and subcutaneous KB-3-1 and KB-V-1 (KB/R) human xenografts, and their activity was compared with that of DOX. Annamycin was clearly superior to DOX in the M5076 and LLC tumor models with survival rates ranging from 39% to 45% for F-Ann and from 63% to 85% for L-Ann, whereas DOX was practically inactive in these models (54). Consistently the liposomal preparations showed higher activity than the free drug.

These experiments well illustrate the potential of the "double-advantage approach" (13), a novel strategy to obtain effective chemotherapeutics, which consists of (a) the design and synthesis of a drug more active than the clinically used parent drug and (b) the use of an appropriate (compatible) drug carrier to enhance the analogue's effectiveness. The double-advantage approach was reviewed recently in detail and therefore will not be discussed here (13).

Interesting in vivo results that confirmed in vitro non-cross-resistant properties of Annamycin were gathered in KB-3-1 and KB-V-1 human xenograft experiments. In sensitive KB-3-1 carcinoma, all drugs were active and showed similar ranges of tumor growth inhibition (TGI), whereas in KB-V-1 MDR tumor, Annamycin and its liposomal preparations displayed significantly higher TGI than did DOX (54).

In summary, Annamycin proved to be non-cross-resistant or partially non-cross-resistant with DOX against MDR cell lines in vitro, and was more active than DOX against different tumor models in vivo, including MDR human xenografts. The pharmacology and organ distribution of Annamycin and its liposomal formulation differ significantly from that of DOX, and it is apparent that the biological properties and antitumor efficacy of Annamycin can be modulated by using different types of liposomes. Annamycin is expected to enter clinical studies in 1994.

2'-Halo Anthracyclines Containing an Amino Group at C-3'

Analogues halogenated at C-2' and containing an amino group at the C-3' position, even though the amino group was significantly less basic than in DOX or daunorubicin, formed ammonium salts easily and were water soluble. Iodine or bromine were introduced in a fashion similar to that in deaminated anthracyclines using glycals as substrates. In the initial synthesis all glycals were prepared from respective amino sugars by the elimination of acetic acid on silica gel (58) and later by using a novel one-pot method based on generation of glycosyl bromide in a mild condition and subsequent elimination reaction by N,N-diisopropylethylamine (59, 60).

2'-Iododaunorubicin and 4'-Epidaunorubicin.   Reaction of daunosaminal (62) with daunomycinone (9) in the presence of NIS gave only a trans addition product having α-L-*talo* configuration.   Subsequent deblocking and

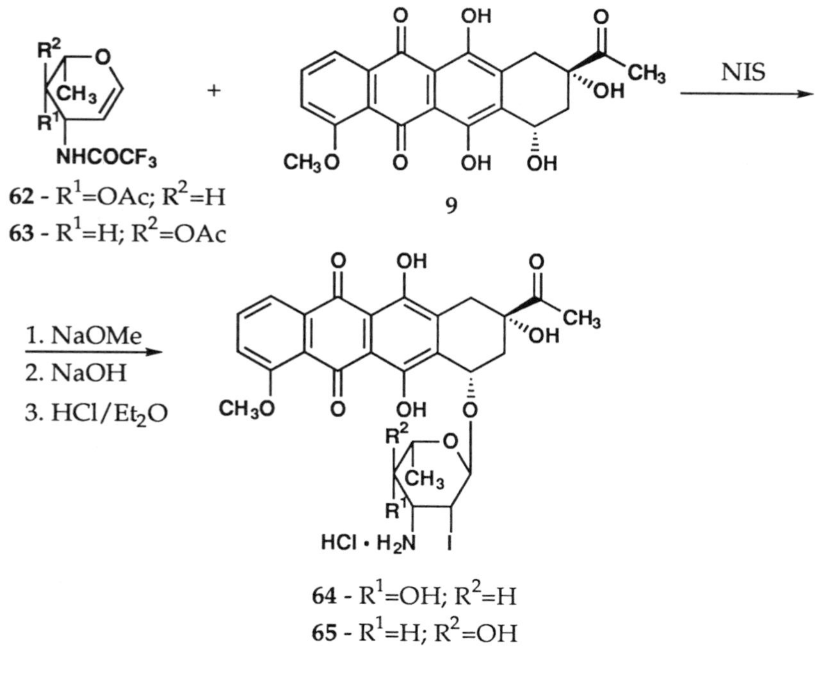

SCHEME 8

precipitation as hydrochloride gave in good yield the 2'-iododaunorubicin 64 (42). Similarly, 2'-iodo-4'-epidaunorubicin 65 was prepared starting from acosaminal 63. Only an α-*manno* isomer was isolated from the reaction mixture. Two-step deblocking and precipitation of final compound **65** as hydrochloride gave a red water-soluble solid (Scheme 8).

Analogue **65** having the α-L-*manno* configuration displayed better activity than analogue **64** having the α-L-*talo* configuration (Table III). In fact analogue **65** showed in vivo activity significantly higher than daunorubicin and comparable with DOX, whereas, α-L-talo analogue **64** showed activity similar to daunorubicin. The similar observation that α-L-talo analogues have activity against sensitive tumors similar to their parent 3'-amino drugs was made for 2'-fluoro daunorubicin (*61*) and carminomycin (*51*). This further confirms the general pattern of SAR established for 2'-halo-3'-hydroxy analogues (Figures 3 and 4), namely, that compounds having α-L-*manno* configuration have superior activity to α-L-*talo* analogues.

Table III. In Vivo Antitumor Activity of 2'-Iodo-3'-amino-daunorubicins 64 and 65 Against P388 Leukemia[a]

Drug	Dose (mg/kg)	%T/C
64	3.12	148
	6.25	153
	12.5	122
	25.0	80
	50.0	55
65	2.5	142
	5.0	175
	10.0	205
	20.0	65
Daunorubicin	6.25	158
	12.5	103
	25.0	45
DOX	10.0	205

[a]Treatment i.p. on day 1;   %T/C = median survival time of treated animals expressed as percent of control.

4'-Epidaunorubicins Brominated at C-2'.   The high activity displayed by analogue **65** indicated that the 2'-bromo analogue in an α-L-*manno* configuration is more promising than the analogue in an α-L-*talo* configuration. Therefore, our initial efforts were directed towards preparation of an α-L-*manno* analogue. Bromination of acosaminal **63** in dichloromethane gave a mixture of glycosyl bromides **66** and **67** epimeric at C-2 with a *manno* isomer as a minor product (ratio of *manno* to *gluco*, 3:7). In separate studies, we have found that iodine added to the bromination reaction alters the stereochemical outcome and that *manno* glycosyl bromide is formed as a main product (ratio of *manno* to *gluco*, 6:4) (*62, 63*). Furthermore, glycosyl bromide in the *gluco* configuration can be removed by crystallization from the reaction mixture. Therefore, coupling reactions can be performed on pure or highly enriched glycosyl bromides.

Table IV. In Vitro Cytotoxicity of 2'-Bromo-daunorubicin Analogues Against L1210 Leukemia and M-5076 Reticulosarcoma[a]

Drug	ID_{50} ($\mu g/ml$)	
	L1210	M5076
WP 401	0.41	0.39
WP 402	2.36	4.72
WP 400	> 25.0	-
DOX	0.52	1.82

[a]MTT assay; 10,000 cells/well; drug incubation: 72 h (L1210), 24 h (M5076); ID_{50} - 50% inhibition dose.

Coupling of *manno* glycosyl bromide **66** was performed in the presence of silver trifluoromethanesulfonate, and one main product having an α-L-*manno* configuration was formed. After two-step deblocking, the 2'-bromodaunorubicin **68** was isolated as the only product in a very good yield (*62, 63*). Coupling of *gluco* glycosyl bromide **67** gave a mixture of α and β anomers. Deacetylation with sodium methoxide and deblocking of the amino group with 1 M sodium hydroxide gave products having α-L-*manno* (**WP401**), α-L-*gluco* (**WP400**), and β-L-*gluco* (**WP402**) configurations (Scheme 9).

Cytotoxic assays (L1210 and M5076 cell lines) indicated that **WP401** was as cytotoxic or more cytotoxic than DOX, whereas **WP400** and **WP402** were significantly less active (Table IV). Again, as for 3'-deamino analogues (Figure 3), the 2'S configuration is responsible for the diminished activity of the **WP400** and **WP402** analogues. A similar deactivating effect of 2'S-oriented fluorine was observed for the daunorubicin analogue (*64, 65*).

Table V. In Vitro Cytotoxicity of WP401 and WP402 Against Sensitive and MDR Human Carcinoma (KB-3-1, KB-V-1) and Childhood Leukemia (CEM, CEM-VBL) Cells[a]

Drug	ID_{50} ($\mu g/ml$)		RI	ID_{50} (ng/ml)		RI
	KB-3-1	KB-V-1		CEM	CEM-VBL	
WP401	0.39	4.75	12	5.0	50	10
WP402	-	-	-	60.0	1,500	25
DOX (1)	4.0	>100	>25	4.5	1,600	356

[a] Resistance index (RI) = ID_{50} for resistant cells/ID_{50} for sensitive cells; MTT assay for KB cells; growth inhibition assay for 8226 and CEM cells.

WP401 and **WP402** were also tested against sensitive CEM and multidrug resistant CEM-VBL cell lines (Table V). **WP401** was significantly more active against the resistant cell line (RI 10), whereas against sensitive

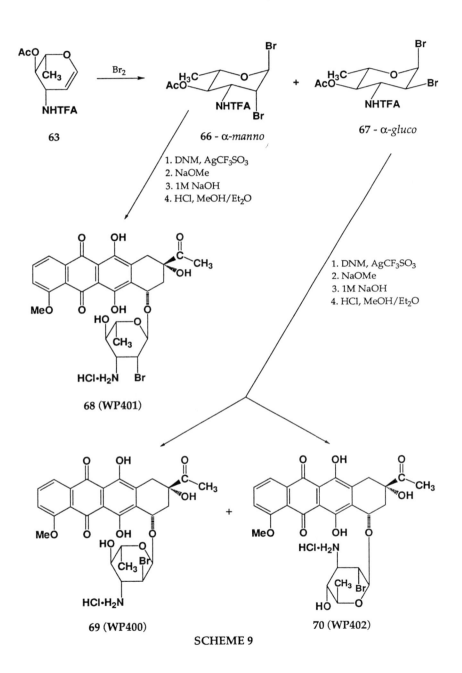

SCHEME 9

CEM cells **WP401** displayed the same potency that DOX did. Interestingly, less potent analogue **WP402** also displayed a significantly lower RI of 25 than DOX, which showed an RI of 356. **WP401** also showed good activity against KB-V-1 human carcinoma (RI 12) and was more then 20-fold more cytotoxic than DOX (Table V).

Anthracyclines Fluorinated at the D Ring

The interesting properties of 4-demethoxy analogues such as idarubicin focused our interest on the development of 4-demethoxy analogues fluorinated at the D ring. Analysis of potential methods for preparation of D-ring fluorinated aglycons indicated that the best method could be a Swenton approach. Collaborative efforts led to synthesis of a series of analogues fluorinated at the D ring, that is, 4-fluoro-daunorubicin (**WP110**), 1,4-difluoro-daunorubicin (**WP133**), and 1-fluoro-daunorubicin (not discussed here) (66-68).

WP110 - R = F
Idarubicin (**IDA**) - R = H

WP133

Table VI. In Vivo Antitumor Activity of 4-Fluoro-daunorubicin (WP110) and 1,4-Difluoro-daunorubicin (WP133) Against P388 Leukemia[a]

Dose (mg/kg)	%T/C					
	Exp. 1		Exp. 2		Exp. 3	
	WP110	IDA	WP110	IDA	WP133	IDA
0.125	-	-	116	134	-	121
0.25	-	-	142	147	-	136
0.5	165	172	153	171	-	162
1.0	102	100	89	89	162	84
2.0	70	85	-	-	182	-
4.0	55	-	-	-	182	-
8.5	-	-	-	-	113	-
17.0	-	-	-	-	89	-

[a]Treatment i.p. on day 1; %T/C = median survival time of treated animals expressed as percent of control.

Comparative studies of 4-fluoro-daunorubicin (**WP110**), 1,4-difluoro-daunorubicin (**WP133**), and idarubicin (**IDA**) in vivo against P388 leukemia indicated that the efficacy as well as potency of 4-fluorodaunorubicin (**WP110**) resembled those of idarubicin (Table VI). Although the potency of 1,4-difluoro-daunorubicin (**WP133**) was lower than that of the 4-fluoro analogue and idarubicin, the **WP133** was as efficacious as idarubicin and **WP110**. Evaluation of the effects of fluorination of the D ring on the biological and pharmacological properties of anthracyclines should be further investigated.

Conclusion

Analogues having an amino group at C-3' replaced with a hydroxyl group were shown to have interesting biological properties. One of the most important effects is non-cross-resistance with DOX. It is not quite clear at this stage of the studies if it is related to direct interaction with P-gp, or to the altered kinetics of passive diffusion through the membranes, or to both. It is also difficult to exclude other possibilities like altered subcellular distribution or interaction with a yet unknown target. Whatever the real explanation, it is clear now that the removal of the basic center alters interactions with biologically important macromolecules including DNA, whereas antitumor properties in vitro and in vivo are preserved.

An important structural factor in our studies is halogen at C-2'. The role of halogen at C-2' in 3'-amino anthracyclines is more apparent as the factor reducing basicity, and as such can be correlated with properties of 3'-deaminated analogues. The effects of 2'-substitution-induced increased stability of the glycosidic bond on properties of anthracyclines is not as apparent and need to be further studied. A strong deactivating effect of equatorial halogen at the 2'S configuration is probably steric in nature. This might offer a good opportunity to dissect steric and electronic effects by careful comparative mechanistic and pharmacological studies of analogues epimeric at C-2' and parent drugs.

Biological evaluations of 2'-halo-substituted anthracyclines indicate that 2'-halo substitution might be a very useful modification. From the perspective of the last 10 years it is a pleasure to observe that the major laboratories involved in synthesizing new anthracycline analogues have adopted our idea of 2'-halo substitution (*46-50*) and use in their work 2'-halo sugars identified by us as important for biological activity.

We have used a combination of four structural changes -- epimerization at C-4', replacement of an amino at C-3' group with hydroxyl, demethoxylation at C-4, and introduction of iodine at C-2' -- to obtain the leading analogue, Annamycin, which has shown promising properties in its preclinical evaluation. These chemical changes balanced well the high biological activity of Annamycin, which is superior to DOX, with the affinity of the drug for liposomes. Annamycin forms stabile liposomes, thus allowing us to take full advantage of this drug delivery system. Use of liposomes to deliver Annamycin has led to increased tumor targeting and reduced acute toxicity of Annamycin. Phase I clinical studies of liposomal Annamycin are scheduled in 1994.

Acknowledgements. This work was supported, in part, by National Institutes of Health grant CA 55320 and by Argus Pharmaceuticals, Inc.

Literature Cited

1. *Cancer. Principles and Practice of Oncology*; DeVita, J. V. T.; Hellman, S.; Rosenberg, S. A., Eds.; J. B. Lippincott Company: Philadelphia, 1993.
2. Bradley, G.; Juranka, P. F.; Ling, V. *Biochem. Biophys. Acta.* **1988**, *948*, 87.
3. *Mechanisms of Drug Resistance in Neoplastic Cells*; Woolley III, P. V.; Tew, K. D., Eds.; Academic Press, Inc.: San Diego, 1988.
4. Kessel, D., *Resistance to Antineoplastic Drugs*; CRC Press, Inc.: Boca Raton, Florida, 1989.
5. *Molecular and Cellular Biology of Multidrug Resistance in Tumor Cells*; Roninson, I. B., Eds.; Plenum Press: New York, 1991.
6. Endicott, J. A.; Ling, V. *Ann. Rev. Biochem.* **1989**, *58*, 137.
7. Pastan, I.; Gottesman, M. *N. Engl. J. Med.* **1987**, *316*, 1388.
8. Priebe, W.; Van, N. T.; Perez-Soler, R. *J. Cancer Res. Clin. Oncol.* **1990**, *116 (Suppl., Part I)*, 439.
9. Priebe, W.; Van, N. T.; Neamati, N.; Grynkiewicz, G.; Perez-Soler, R. *Third Internat. Symp. on Molecular Aspects of Chemother.*, Gdansk, Poland, **1991**.
10. Priebe, W.; Van, N. T.; Perez-Soler, R. *Proc. Am. Assoc. Cancer Res.*, 2245, **1991**.
11. Priebe, W.; Van, N. T.; Perez-Soler, R. *Proc. Am. Chem. Soc. Meet.*, MEDI-40, San Francisco, **1992**.
12. Priebe, W.; Van, N. T.; Burke, T. G.; Perez-Soler, R. *Anti-Cancer Drugs* **1993**, *4*, 37.
13. Priebe, W.; Perez-Soler, R. *Pharmacol. Ther.* **1993**, *60*, 215.
14. Umezawa, K.; Haresaku, M.; Muramatsu, M.; Matsushima, T. *Biomed. Pharmacother.* **1987**, *41*, 214.
15. El Khadem, H. S.; Swartz, D. L.; Cermak, R. C. *J. Med. Chem.* **1977**, *20*, 957.
16. Fuchs, E.-F.; Horton, D.; Weckerle, W. *Carbohydr. Res.* **1977**, *57*, C36.
17. El Khadem, H. S.; Swartz, D. L. *Carbohydr. Res.* **1978**, *65*, C1.
18. El Khadem, H. S.; Swartz, D. L. *J. Med. Chem.* **1981**, *24*, 112.
19. Fuchs, E.-F.; Horton, D.; Weckerle, W.; Winter-Mihaly, E. *J. Med. Chem.* **1979**, *22*, 406.
20. Horton, D.; Priebe, W.; Turner, W. R. *Carbohydr. Res.* **1981**, *94*, 11.
21. Horton, D.; Priebe, W.; Varela, O. *J. Antibiot.* **1984**, *37*, 853.
22. Horton, D.; Priebe, W.; Varela, O. *J. Antibiot.* **1984**, *37*, 1635.
23. Priebe, W.; Neamati, N. *Abstr. Pap. XIV IUPAC Internat. Carbohydr. Symp.* B-64, Stockholm, **1988**.

24. Priebe, W.; Neamati, N.; Perez-Soler, R. *J. Antibiot.* **1990**, *43*, 838.
25. Perez-Soler, R.; Priebe, W. *Cancer Chemother. Pharmacol.* **1992**, *30*, 267.
26. Solary, E.; Ling, Y.-H.; Perez-Soler, R.; Priebe, W.; Pommier, Y. *Int. J. Cancer*, **1994**, *58*, 85-94.
27. Chaires, J. B.; Priebe, W.; Graves, D. E.; Burke, T. G. *J. Am. Chem. Soc.* **1993**, *115*, 5360.
28. Friedman, R. A. G.; Manning, G. *Biopolymers* **1984**, *23*, 2671.
29. Lothstein, L.; Wright, H. M.; Sweatman, T. W.; Israel, M. *Oncol. Res.* **1992**, *4*, 343.
30. Lothstein, L.; Sweatman, T. W.; Docter, M. E.; Israel, M. *Cancer Res.* **1992**, *52*, 3409.
31. Skibicki, P.; Perez-Soler, R.; Burke, T. G.; Priebe, W. *Proc. Am. Chem. Soc. Meet.*, CARB-30, Denver, **1993**.
32. Lothstein, L.; Hosey, M.; Sweatman, T. W.; Koseki, Y.; Dockter, M.; Priebe, W. *Oncol. Res.* **1993**, *5*, 229.
33. Horton, D.; Priebe, W., In *Anthracycline Antibiotics*; El Khadem, H. S., Ed.; Academic Press, Inc.: New York, 1982, p. 197.
34. Priebe, W.; Zamojski, A. *Tetrahedron* **1980**, *36*, 287.
35. Priebe, W.; Grynkiewicz, G.; Neamati, N.; Perez-Soler, R. *Tetrahedron Lett.* **1991**, *32*, 3313.
36. Horton, D.; Priebe, W. *Proc. Am. Chem. Soc. Meet.*, CARB-22, San Francisco, **1980**.
37. Horton, D.; Priebe, W. *Proc. Am. Chem. Soc. Meet.*, CARB-11, New York, **1981**.
38. Horton, D.; Priebe, W.; Varela, O. *Abstr. Pap. XIth Internat. Carbohydr. Symp.* Vancouver, **1982**.
39. Naff, M. B.; Plowman, J.; Narayanan, V. L., In *Anthracycline Antibiotics*; El Khadem, H. S., Ed.; Academic Press, Inc: New York, 1982, p. 1.
40. Priebe, W.; Horton, D.; Wolgemuth, R. L. *Eur. Pat.* **1984**, *EP116,222*, August 22.
41. Horton, D.; Priebe, W. *U.S. Patent.* **1984**, *4,427,664*, January 24.
42. Horton, D.; Priebe, W. *U.S. Patent.* **1985**, *4,562,177*, December 31.
43. Horton, D.; Priebe, W. *U.S. Patent.* **1985**, *4,537,882*, August 27.
44. Horton, D.; Priebe, W. *Carbohydr. Res.* **1985**, *136*, 391.
45. Horton, D.; Priebe, W.; Varela, O. *Carbohydr. Res.* **1985**, *144*, 305.
46. Tsuchiya, T.; Takagi, Y.; Ok, K. D.; Umezawa, S.; Takeuchi, T.; Wako, N.; Umezawa, H. *J. Antibiot.* **1986**, *39*, 731.
47. Ok, K. D.; Takagi, Y.; Tsuchiya, T.; Umezawa, S.; Umezawa, H. *Carbohydr. Res.* **1987**, *169*, 69.
48. Tsuchiya, T.; Takagi, Y.; Umezawa, S.; Takeuchi, T.; Komuro, K.; Chisato, N.; Umezawa, H.; Fukatsu, S.; Yoneta, T. *J. Antibiot.* **1988**, *41*, 988.
49. Gerken, M.; Blank, S.; Kolar, C.; Hermentin, P. *J. Carbohydr. Chem.* **1989**, *8*, 247.
50. Florent, J. C.; Genit, A.; Monneret, C. *J. Antibiot.* **1989**, *12*, 1823.

51.   Baer, H. H.; Mateo, F. H. *Can. J. Chem.* **1990**, *68*, 2055.
52.   Umezawa, K.; Tsuchiya, T.; Takagi, Y.; Sohtome, H.; Chang, M. S.; Kobayashi, N.; Tukuoka, Y.; Takeuchi, T. *Abstr. Pap. XV Internat. Carbohydr. Symp.*, D037, **1990**.
53.   Horton, D.; Priebe, W.; Wolgemuth, R. L. *U.S. Patent.* **1988**, *4,772,688*, September 20.
54.   Zou, Y.; Ling, Y. H.; Van, N. T.; Priebe, W.; Perez-Soler, R. *Cancer Res.* **1994**, *54*, 1479.
55.   Zou, Y.; Priebe, W.; Ling, Y. H.; Perez-Soler, R. *Cancer Chemother. Pharmacol.* **1993**, *32*, 190.
56.   Ling, Y. H.; Priebe, W.; Perez-Soler, R. *Cancer Res.* **1993**, *53*, 1845.
57.   Ling, Y. H.; Priebe, W.; Yang, L. Y.; Burke, T. G.; Pommier, Y.; Perez-Soler, R. *Cancer Res.* **1993**, *53*, 1583.
58.   Horton, D.; Priebe, W.; Sznaidman, M. *Carbohydr. Res.* **1989**, *187*, 145.
59.   Priebe, W.; Grynkiewicz, G.; Krawczyk, M.; Fokt, I. *Abstr. Pap. XVII Internat. Carbohydr. Symp.* Ottawa, **1994**, B2.55.
60.   Priebe, W.; Grynkiewicz, G.; Krawczyk, M.; Fokt, I. *Proc. Am. Chem. Soc. Meet.* 207, CARB-78, **1994.**
61.   Takagi, Y.; Park, H.; Tsuchiya, T.; Umezawa, S.; Takeuchi, T.; Komuro, K.; Nosaka, C. *J. Antibiot.* **1989**, *42*, 1315.
62.   Priebe, W.; Neamati, N.; Grynkiewicz, G.; Perez-Soler, R. *Proc. Am. Chem. Soc. Meet.,* CARB-11, Atlanta, **1991**.
63.   Priebe, W.; Neamati, N.; Grynkiewicz, G.; Van, N. T.; Burke, T. G.; Perez-Soler, R. *Proc. Am. Assoc. Cancer Res.* 3332, **1992**.
64.   Castillon, S.; Dessinges, A.; Faghih, R.; Lukacs, G.; Olesker, A.; Ton, T. T. *J. Org. Chem.* **1985**, *50*, 4913.
65.   Baer, H. H.; Siemsen, L. *Can. J. Chem.* **1988**, *66*, 187.
66.   Morrow, G. W.; Swenton, J.; Filppi, J. A.; Wolgemuth, R. L. *J. Org. Chem.* **1987**, *52*, 713.
67.   Swenton, J.; Horton, D.; Priebe, W.; Morrow, G. W. *U.S. Patent.* **1987**, *4,663,445*, May 5.
68.   Swenton, J.; Morrow, G. W.; Priebe, W. *U.S. Patent.* **1987**, *4,697,005*,

RECEIVED August 16, 1994

Chapter 3

Fluorinated Anthracyclinones and Their Glycosylated Products

Antonio Guidi, Franca Canfarini, Alessandro Giolitti, Franco Pasqui, Vittorio Pestellini, and Federico Arcamone

Laboratori di Ricerca Chimica, A. Menarini Industrie Farmaceutiche Riunite S.r.l., via Sette Santi 3, 50131, Florence, Italy

Fluorine substitution on the aglycone moiety of antitumor anthracyclines has been object of study by different groups in the last ten years. Doxorubicin derivatives bearing one or two fluorine atoms onto the ring D were obtained by total synthesis, side chain and ring A analogues were synthesized either by total synthesis or following a semisynthetic approach. Compounds substituted at C-8 onto the saturated ring of idarubicin and daunorubicin have been designed with the aim at improving the affinity for guanine residues, typical of clinically useful anthracyclines.

The general interest of medicinal chemists in organofluorine compounds arose in the 1950s when the synthesis of 9α-fluorohydrocortisone acetate, the first fluorine-containing drug, was published and its improved glucocorticoid activity, with respect to cortisol acetate, as well as that of the other 9α-halogen-substituted derivatives was demonstrated (1). Since then, the role of fluoroorganic compounds in the biological and related sciences has increased enormously and, in the last years, we have observed a real boom in this area. This was of such importance that even a new term (*flustrates*: fluorine-containing substrates) was coined to indicate fluorine-containing compounds having a role in some branch of biological chemistry (2). In fact, fluorine substitution can modify the peculiar features of organic compounds in several ways: in a molecule, the replacement of a hydrogen atom with a fluorine can strongly affect the properties of a neighboring functional group, inhibit oxidative metabolism, induce the stabilization of particular conformers, and increase the lipophilicity without altering its steric bulk.

The anthracycline antibiotics, the most important one being doxorubicin, are well known chemotherapeutic agents against human cancers. The discovery and structure elucidation of doxorubicin was reported in 1969 (3) and its pharmacological and curative properties against a number of human cancers were soon recognized as a major improvement over already known anti cancer compounds. However, although doxorubicin is still today considered the antineoplastic drug possessing the broadest spectrum of antitumor activity, several undesirable side effects, the most serious of which is the cumulative dose-limiting cardiotoxicity, and the appearance of resistance phenomena (not to say the lack of

0097–6156/95/0574–0047$08.00/0

activity in important tumor types) severely limit its clinical use. Since the discovery of doxorubicin, the chemists have probably synthesized more than 2000 analogues, in order to enlarge the spectrum of antitumor activity and to overcome the pharmaceutical and toxicological limitations (4). The title of the paper just cited, "The Anthracyclines: Will We Ever Find a Better Doxorubicin?" witnesses the scepticism that is starting to pervade the relevant scientific quarters with respect to the likelihood of further progress in this specific area.

Here, an account of the synthesis of anthracycline analogues carrying one or more fluorine atoms on the aglycone moiety will be given. We have been interested in this type of substitution as a means to obtain new daunorubicin (**1a**) and doxorubicin (**1b**) analogues and possibly to give a positive answer to Weiss' question cited above. On the other hand, the introduction of a fluorine onto the tetracyclic system bearing the quinone functionality of the anthracyclines of the daunorubicin-doxorubicin type has also been pursued by others. So, in this report we shall mention the synthetic work done by Swenton et al. at Ohio State University, by Terashima et al. at Sagami Chemical Research Center, by Russel et al. at the University College, U.N.S.W. and at the National Australian University, and describe the results of recent studies performed in our laboratory.

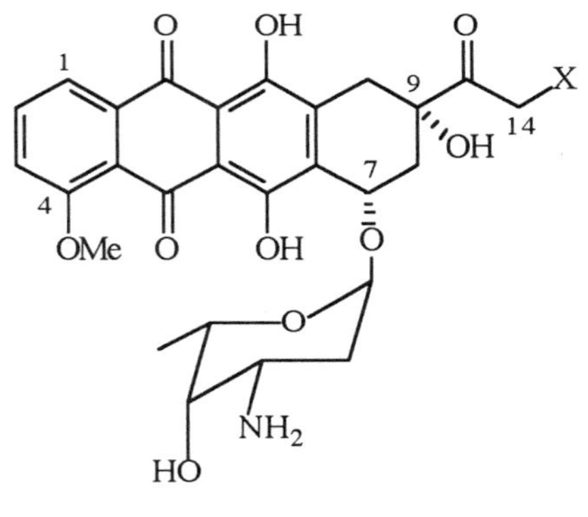

1a: X = H

1b: X = OH

Ring-D and Side Chain Fluorinated Analogues.

Aside from the motivation for evaluating the effect of the substitution of an hydrogen atom on ring D of the antitumor anthracyclines with fluorine, aromatic nucleophilic substitution of fluorine in the presence of electron withdrawing groups was considered as a property potentially inducing biochemical consequences by Morrow et al. (5-7). Starting from the appropriate cyanofluorophthalides, Morrow et al. prepared 1-fluoro, 4-fluoro, and 1,4-difluoro-4-demethoxydaunorubicinones using a regioselective base-promoted annelation reaction of the said precursors to a quinone monoketal previously developed in Swenton's laboratory for the synthesis of daunomycinone. 4-Demethoxydaunorubicin (idarubicin) derivatives **2a-2c** were eventually obtained and tested for bioactivity. It was also shown that the fluoro aglycones reacted readily with a nucleophile such as 2-/(2-aminoethyl) amino/ethanol to give the expected aromatic substitution and to introduce the mitoxantrone side chain onto the anthracyclinone chromophore.

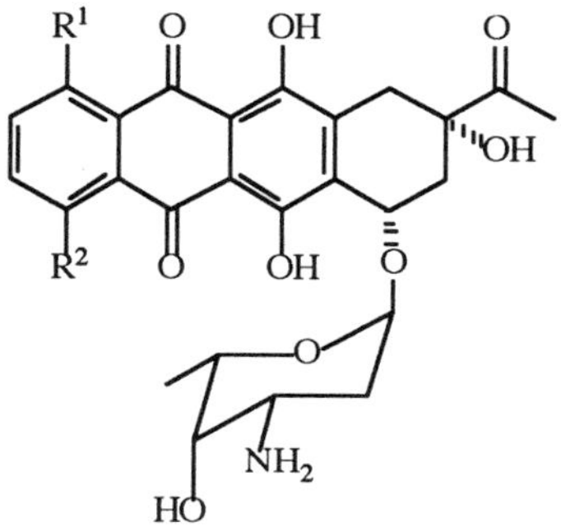

2a: $R^1 = F$, $R^2 = H$

2b: $R^2 = F$, $R^1 = H$

2c: $R^1 = R^2 = F$

The required fluorocyanophtalides **3a-c** were obtained by a metalation/functionalization route from the appropriate commercially available fluoro-substituted aromatic compounds. In the annelation reaction, compound **4**, in the racemic form, was used as an AB building block to give, in three steps, the aglycones **5a**, **5b**, and **5c**, in 70%, 41%, and 52% overall yields, respectively. In addition, compounds **5b** and **5c** were also prepared in enantiomeric pure form, starting from optically active **4**, which was available on a multigram scale (*8*).

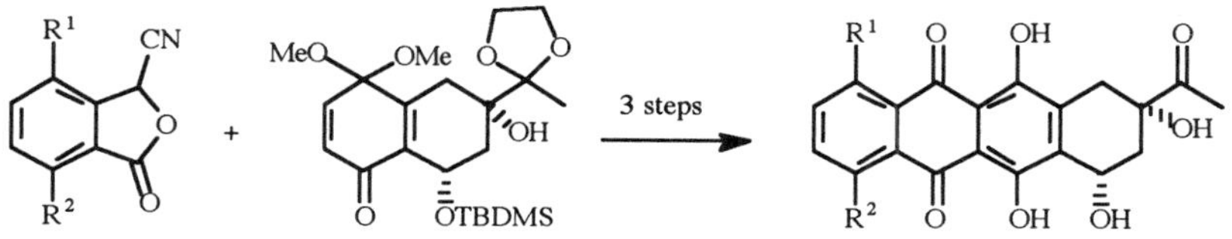

3a: $R^1 = F$, $R^2 = H$ **4**   **5a**: $R^1 = F$, $R^2 = H$

3b: $R^1 = H$, $R^2 = F$ **5b**: $R^1 = H$, $R^2 = F$

3c: $R^1 = R^2 = F$ **5c**: $R^1 = R^2 = F$

As part of a program to employ heteronuclear NMR to probe the interaction of DNA with intercalating agents, Irvine et al. (9) have prepared 2-fluoroidarubicin **9a** and 3-fluoroidarubicin **9b**. The cyanophtalides **6a** and **6b**, obtained via the fluorinating deamination of the corresponding aminophtalates, were annelated to compound **7**, available only as a racemic product, to give, in two steps, racemic aglycones **8a** and **8b**. Glycosidation of the aglycones followed by separation of the diastereomeric products and deprotection afforded the target compounds.

6a: $R^1 = F, R^2 = H$ **7**   **8a**: $R^1 = F, R^2 = H, R^3 = H$

6b: $R^1 = H, R^2 = F$ **8b**: $R^1 = H, R^2 = F, R^3 = H$

                                                                          9a: $R^1 = F, R^2 = H, R^3 = $ daunosaminyl

                                                                          9b: $R^1 = H, R^2 = F, R^3 = $ daunosaminyl

Terashima and coworkers (10) started by taking into account the notable difference in the pharmacological behaviors of daunorubicin and doxorubicin, and hypothesized that analogs containing fluorine atoms at C-14 could be of great interest in terms of structure-activity relationships. In 1988, Terashima and al. published the synthesis of different 14-fluoroanthracyclines, namely 14-fluoroidarubicin (**10a**) and 14-fluorodaunorubicin (**10b**), together with four glycosidic analogs, namely the 2-deoxy-β-D-ribopyranosides (**11a** and **11b** respectively) and the 2-deoxy-α-L-fucopyranosides (**12a** and **12b** respectively) of 14-fluoroidarubicinone (14-fluoro-4-demethoxydaunomycinone) and of 14-fluorodaunomycinone. The 14-fluoroaglycones were synthesized starting from 7-deoxyidarubicinone and 7-deoxydaunorubicinone, respectively. After bromination of the acetyl chain, the bromine atom was exchanged with a fluorine one, using 6 equivs. of tetrabutylammonium fluoride and 3 equivs. of *p*-toluenesulfonic acid at room temperature in anhydrous tetrahydrofurane (any attempt to adapt this fluorination procedure to 4-demethoxy-14-bromoidarubicinone or 14-bromodaunorubicin failed). The hydroxylation at C-7 was then carried out, essentially by the classical benzylic bromination route. Glycosidation of the aglycones with the appropriate sugar reagent was performed following a method based on the use of trimethylsilyltriflate as condensing agent, under strictly controlled reaction conditions (11).

In a more recent paper, Terashima and coworkers reported the synthesis of difluoro derivative **13a**. The key step in the synthesis of compound **13a** (14,14-difluoroidarubicin) was the Reformatsky reaction of ethyl bromodifluoroacetate with aldehyde **13b**. The following reoxidation step required the Dess-Martin periodinane reagent, since the use of other oxidizing systems resulted in complete recovery of starting material (12). Finally, the patent literature reports the synthesis of a trifluoroderivative of structure **13c**. This compound was obtained by allowing carboxylic acid **13d**, previously activated with carbodiimidazole, to react with 3,3,3-trifluoropropyl-magnesium bromide, in a THF-HMPA mixture, at low temperature and in the presence of trimethylsilyltriflate (13)

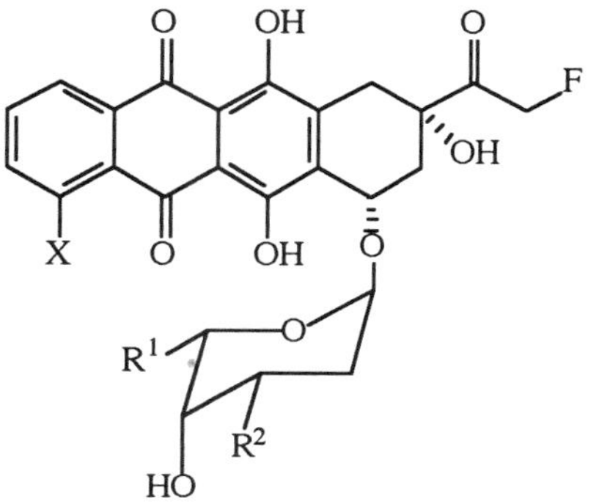

10a: X = H, R^1 = Me, R^2 = NH$_2$

10b: X = OMe, R^1 = Me, R^2 = NH$_2$

11a: X = H, R^1 = H, R^2 = OH

11b: X = OMe, R^1 = H, R^2 = OH

12a: X = H, R^1 = Me, R^2 = OH

12b: X = OMe, R^1 = Me, R^2 = OH

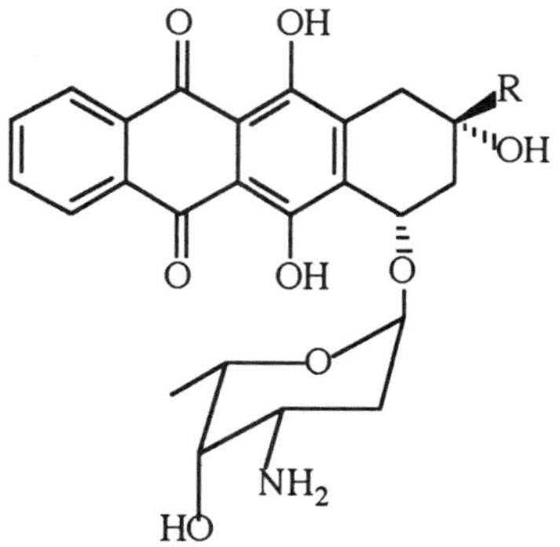

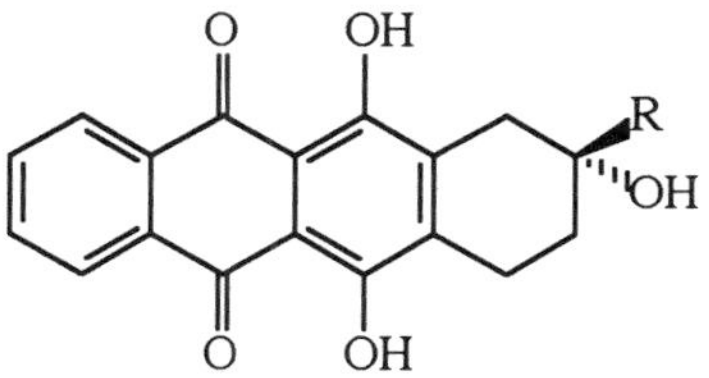

13a: R = COCHF$_2$

13c: R = CO(CH$_2$)$_2$CF$_3$

13b: R = CHO

13d: R = COOH

Ring-A Fluorinated Anthracyclines

Our own work deals with the synthesis of anthracyclines in which one of the hydrogen atoms at position 8 on ring A is substituted with a fluorine. This work is part of an ongoing program carried out with the collaboration of Bristol Myers-Squibb S.p.A. (Rome, Italy) and aimed at the preparation of compounds endowed with a higher affinity or improved selectivity of binding to double-stranded DNA. In fact, although the ultimate molecular event responsible for the antiproliferative effects of the antitumor anthracyclines is still not completely understood, cell DNA is the currently accepted target, and other hypotheses involving the formation of radical species or effects at the membrane levels can be considered unlikely because of structure-activity considerations and the absence of pharmacological evidence (*14*). When clinically useful anthracyclines were added to the culture medium of K542 cells (a human leukemia cell line) a reduction to 50% in the doubling number of the cells in exponential phase was observed at the same intranuclear DNA-bound drug concentrations independently of cellular phenotype, whether sensitive or resistant, and also independently of the C-4' substitution. The intranuclear concentrations were determined according to a non destructive quantitative microspectrofluorometric technique (*15*). It appears therefore that intranuclear DNA-bound drug accumulation is responsible for the cytotoxicity within this class of drugs. In addition, in recent years considerable attention has been paid to an interference with the topoisomerase II reaction (*16*); in fact, the extent of DNA breaks, due to the interference of anthracyclines with the DNA-topoisomerase II reaction, together with the persistence of the same, are well correlated with the cytotoxic effects of **1b** and some strictly related analogs (*17*).

Finally, these conclusions are in agreement with the relationship between affinity for double helical B-DNA and the level of optimal therapeutic doses in tumor bearing mice of different anthracyclines. The compounds that have reached the clinical stage after careful pharmacological and toxicological selection belong to the group showing the highest affinity (*18*). However, the value of the DNA binding constant is not always related to cytotoxicity. In fact a high correlation is found between the cytotoxic activity of different anthracycline glycosides and lipophilicity and binding affinity to DNA if both parameters are taken into account simultaneously (*19*). The case of 9-deoxydoxorubicin (**14a**), a compound with a high binding constant, is an exception because the pharmacological data indicate that this compound is distinctly less active than **1b** (*20*). It appears therefore that the presence of the 9-OH determines a peculiar type of interaction important for the exhibition of bioactivity. The importance of a hydrogen bond donating substituent at C-9 is shown also by the lack of antitumor activity of 9-deoxy-9-methylidarubicin (**14b**) and of 9,10-anhydrodaunorubicin (**15**) (*20*).

As for the mode of DNA binding, X-ray diffraction studies (*21, 22*) allowed the definition of structural features and bonding interactions of the complexes of daunorubicin with the hexanucleotide duplexes d(CGTACG) and d(CGATCG) showing, inter alia, the important hydrogen bond between the 9-hydroxyl and a guanine residue at the intercalation site (Figure 1). In fact, the crystal structures indicate that two drug molecules are intercalated in the d(CpG) sequences at either end of the right-handed double helix. The conformation of the drug molecule in the complex is somewhat modified as compared with that present in the crystals of the free drug. Ring A appears in a different conformation, and also the torsion angles around the glycosidic linkage are varied by 20°. The distance between O-9 and N-3 of the guanine at the intercalation site indicates a strong hydrogen bond, while an additional hydrogen bond is deduced for the amino group at N-2 in the said guanine and the O-9. On the other side of the aglycone ring, the C-9 side chain carbonyl oxygen is bound to a water molecule bridging through another hydrogen bond towards the O-2 on the cytosine above the intercalator.

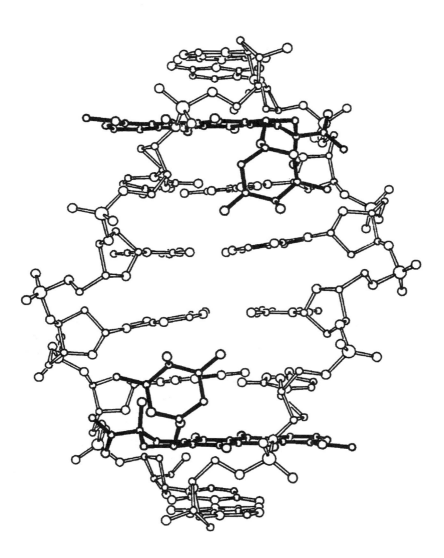

Figure 1. Graphic showing the intercalation complex between d(CGTACG) and daunorubicin as deduced from X-ray analysis. By courtesy of Dr S. Ughetto, CNR laboratory, Montelibretti, Rome.

14a: R^1 = OMe, R^2 = H, R^3 = OH   **15**

14b: R^1 = H, R^2 = Me, R^3 = H

Taken as a whole, these observations point out the importance of ring A substitution for the geometry of the DNA complex at this specific site. Other interactions involving hydrogen bonds are those displayed by the charged amino group on the sugar moiety to the adenine, through a molecule of water, and by the C-4 and C-5 oxygens on the other groove of the helix. The interaction of different anthracyclines, including doxorubicin, and 9-deoxydoxorubicin with the d(CGTACG) duplex, identical to that used in the X-ray diffraction study of Wang et al. (*21*) and with the d(CGCGCG) duplex was studied taking into account the different fluorescence yields of anthracyclines intercalated at the CpG site (complete quenching) with respect to ApT sites (large residual fluorescence). The analysis indicated preferential intercalation of the first two compounds at CpG site in d(CGTACG), as compared with the same sites in d(CGCGCG), whereas the 9-deoxy derivative showed a lower affinity in both cases. These experimental results are in agreement with theoretical computations (*23, 24*). It appeared of interest, therefore, to synthesize analogs of the clinically useful anthracyclines carrying a fluorine atom adjacent to the 9 position. This was done, and hereinafter the synthesis of two of these derivatives containing a fluorine atom at position 8 is outlined.

The first compound belongs to the 4-demethoxy (idarubicin) series. Idarubicin itself is a powerful antileukemic anthracycline recently introduced into therapeutic use worldwide (*4*). The fluorine atom was introduced onto the idarubicin aglycone, idarubicinone, starting from epoxyketone **16a** (as a racemic compound). Similarly, the corresponding derivative in the natural (daunorubicin) series was obtained starting from **16b** (in the optically active form). The two compounds were therefore chosen as intermediates for the synthesis of 8-(S)-fluoro-anthracyclines **17a** and **17b**.

Compound **16a** was synthesized in four steps (Figure 2), at a 52% overall yield, from the readily available 1,4-dimethoxy-2,3-bis-(bromomethyl)-anthraquinone, following an adaptation of Cava's route to anthracyclinones (*25*). The introduction of fluorine was then accomplished by means of the Olah's reagent and the desired fluorohydrin (Figure 3) was obtained in 60% yield (*26*). Conversely,

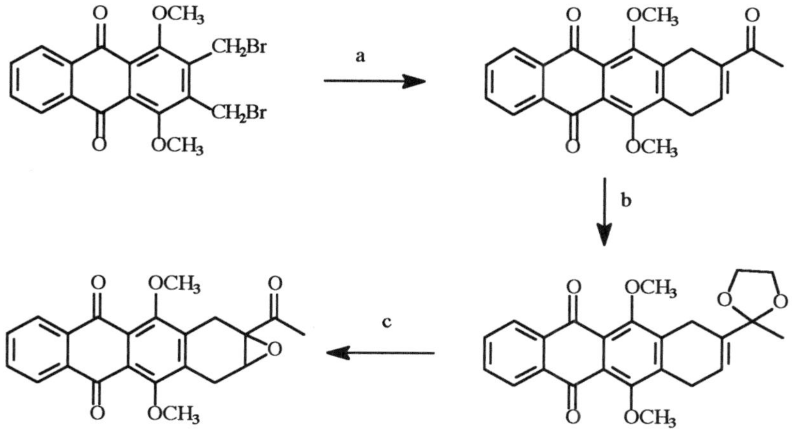

Figure 2. Synthesis of epoxide **16a.** (a) 3-butyn-2-one, NaI, DMA, 60°C, 80%; (b) collidinium tosylate cat., ethylene glycol, reflux.; (c) mCPBA, CHCl3, r.t., then TFA, total yield of 65% for steps b and c.

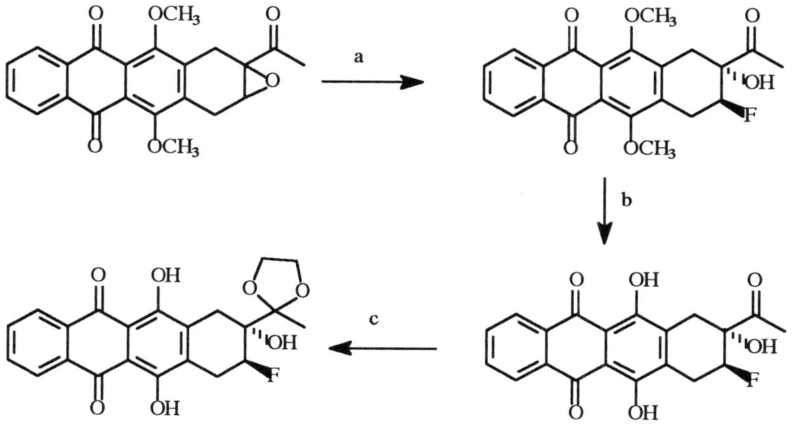

Figure 3. Synthesis of 7-deoxy-8-fluoroidarubicinone protected at C-13 as ethylene ketal. (a) HF/py 70%, 1 h, r.t., 60%; (b) BCl3, -60°C; (c) TsOH cat., ethylene glycol, reflux, total yield of 85% for steps b and c.

the known C-13 epimeric mixture of epoxides **18** (*25*) was used as starting material for the preparation of compound **16b** (75% overall yield in four steps). In the case of **16b**, anhydrous hydrogen fluoride at low temperature was used for the epoxide opening reaction and the yield was nearly quantitative.

16a: R^1 = H, R^2 = Me

16b: R^1 = OMe, R^2 = H

17a: R^1 = H, R^3 = COCF$_3$

17b: R^1 = OMe, R^3 = H

The following hydroxylation at position 7 by the commonly used benzylic bromination route manifested regioselectivity problems for both compounds and resulted in lower yields as compared with those usually obtained with the nonfluorinated compounds. Finally, racemic 8-fluoroidarubicinone, and 8-(S)-fluorodaunorubicinone were coupled with daunosamine, following Terashima's method (*11*), to give fluoroanthracyclines **17a** and **17b**, after the necessary deprotection steps and chromatographic separations. On the other hand, an alternative short synthesis of 8-(S)-fluorodaunorubicinone was also developed. According to this procedure, the known C-13 epimeric mix ture of alcohols **19**, deriving from the regio- and stereoselective methanolisis of compound **18** (*27*) gave a fluorodehydroxylation reaction with retention of configuration when treated with DAST in anhydrous THF. The desired aglycone 8-(S)-fluorodaunomycinone, was obtained from **20** after only a further, one-pot, cumulative deprotection step (*28*).

Conclusions

Fluorine substitution in the tetracyclic aglycones of antitumor anthracyclines has been pursued in the last 10 years by different groups with different motivations. These studies have contributed to a better knowledge of anthracycline synthesis and chemical behavior. It appears that the preparation of ring A fluoro derivatives can be carried out both by total synthesis and by transformation of the natural compounds. This work, therefore, opens the way to a novel class of anthracycline aminoglycosides designed for an improvement of the GC base-pair binding affinity. As for the pharmacological aspects, the compounds whose bioactivity is mentioned in the literature cited as well as 8-(S)-fluoroidarubicin, obtained upon detrifluoroacetylation of **17a** (*29*), exhibit cytotoxic effects in cell cultures and antitumor activity against human tumors in immunodepressed mice that are in the range of those shown by the parent, nonfluorinated, anthracyclines. Published data are not adequate for the comparative evaluation of potential antitumor efficacy of different compounds. Complete biological evaluation of the 8-(S)-fluoroderivatives is still under way and will be reported in due time.

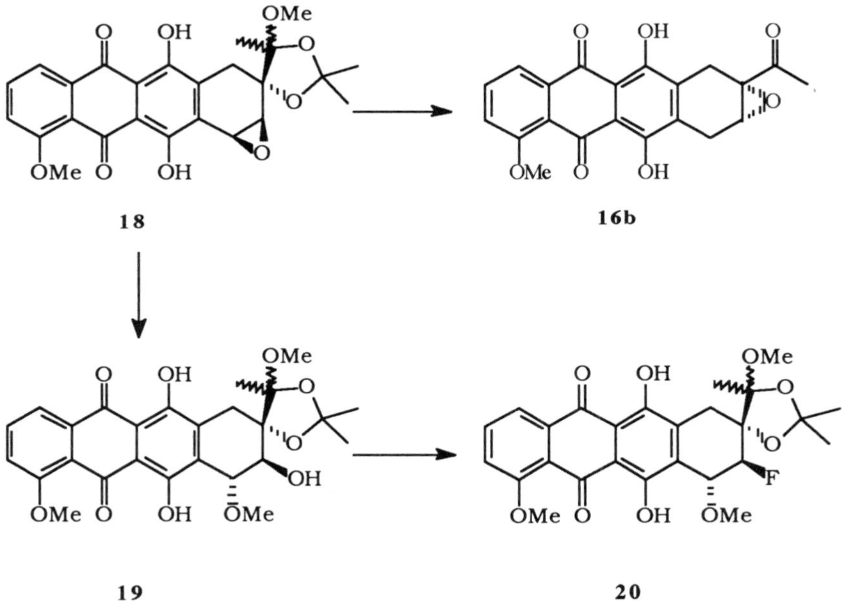

18 **16b**

19 **20**

Literature Cited

1. Fried, J. H.; Sabo, E. F. *J. Am. Chem. Soc.* **1954**, *76*, 1455.
2. Seebach, D., *Angew. Chem. Int. Ed. Engl.* **1990**, *29*, 1320. And reference therein quoted.
3. Arcamone, F., Franceschi, G.; Penco, S; Selva, A. *Tetrahedron Lett.* **1969**, 1007.
4. Weiss, R. B. *Semin. Oncol.* **1992**, 19, 670.
5. Morrow, G. W.; Swenton, J. S.; Filppi, J. A; Wolgemuth; R. L. *J. Org. Chem.* **1987**, *52*, 713.
6. Swenton, J.; Morrow, G. W.; Priebe, W. U. S. Patent 4,697,005 (29 Sept 1987).
7. Swenton, J.; Horton, D.; Priebe, W.; Morrow, G. W. U. S. Patent 4,663,445 (5 May 1987).
8. Swenton, J. S.; Freskos, J. N.; Morrow, G. W.; Sercel, A. D. *Tetrahedron* **1984**, *40*, 4625.

9. Irvine, R. W.; Kinloch, S. A.; McCormick, A. S.; Russell, R. A.; Warrener, R. N. *Tetrahedron* **1988**, *44*, 4591.
10. Matsumoto, T.; Ohsaki, M.; Yamada, K.; Matsuda, F.; Terashima, S. *Chem. Pharm. Bull.* **1988**, *36*, 3793.
11. Kimura, Y.; Suzuki, M.; Matsumoto, T.; Abe, R.; Terashima, S. *Bull. Chem. Soc. Jpn.* **1986**, *59*, 423.
12. Matsuda, F.; Matsumoto, T.; Ohsaki, M.; Terashima, S. *Bull. Chem. Soc. Jpn.* **1991**, *64*, 2983.
13. Terajima, A.; Suzuki, M.; Matsuda, F.; Matsumoto, T.; Osaki, M.; Yamada, K. *Jpn Tokkyo Koho*, JP 63,156,742 (29 Jun 1988); C. A. **1989**, 110, 23632.
14. *Antitumor Natural Products, Basic and Clinical Research, Gann Monograph on Cancer Research No. 36* Takeuchi,T.; Nitta, K.; Tanaka, N. Eds.: Japan Sci. Soc. Press, Tokyo and Taylor & Francis Ltd. London and Bristol, **1989**, 81.
15. Gigli, M.; Rasonaivo, T. W. D.; Millot, J.-M.; Jeannesson, P.; Rizzo, V.; Jardillier J.-C.; Arcamone; F. and Manfait, M. *Cancer Res.* **1989**, *49*, 560-564.
16. Glisson, B. S. and Ross, W. E. *Pharmacol. Ther.* **1987**, *32*, 89.
17. Capranico, G.; De Isabella, P., Penco, S., Tinelli, S., and Zunino, F. *Cancer Res.* **1989**, *49*, 2022.
18. Valentini, L.; Nicolella, V.; Vannini, E.; Menozzi, M.; Penco, S. and Arcamone, F. *Il Farmaco Ed. Sci.* **1985**, *40*, 377.
19. Hoffman, D.; Berscheid, H. G.; Boettger, D.; Hermentin, P.; Sedlacek, H. H. and Kraemer, H. P. *J. Med. Chem.* **1990**, *33*, 166.
20. Arcamone, F. *Doxorubicin Anticancer Antibiotics*, Academic Press, New York, 1981.
21. Wang, A. H.-J.; Ughetto G.; Quigley G. J.; Rich, A. *Biochemistry* **1987**. *26*, 1152.
22. Moore, M. H.; Hunter, W. N.; Langlois d'Estainot, B.; Kennard, O. *J. Mol. Biol.* **1989**, *206*, 693.
23. Rizzo, V.; Battistini, C.; Vigevani, A.; Sacchi, N.; Razzano, G.; Arcamone, F.; Garbesi, A.; Colonna, F. P.; Capobianco, M.;Tondelli, L. *J. Mol. Recognit.* **1989**, 2, 132.
24. Gresh, N.; Pullman, B.; Arcamone, F.; Menozzi, M.; Tonani, R. *Mol. Pharmacol.* **1989**, *35*, 251.
25. Kerdesky, F. A. J.; Cava, M. P. *J. Am. Chem. Soc.* **1981**, *103*, 1992.
26. Giolitti, A.; Guidi, A.; Pasqui, F.; Pestellini, V.; Arcamone, F. M. *Tetrahedron Lett.* **1992**, *33*, 1637.
27. Penco, S.; Angelucci, F.; Ballabio, M.; Vigevani, A.; Arcamone, F. *Tetrahedron Lett.* **1980**, *21*, 2253.
28. Canfarini, F; Guidi, A.; Pasqui, F.; Pestellini, V.; Arcamone, F. M. *Tetrahedron Lett.* **1993**, *34*, 4697.
29. F. Animati, F. Arcamone, P. Lombardi, M. Bigioni, G. Pratesi, and F. Zunino. *Proc. Am. Assoc. Cancer Res.* **1993**, *34*, 374.

RECEIVED June 3, 1994

Chapter 4

Semisynthetic Rhodomycins and Anthracycline Prodrugs

Cenek Kolar[1], Klaus Bosslet, Jörg Czech, Manfred Gerken,
Peter Hermentin, Dieter Hoffmann, and Hans-Harold Sedlacek

Research Laboratories of Behringwerke AG, 35001 Marburg, Germany

A series of β- and ε-(iso)rhodomycins and anthracycline prodrugs have been synthesized. The range of structural variants in the carbohydrate portion of the rhodomycins was designed to furnish a suitable comprehensive series of stereochemical and substitutional variants in the daunosamine moiety and provide anthracycline glycosides synthesized in sufficient quantity for antitumor evaluation in an *in vitro* and *in vivo* test system. One approach to improve the efficiency of drug action and the selectivity of drug delivery is to prepare anthracyline prodrugs which itself pharmacologicaly inactive, but which becomes activated in vivo to liberate the parent drug by enzymatic attack. The new potentially useful prodrugs obtained along this line were glycosyl-anthracyclines, which can be activated by the tumor-specific immunoenzymatic conjugates.

Anthracyclines of the doxorubicin-daunorubicin group play a major role in the effective treatment of a number of neoplastic diseases (*1,2*). However, their use is restricted by cardiotoxic and other undesirable side effects involving the emergence of multiresistant cancer cells (*3*). In order to improve tumor therapy, we have concentrated on obtaining to new anthracycline compounds via modification of microbial β-rhodomycins, a class of anthracyclines first described by Brockmann (*4,5*), and on developing anthracycline prodrugs for fusion protein mediated prodrug activation (FMPA) (*6,7*).

β-Rhodomycins

Degradation and Isolation of β-Rhodomycins. The rhodomycins produced by *Streptomyces purpurascens* were the first anthracycline compounds to be isolated and studied (*8*). In his review published in 1963, Brockmann described aglycons and glycosides that belong to the rhodomycin-isorhodomycin family, namely rhodomycin A **1** and rhodomycin B **2**, whose component was β-rhodomycinone (**3**, β-RMN) (*9*). The structure of rhodomycin A, isorhodomycin A ($R^1 = OH$), and other aminoglycosides was also clarified by Brockmann by various methods (e.g., hydrolysis, hydrogenation, and later, NMR spectroscopy) (*10*).   Structures of other

---

[1]Current address: Glycon Biochemicals, Deutschhaus Strasse 20, P.O. Box 1530, D–35005 Marburg, Germany

0097–6156/95/0574–0059$08.00/0

β-rhodomycinone mono- or bis-glycosides containing complex oligosaccharide chains have been described by Oki et al. (*11*) and other authors (*12,13*).

We have established an alternative route (Figure 1) for the preparation of rhodomycin A and rhodomycin B, β-RMN (**3**), and L-rhodosamine (**4**, DauMe$_2$) (*10*). A mixture of β-(iso)rhodomycins containing different glycosidic side chains, with α-L-rhodosamine as the first sugar moiety, at positions 7 and 10 of the aglycon, was hydrolyzed in acetic acid/water (1:10) at 85°C for 72 h. First, **1** (7,10-bis-*O*-(α-L-DauMe$_2$)-β-RMN) was obtained, which was further hydrolyzed (addition of HCl to pH 0.5 at 35°C) to **2** (7-*O*-α-L-DauMe$_2$-β-RMN). This important intermediate can be isolated from the crude rhodomycin hydrolyzate using chromatography (silica gel 60; 0.040-0.063 mm; Merck Nr. 9385; solvent: dichloromethane/methanol 4:1). An alternative route for the synthesis of **2**, based on the glycosylation of 10-*O*-trifluoroacetyl-β-RMN with the readily accessible 1,4-di-*O*-acetyl-L-rhodosamine using trimethylsilyl triflate (TMSOTf), has also been developed (see section on glycosylation below).

Modification of Rhodomycin B. Rhodosaminyl type anthracyclines are effectively *N*-monodemethylated upon irradiation with visible light, leading to 3'-*N*-methyl-α-daunosaminyl anthracyclines (*14,15*). These compounds have turned out to be interesting intermediates for the preparation of mixed dialkylamino derivatives. Thus, rhodomycin B **2** was converted by photolytic demethylation into 3'-*N*-methyl-daunosaminide **5** in about 60-65% yield (*16*). For large scale photolysis, a special recycling apparatus was developed, allowing demethylation of 10-20 g of **2** within a few hours, using a mercury diving-lamp (*16*) in the presence of dichloroethane as a solvent at 25-30°C (Figure 2).

N-Alkyl-N-methyl-daunosaminyl Rhodomycins. A distinct class of *N*-alkyl-*N*-methyl analogues has been generated by reductive alkylation of the *N*-methyl-daunosaminide **5** with aldehydes. For example, the introduction of alkyl residues [CH$_3$(CH$_2$)$_n$ with n=1-7] continuously increases the lipophilicity of the drug (Figure 3), which in turn influences the drug's uptake into the cell and hence its cytotoxic effect (*17*). The *N*-alkyl (n > 4) as well as the *N*-benzyl (IC$_{50}$ (1 h) = 0.32 µg/mL) derivatives proved to be too lipophilic for the uptake into the cell. As expected, the lipophilicity of such compounds can be reduced upon the introduction of heteroatoms such as oxygen or nitrogen. Thus, the picolyl, furfuryl, and thenyl derivatives of **5** provided improved solubility and cytotoxicity against L1210 cells in vitro (*18*). However, the most promising derivatives gained so far have been obtained upon the introduction of a glycidyl residue (*18*).

B 880308 Bifunctional Cytostatic. Today only 6-7% of all tumors can be cured by cytostatic agents. In about 40% of tumor types, treatment with cytostatic agens leads to a transient tumor regression and/or to an increase in survival time, which is limited by development of resistance of the tumor against any further treatment. Our goal was the improvement of cytostatic agents by using the synthetic combination of two structures with proven antitumor activity (*19*). One of the compounds synthesized, which contains β-RMN aglycon as intercalating moiety and glycidyl residue as alkylating moiety, is 3'-*N*-glycidyl-3'-*N*-methyl-oxaunomycin (**7**; B 880308) (*20*).

In the synthesis of B 880308 we applied the following two-step strategy: monodemethylation of 7-α-DauMe$_2$-β-RMN **2** to the 7-*O*-(3-*N*-methyl-α-L-daunosaminyl)-β-RMN followed by alkylation of the methylamino group with epichlorohydrin (*20*) or glycidyl tosylate (*21,22*). The key reaction in the synthesis of B 880308 is no doubt the *N*-alkylation step. The reaction of 7-*O*-(3-*N*-methyl-α-L-daunosaminyl)-β-RMN (**5**) with epichlorohydrin provides 3-*N*-(*R*)- (**7**) and

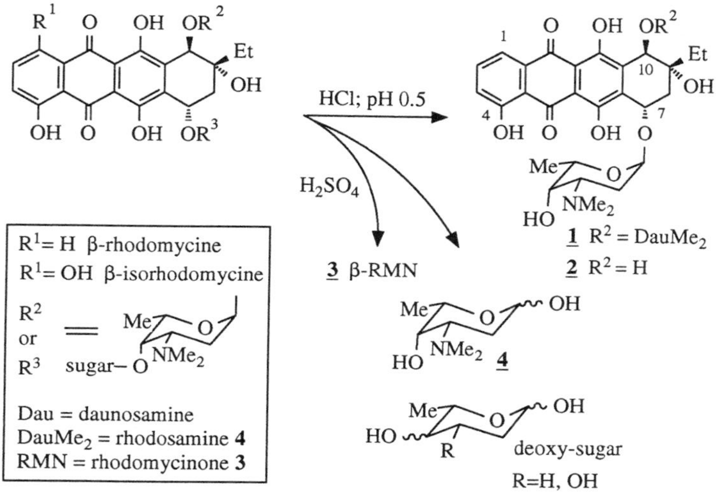

Figure 1. Degradation of Rhodomycins.

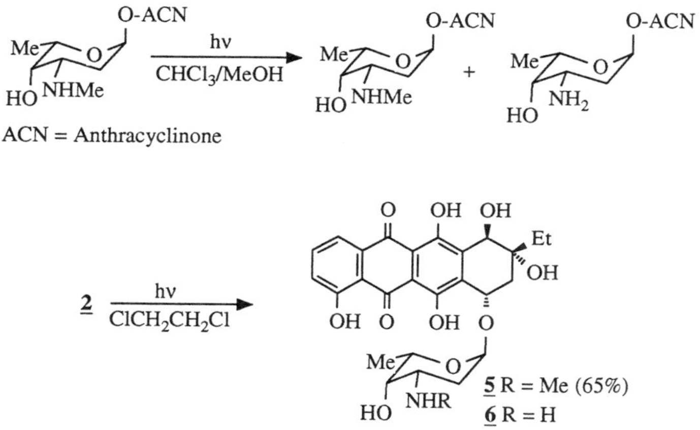

Figure 2. Demethylation of 7-*O*-Rhodosaminyl-Anthracyclines.

(S)-glycidyl diastereomer (**8**). The (R)- and (S)-diastereodiastereomers were characterized as 2,3-(R)- or (S)-dihydroxypropyl derivatives. which were obtained by stereoselective acidic or basic cleavage of the glycidyl moiety. The absolute configuration of the obtained (R)- or (S)-diol derivatives is based on D-glyceraldehyde, which was linked to **5** by reductive alkylation.

Facing the antitumor activity of the (R)-glycidyl compound (**7**), a new stereoselective approach for the introduction of the glycidyl residue was developed (*21*). Thus, the use of (R)-glycidyl-tosylate for the N-alkylation of **5** in the presence of potassium carbonate and dry DMF as a solvent at 85°C gives B 880308 as a main product, which after chromatography on silica gel (Machery and Nagel; Nucleosil 100 - 1525; desactivated with a mixture of H_2O/Et_3N 100:1 and concentrated phosphoric acid [until pH 3]; solvent: dichloromethane/isopropanol/acetonitrile, 80:12:8) gave B 880308 at a purity > 98% (Figure 4).

B 880308 demonstrates activity towards tumor cells (Table 1), which are sensitive or are made resistant to standard cytostatic drugs (*23*). On human tumors transplanted into immunodeficient (*nu/nu*) mice, B 880308 was revealed to be at least as strong as or stronger than all standard anthracyclines (Table 2). In all human ovarian carcinomas tested till now, B 880308 induced either partial or even complete responses and seems to be, at least experimentally, the most effective cytostatic compound for this tumor type. Bone marrow toxicity seems to be the dose limiting toxicity (*24*).

Synthesis of β-Rhodomycins with Mono- and Oligosaccharide Moiety.

Synthesis of Glycosyl Donors. The most important step in the synthesis of anthracyclines is glycosylation. The glycosyl donors of the 3-amino-2,3,6-tri-deoxy-L-hexopyranose type used in the condensation step are typically 4-O-p-nitrobenzoyl (pNBz) or 4-O-trifluoroacetyl (TFAc) and 3-N-trifluoroacetyl derivatives. The amino sugar is activated for coupling either via the glycosyl halide, glycal, or p-nitrobenzoate (Figure 5) (*25*).

An alternative route for the synthesis of the *arabino*- and *lyxo*-hexopyranosyl donors has been developed, based on readily accessible 4-O-acetyl-3-azido-2,3,6-trideoxy-L-*arabino*-hexopyranose **9** (Figure 6) (*26,27*). Thus, silylation of the *arabino*-hexopyranose **9** with ^{t}BuMe$_2$SiCl gave with high selectivity the β-anomer **10** in the presence of imidazole and dichloroethane. After deacetylation of **10** by the method of Zemplén, compound **11** was obtained and then treated with trifluoromethanesulfonic anhydride in pyridine-dichloromethane at -30°C to give triflate **12**. The C-4 epimerization of **12** by treatment with 10 equiv. NaNO$_2$/dimethylaminopyridine (pH 7.5) in DMF at 0°C gave the *lyxo*-compound **13**. The hydrogenolysis of the azide **13** gave the amine **14**, which was N-trifluoroacetylated to **15**, followed by O-acylation of **15** to the acetate **16** or p-nitrobenzoate **17** (Figure 7) (*27*).

In order to synthesize the intermediate **5** for the preparation of B 880308, the rhodosamine glycosyl donor **19** was also prepared (Figure 8) (*28*). We have described the use of β-glycosyloxy-*tert*.-butyl-dimethylsilanes as glycosyl donors for the preparation of α-glycosides (*29,30*). Most interesting is the fact that, in such glycosyloxysilanes the *tert*.-butyldimethylsilyloxy group is the leaving group, and for this reason the formation of the glycoside takes place at the anomeric carbon.

The main problem in the synthesis of anthracycline oligosaccharides is considered to be the condensation step (*31*), in which the unreactive axial OH-4 of the daunosamine acceptor should react with the "2-deoxy-L-fucose" or "L-daunosamine" donors. We have described the use of such donors for the synthesis of 7-O-(2,6-dideoxy-α-L-*lyxo*- and *arabino*-hexopyranosyl)-RMNs (*30,32*) and for the preparation of di- and trisaccharide units of anthracyclines (*27,32*).

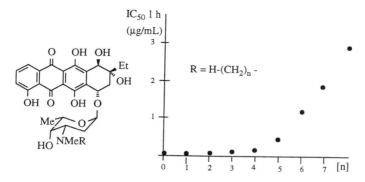

Figure 3. Structure Activity Relationship of 3'-*N*-Alkyl-rhodomycins.

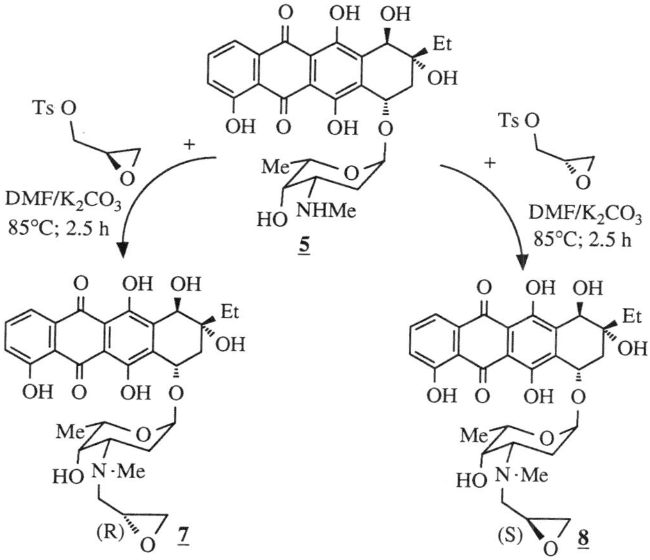

Figure 4. Synthesis of B 880308.

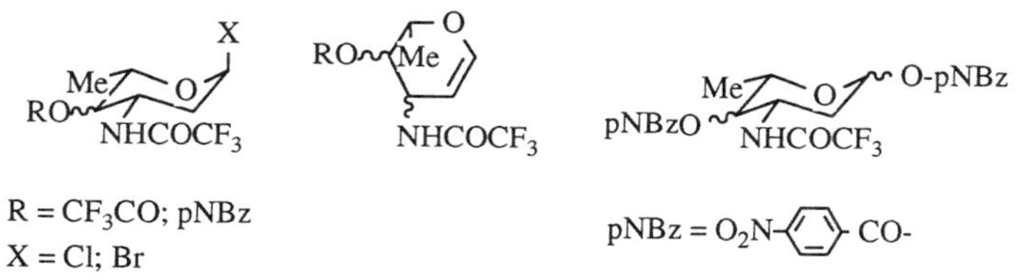

R = CF$_3$CO; pNBz
X = Cl; Br

pNBz = O$_2$N-⟨⟩-CO-

Figure 5. Glycosyl donors for glycosylation of anthracyclinones.

Table 1. Antitumor Activity of B 880308 *in vitro*

cytotoxicity	cross-resistance
IC_{50} = 0.001 mg/kg (10 x stronger than doxorubicin)	no cross-resistance to: - Doxorubicin (intercalation) - Etoposide (topoisomerase inhib.) - Vinblastine (spindle toxin) - Cisplatin (DNA cross link) - Melphalan

Table 2. Antitumor Activity of B 880308 against Human Tumors *in vivo*[a]

Human Tumors[b] Transplanted into Kidneys of Nude Mice		Regression of Human Tumors			
		>80%	>50%	<50%	no effect
bronchial tumors	B 880308	1	1	5	-
	DXR[c]	1	3	2	-
ovarian tumors	B 880308	2	1	-	-
	DXR	-	-	2	1
Transplanted Subcutaneously into Nude Mice					
bronchial tumors	B 880308	1	1	1	-
	DXR	1	-	2	-
ovarian tumors	B 880308	1	4	-	-
	DXR	-	-	5	-

[a]Adapted from ref. 24.
[b]Murine tumors: activity similar to DXR in leukemia, melanoma,
  ovarian and colon carcinoma.
[c]DXR = doxorubicin.

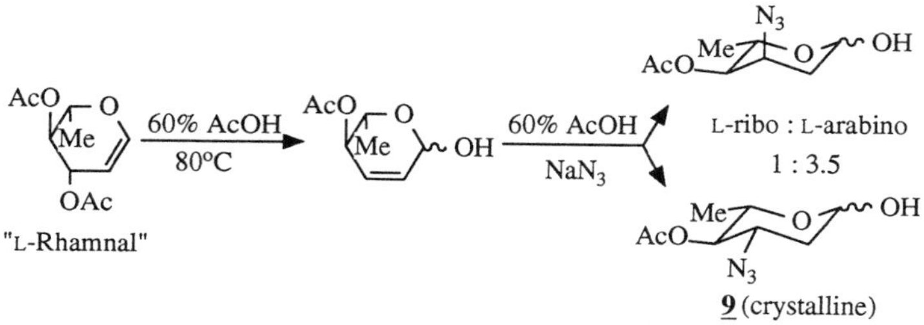

Figure 6. Synthesis of 3-Azido-2,3,6-trideoxy-arabinose.

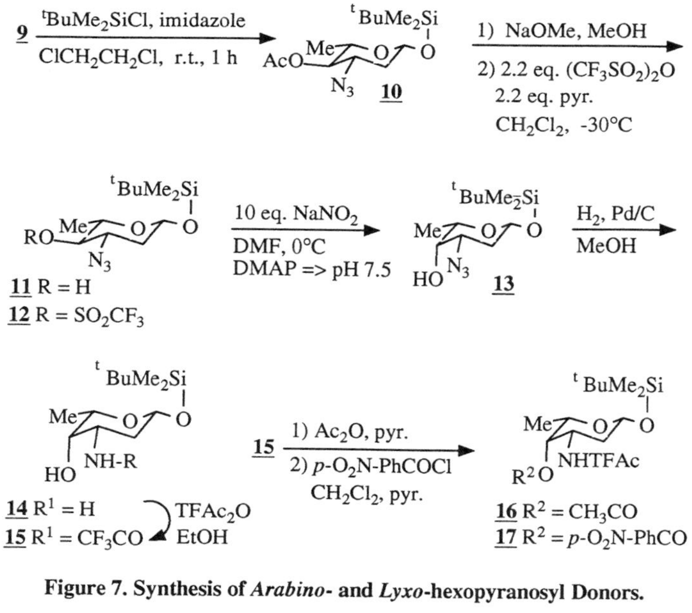

Figure 7. Synthesis of *Arabino-* and *Lyxo*-hexopyranosyl Donors.

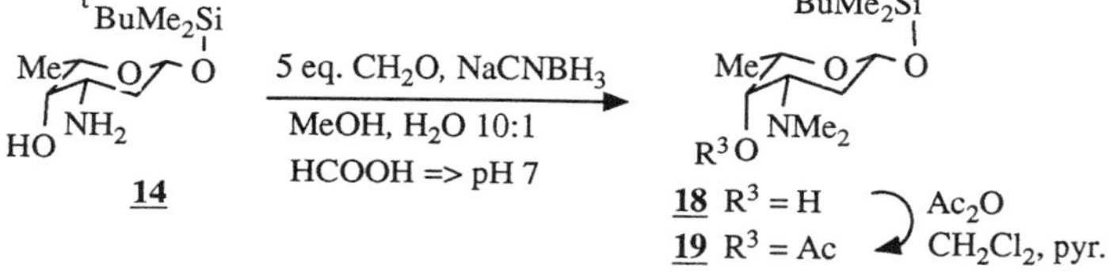

Figure 8. Synthesis of Rhodosamine Glycosyl Donor.

Thus far, L-daunosaminyloxysilane (16), 2-deoxy-L-fucosyloxysilane (23) and L-fucal (24) donors have been used for glycosylation of benzyl-N-trifluoroacetyl-L-daunosaminide (21), which provide α-linked disaccharides (22) and (25) in the presence of TMSOTf-Et₃N (5:1) and 4Å molecular sieves in dichloromethane at -65°C in yields of 75-80% (Figure 9) (27). Hydrogenation (Pd/C, EtOAc) of disaccharides 22 and 25 gave suitable starting compounds 26 and 27 for the preparation of glycosyl donors. For comparative purposes, further modification of 26 and 27, especially the introduction of p-nitrobenzoyl (28 and 29) or a tert.-butyldimethylsilyl group or glycal function, have been studied in order to improve the accessibility of disaccharidyl-RMNs.

Synthesis of Glycosyl Acceptor and Protection of β-Rhodomycinone. For the selective glycosylation of the β-RMN 3 at 7-OH, it is necessary to protect 10-OH because the reactivity of the alcoholic hydroxy groups decreases from 7-OH to 10-OH to 9-OH (32,33). The procedure (34) used was as follows: in the first step 3 was converted into 7,9-O,O'-phenylboronate 30 by phenylboronic acid and glacial acetic acid as a catalyst. The free 10-OH group was then acylated with trifluoroacetic anhydride-Et₃N to give 31. Esterification of the phenolic hydroxy group was not observed in the reaction, which is in contrast to the reaction with acetic anhydride. In the next step, the phenylborylene protecting group in 31 was cleaved by treatment with 2-methylpentan-2,4-diol using the method described by Broadhurst et al. (35). The 10-O-trifluoroacetylated product 32 was stable in the acidic pH range and could easily be purified on silica gel (Figure 10).

Glycosylation of Anthracyclinones and Deprotection Reactions. For the glycosylation of rhodomycinones, we used both a slight modification of Terashima's TMSOTf (36) method and our glycosyloxysilane method (29). Either method can be used successfully for the glycosylation of β- and ε-rhodomycinones with 3-N-trifluoroacetylated 2,3,6-trideoxyhexopyranoses.

We also succeeded in developing a direct glycosylation method for the synthesis of 7-O-rhodosaminyl-β-rhodomycinone, which is an important intermediate for the preparation of B 880308 (Figure 11). In the following glycoside synthesis step, 19 or 1,4-di-O-acetyl-L-rhodosamine 33 (37) was reacted with 32 to give the rhodosaminide 34 in the presence of TMSOTf, which was further deprotected to rhodomycin 2.

After several preliminary experiments, disaccharide p-nitrobenzoates 28 and 29, especially of the β-anomers, were chosen as the most promising donors for the glycosylation of aglycon 32. Under Terashima conditions, the reaction gave about 70% of rhodomycins 35 and 36. During processing, both α- and β-p-nitrobenzoates reacted with the β-RMN acceptor; however, the glycosylation with the α-anomers required longer reaction time and higher reaction temperature (about -40°C). The use of disaccharidyloxysilanes and glycals for the glycosylation of 32 in the presence of TMSOTf/Et₃N also gave 7-O-glycosyl-β-RMNs, although this required an excess of glycosyl donors (Figure 12).

Rhodomycins 35 and 36 could be deprotected either partially by cleaving the O-trifluoroacetyl group with amino-silica gel or completely using 1 M NaOH to provide 37 and 38. In order to obtain pure compounds, stepwise deprotection was employed. ¹H,¹H-Cosy experiments were used to assign the disaccharide moiety and β-RMN ring-D protons in the ¹H-NMR spectra of 37 and 38. As expected, the lyxo-hexopyranosyl moieties in these compounds were in the ¹C₄-conformation, as in the daunosaminyl or 2-deoxy-fucosyl moiety in related analogues (27).

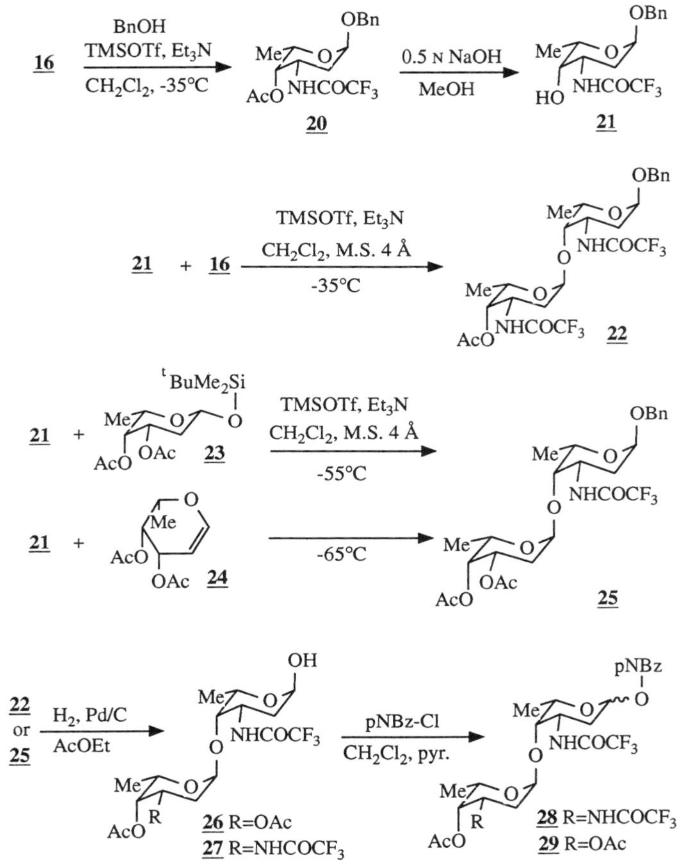

Figure 9. Synthesis of Disaccharides α-dFuc(1→4)-α-Dau and α-Dau(1→4)-α-Dau.

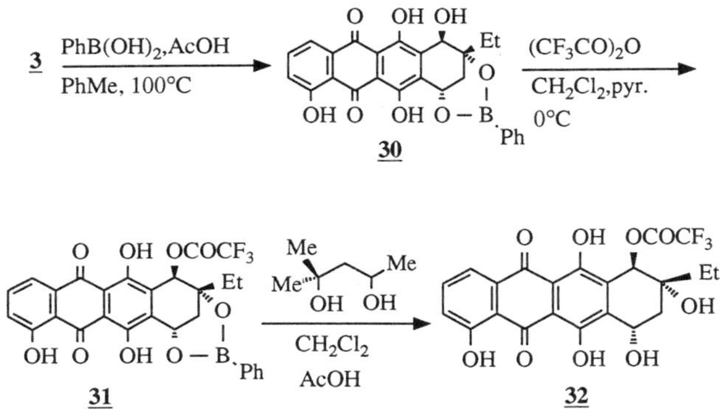

Figure 10. Protection of β-Rhodomycinone.

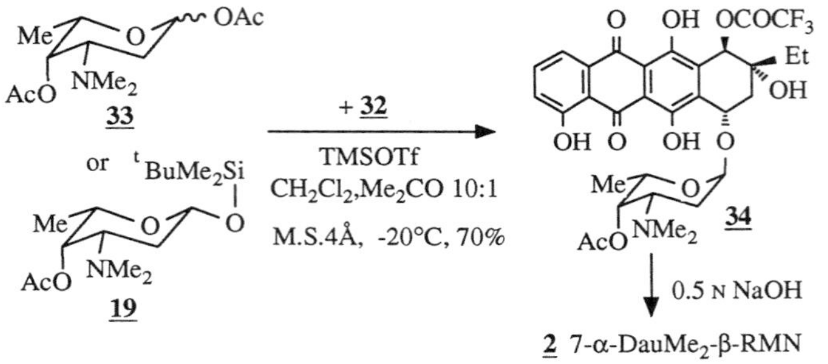

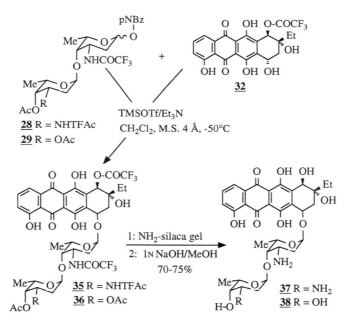

Figure 11. Glycosylation with Rhodosaminyl Donors.

Figure 12. Glycosylation and Deprotection Reactions.

Structure-Activity Relationship of Some Rhodomycins.

The cytostatic activity of the semisynthetic rhodomycins was tested on L1210 mouse leukemia cells in a clonogenic assay (*38*). This method is used to detect the effect of the test substances on growth behavior of cells over 1 h or 7 days (about 14 consecutive generations with a cell cycle lasting 10-12 h). The results are stated as percentages of the surviving colonies in the treated group versus the untreated groups. The cytotoxicity (IC_{50}, μg/mL) for continuous and for 1-h incubation were determined from the dose-effect curve (Table 3). For comparative purposes, table 3 shows antitumor assay data for doxorubicin, 4-*O*-methyl-rhodomycins and β-rhodomycins containing other sugar moieties (*8,29*).

   The resulting data lead us to the following conclusions concerning the structure-activity relationship:

(a) the cytotoxicity of 7-*O*-glycosyl-β-RMNs is decreased by alkylation of the 3'-amino group in the sequence NH_2 > NHMe > NMe_2;

(b) the disposition of the equatorial group is essential for the development of the cytotoxic effect;

(c) β-RMNs containing the 4-*O*-glycosyl-daunosamine or -acosamine moiety are less potent in vitro than the monosaccharide derivatives.

Fusion Protein-mediated Prodrug Activation (FMPA)

One of our projects is the evaluation of the therapeutic possibility of monoclonal antibodies (MAbs) (*6,7*). As we know that the amount of antibody localizing within the tumor is very low (~ 0.01% of the applied MAbs per gram tumor) (*38*), any approach to tumor therapy with MAbs has a chance to be successful only if we can introduce an amplification step between antibodies and the cytostatic molecules. (Such an amplification system already exists in in vitro immunodiagnostic techniques using antibody-enzyme conjugates and chromogenic substrates). An additional prerequisite is a high concentration of the cytotoxic compound at the tumor site but a low one in normal tissue. To fulfill both requirements, the amplification step of cleavage of <u>many</u> substrate molecules by <u>one</u> enzyme molecule was incorporated into the biphasic antibody-mediated treatment of tumors, called antibody-"dependent" enzyme-mediated prodrug therapy (ADEPT) (*39,40*). ADEPT consists of an antibody linked to an enzyme and an (nontoxic) low molecular weight prodrug that is cleaved by the enzyme into a toxic drug. A diagram demonstrating the principles of ADEPT is shown in Figure 13.

   The principle of biphasic treatment with enzyme linked to antibody for converting prodrugs into cytostatic drugs was first elaborated by Philpott et al. (*41,42*)for killing tumor cells, by Knowles et al.(*43*) for killing bacteria, and by Bode et al.(*44*) lysing fibrin clots. Bagshawe et al. (*40*) were the first to synthesize a considerably less toxic prodrug of cytostatic agents, which was cleaved through the selected enzyme into the cytotoxic drug. Moreover, they used F(ab')$_2$ fragments of tumor-selective MAbs, to which they chemically linked bacterial carboxy peptidase. In subsequent experiments Senter et al.(*45*) and numerous other investigators broadened the ADEPT system by using additional and different xenogeneic enzymes chemically linked to MAbs and synthesizing suitable prodrugs of the main cytostatics clinically used in tumor therapy. Bagshawe et al.(*40*) and subsequently Senter et al.(*43*) were also the first to show tumor-specific activity of the ADEPT system in vivo on tumors xenografted into nude mice. The first proof that antigen-negative tumor cells in the neighborhood of antigen positive tumor cells are indeed killed through the ADEPT process was supplied in vitro by Sahin et al. (*46*).

   Based on our knowledge of MAbs and antibody engineering as well as on our experience in the chemistry of antitumor drugs, especially anthracyclines, we have defined the conditions that must be fulfilled for further improvement of ADEPT (Table 4) (*6*). Essential are the antibody specificity and the enzyme activity

Table 3. Structure Activity Relationship of β-Rhodomycins

compound	(structure 1: O OH OH Et OH Me O O OH O–Sugar)		(structure 2: O OH OH Et OH OH O OH O · Sugar)	
sugar	IC_{50}(L 1210) [µg/mL] 1 h 7 d	LD_{50} [mg/kg]	IC_{50}(L 1210) [µg/mL] 1 h 7 d	LD_{50} [mg/kg]
α-Dau	0.029 0.012	1.0 (3xip)	0.005 0.002	<0.3 (3xip)
α-Aco	0.12 0.04	7.0 (3xip)	0.007 0.004	<0.3 (3xip)
α-dF	0.4 0.3	10.0 (3xip)	0.08 0.04	1.0 (3xip)
α-Dau(1-4)-α-Dau			0.26 0.01	
α-dF(1-4)-α-Dau			0.72 0.02	
α-Aco(1-4)-α-Dau			0.46 0.01	
α-dF(1-4)-α-Aco			1.8 0.5	
α-Dau-ADN	0.04 0.02			

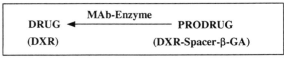

DRUG ← MAb-Enzyme —— PRODRUG
(DXR)   (DXR-Spacer-β-GA)

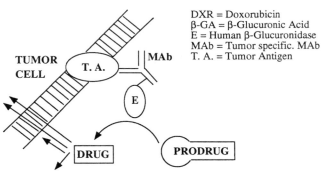

DXR = Doxorubicin
β-GA = β-Glucuronic Acid
E = Human β-Glucuronidase
MAb = Tumor specific. MAb
T. A. = Tumor Antigen

Figure 13. Antibody-Directed Enzyme-mediated Prodrug Therapy.

of the antibody enzyme conjugate, its unreduced localization rate and long retention in the target tissue, its quick metabolization in normal tissue, and its lack of immunogenicity. Our approach to FMPA concept is based on the availability of MAbs binding to CEA with high avidity (10^{10} L/mol) and their humanized version (*47*) and on access to cloned and expressed cDNA for human placental β-glucuronidase (*49*) and other lysosomal endogenous enzymes (*50*).

Based on DNA sequence of the humanized anti CEA MAb and human β-glucuronidase, a gene for a fusion protein was constructed (Figure 14) (*6,7,51*). The resulting fusion protein maintained all functions of the parental proteins. For the design of prodrug/drug system for FMPA we used some experience in designing prodrugs (*52-55*).

Our first approach in the preparation of prodrugs used *O*-glycosylated cytostatic agents like etoposide (*56*), oxaunomycin (*13*) and adriamycin (*8*). Unfortunatly our enzymatic studies revealed, that prodrugs like 14-*O*-(α-D-galacto-pyranosyl)- or 14-*O*-(β-D-glucuronyl)-adriamycin could not sufficiently be cleaved by the corresponding enzyme or fusion protein with the required high turnover rate (*57*).

Our second approach was based on the report of Katzenellenbogen et al. (*58*) and on our knowledge of the serum stability of anthracycline urethanes (59). Katzenellenbogen et al. (*58,60*) described a model system for a tripartite prodrug where chromophore (in place of a drug) and specifier (2-*N*-BOC-L-Lys) are linked by a 4-aminobenzyloxycarbonyl moiety (Figure 15). In the presence of trypsin, the model compound undergoes rapid hydrolysis to release first the amino acid and then the chromophore because of activation of the linker.

First, we elaborated structural requirements for different substrates by considering the pH dependence of the enzymatic cleavage and turnover rate under physiological conditions. On the basis of those results, we developed a new model of the glycosyl-spacer-drug system (*6*). In this model system, the glycosyl moiety represents α-D-galactopyranosyl or β-D-glucuronyl residue; the spacer moiety represents hydroxybenzyloxycarbonyl derivative; and drug represents anthracycline. Such types of prodrugs were developed and synthesized in cooperation with different French partners (Figure 16) (*61*).

The glycosyl-anthracycline prodrugs were screened for their cytotoxic activity against L1210 cells and also for their cleavability by enzymes. As expected, all the tested prodrugs were less cytotoxic than the free anthracycline drugs. Cleavability of the β-glucuronyl or α-galactopyranosyl moiety and subsequently the spontaneous cleavage of the spacer depends on the position of the substituents in the phenyl ring of the spacer (Table 5). The linkage of the glycosyloxy group in position C-4 of the benzyl group and an electron-withdrawing substituent in position C-5 speed up the turnover rate of the prodrug cleavage (*61-63*).

Our biphasic treatment schedule with FMPA is shown in Figure 17. The fusion protein, consisting of a humanized anti-CEA monoclonal antibody and human β-glucuronidase (*64*), was used as tumor-specific prodrug-activating moiety. We have demonstrated CEA-antigen specific binding of the fusion protein to cells from several tumor lines through activation of a substrate (4-methyl-umbelliferyl-β-glucuronide) for β-glucuronidase (*64*).

In the biphasic therapy approach, first the tumor-specific antibody-enzyme conjugate is injected. There, after a so-called localization phase (about 7 days), in which the conjugate is cleared out of the blood and other organs, but still remains at the tumor site, the prodrug is administered and should be specifically cleaved by the localized fusion protein at the tumor site. We used the corresponding substrate 4-methylumbelliferyl glycoside to prove that the prodrug would not be cleaved unspecifically in blood or the organs (*64*).

Furthermore, we have tested a doxorubicin glucuronide prodrug in vitro and

Table 4. Parameters Critical for Effective ADEPT and FMPA

Components	*Parameters and goals*
Monoclonal antibodies	- specificity for tumor antigens - number of epitopes ($>10^{4-5}$/cell) - high affinity - no immunogenicity
Enzyme	- not present in blood or interstitium - no immunogenicity
MAb enzyme conjugate	- low molecular weight - localization rate equal to that of MAb - no immunogenicity
Prodrug	- stable in vivo - low molar toxicity - hydrophilic molecule - long half-life
Drug	- hydrophobic molecule - high molar toxicity - short half-life - cytotoxicity not dependent on cell cycle phase

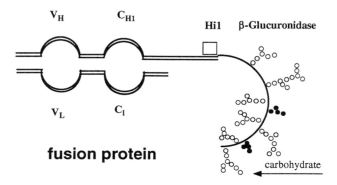

Figure 14. A Schematic Diagram of the Monovalent Fusion Protein
Detected in Transfectoma Supernatants under
Denaturing Conditions.

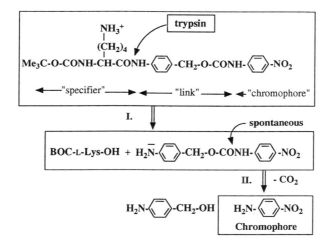

Figure 15. Tripartite Prodrug.

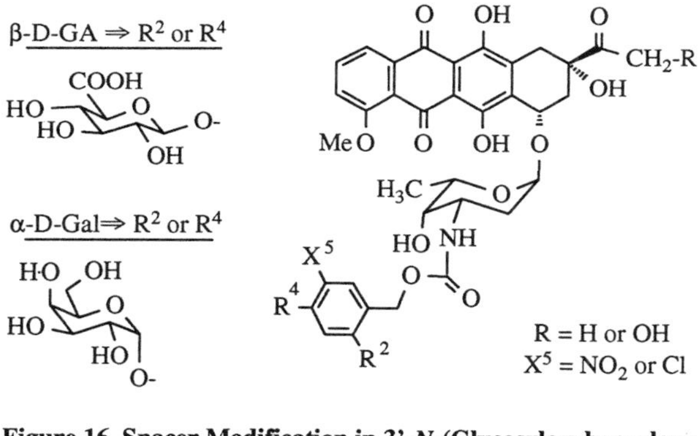

Figure 16. Spacer Modification in 3'-N-(Glycosyloxybenzyloxy-carbonyl)-Anthracycline Derivatives.

Table 5. Relationship between the Position of the Substituents in the Spacer and the Cleavability of the Prodrug

anthracycline				Cleavage of the Glycoside/Spacer		*L1210*
	R^2	R^4	R^5	$t_{1/2}$ (h)	$t_{1/2}$ (h)	$IC_{50}(\mu g/mL)$
R = H daunomycin	β-GA-O	OMe	NO_2	1.88	0.75	>1
	β-GA-O	H	NO_2	0.69	5.30	>1
	β-GA-O	H	Cl	4.06	12.00	>1
	H	β-GA-O	NO_2	0.42	< 0.08	>1
	H	α-Gal-O	Cl	0.02	< 0.08	>1
R = OH doxorubicin	β-GA-O	H	NO_2	0.92	6.00	>1
	H	β-GA-O	NO_2	0.40	< 0.08	>1

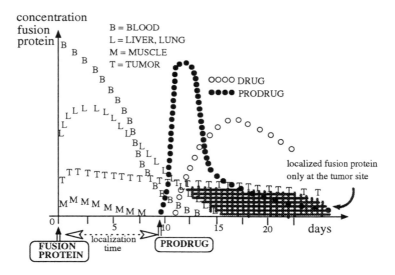

Figure 17. ADEPT: Biphasic Tumor Therapy.

observed detoxification of at least 100-fold when compared with drug (*61,63*). Toxicity toward antigen-positive tumor cells could partially be restored by preincubation with fusion protein. Since CEA expression is more pronounced in human tumor xenografts than in in vitro cell culture, we expected antitumoral efficacy in vivo. In nude mice bearing established human colon carcinoma xenografts, it could be proven that the combination of fusion protein and prodrug is effective for tumor therapy (*65*). Fusion protein (400 µg) was injected iv. Seven days later, when the fusion protein concentration was < 1 ng/mL in plasma and 200-400 ng/g of tumor, 500 mg/kg of prodrug (3'-*N*-[4-(β-D-glucuronyl-oxy)-3-nitro-benzyloxycarbonyl]-doxorubicin (*63*) was infused. In the group wherein fusion protein was combined with prodrug, tumor growth was strongly inhibited and resulted in regression (72%). Tumor growth in the prodrug group as well as the doxorubicin group was not significantly inhibited (20%) compared with that in the untreated control. The superior effects can be explained by the at least 10-fold higher concentrations of doxorubicin found in the tumors of animals treated with fusion protein and prodrug compared with those receiving the maximal tolerable dose of doxorubicin (61,65).

Literature

1. Arcamone, F. *Cancer Res.* **1985**, 45, 5995-5999.
2. Weiss, R. B.; Sarosy, G.; Clagett-Carr, K.; Russo, M.; Leyland-Jones, B. *Cancer Chemother. Pharmacol.* **1986**, 18, 185-197.
3. *Antitumor Drug Resistance*, Fox, B. B. W.; Fox, M., Eds.; Springer-Verlag; **1984**.
4. Brockmann, H.; Patt, P. *Chem. Ber.* **1955**, 88, 1455-1458.
5. Brockmann, H.; Waehneldt, T.; Niemeyer, J. *Tetrahedr. Lett*, **1969**, 6, 415-417.
6. *Antibodies as Carriers of Cytotoxicity*; Sedlacek, H. H.; Seemann, G.; Hoffmann, D.; Czech, J,; Lorenz, P.; Kolar, C.; Bosslet, K. Eds.; Contributions to Oncology; Karger: München, Germany, **1992**; Vol. 43.
7. Sedlacek, H. H.; Hoffmann, D.; Czech, J.; Kolar, C.; Seemann, G.; Güssov, G.; Bosslet, K. *Chimia*, **1991**, 45, 311-316.
8. Arcamone, F. In *Doxorubicin*; Medicinal Chemistry; Academic Press: New York, USA, **1981**; Vol 17.
9. Brockmann, H. *Fortschr. Chem. Org. Naturst.* **1963**, 21, 121-123.
10. Brockmann, H.; Niemeyer, J. *Chem. Ber.* **1967**, 100, 3578-3583; ibid. **1968**, 101, 1341-1347.
11. Oki, T. In *Anthracycline Antibiotics*, El Khadem, H. S. Ed.; Academic press, **1982**, 75-96.
12. Ihn, W.; Schlegel, B.; Fleck, W. F.; Tresselt, D.; Gutsche, W.; Sedmera, P.; Vokoun, J. *Pharmazie*, **1984**, 39, 176-180.
13. Yoshimoto, A,; Fujii, S.; Johdo, O.; Kubo, K.; Ishikura, T.; Naganawa, H.; Sawa, T.; Takeuchi, T.; Umezawa, H. *J. Antibiotics.* **1986**, 39, 902-907.
14. Oki T. *J. Antibiotic*, **1979**, 32, 801-805; ibid. **1982**, 35, 312-316.
15. Kolar, C.; Paal. M.; Hermentin, P.; Gerken, M. *J. Carbohydr. Res.* **1989**, 8, 295-382.
16. Hermentin, P.; Paal, M.; Kraemer, H. P.; Kolar, C.; Hoffmann, D.; Gerken, M. *J. Carbohydr. Res.* **1990**, 9, 235-243.
17. Hoffmann, D.; Berscheid, H. G.; Boettger, D.; Hermentin, P.; Sedlacek, H. H.; Kraemer, H. P. *J. Med. Chem.* **1990**, 33, 166-171.
18. Hermentin, P.; Paal, M.; Kraemer, H. P.; Kolar, C.; Hoffmann, D.; Gerken, M. EP 0270992, **1987**.
19. Stache, U.; Sedlacek, H. H.; Hoffmann, D.; Kraemer, H. P. *J. Cancer Res. Clin. Oncol.* **1990**, 116, suppl. part I, 439-441.

20. Hermentin, P.; Kraemer, H. P.; Kolar, C.; Hoffmann, D.; Gerken, M.; Stache, U. EP 0345598, **1989**.
21. Gerken, M.; Grim, M.; Raab, E.; Hoffmann, D.; Straub, R. DE 4036155 A1, **1990**.
22. Klunder, M. J.; So, Y. S.; Shapless, K. B. *J. Org. Chem.* **1986**, 51, 3710-3712.
23. Hoffmann, D.; Gerken, M.; Hermentin, P.; Sedlacek, H. H. *82nd Annual Meeting of the AACR for Cancer research*, Housten, **1991**.
24. Sedlacek, H. H.; Hoffmann, D.; Czech, J.; Kolar, C.; Seemann, G.; Güssow, D.; Bosslet, K.; *Chimia* , **1991**, 45, 311-316.
25. Acton, E. M. *J. Med. Chem.* **1974**, 17, 659-664; Arcamone, F. *Cancer Chemother. Rep.* **1975**, Part 3, 6, 123-127; Arcamone, F. US 4020270 **1977**; Umezawa. H. *J. Antibiot.* **1980**, 33, 1581-1584; Boivin, J. *Carbohydr. Res.* **1980**, 79, 193-196; Kimura, Y. *Chem. Lett.* **1984**, 501.
26. Florent, J. C.; Monneret, C. *J. Chem. Soc., Commun.* **1987**, 1171-1174.
27. Kolar, C.; Kneissl, G. *Angew. Chem.*, **1990**, 102, 827-828; Kolar, C. CA **1991**, 115(13):136652; *Ger. Offen.* DE 3943029, **1991**.
28. Kolar, C.; Dehmel, K.; Knoedler, U.; Paal, M.; Hermentin, P.; Gerken, M. *J. Carbohydr. Chem.* **1989**, 8, 295-305.
28. Kolar, C.; Kneissl, G.; Knödler, U.; Dehmel, K. *Carbohydr. Res.* **1991**, 209, 89-100.
29. Kolar, C.; Dehmel, K.; Moldenhauer, H.; Gerken, M. *J. Carbohydr. Chem.* **1990**, 9, 873-890.
30. Kolar, C.; Kneissl, G.; Wolf, H.; Kämpchen, T. *Carbohydr. Res.* **1990**, 208, 111-116.
31. Thiem, J.; Karl, H.; Schwentner, J. *Synthesis* **1978**, 696-701; Martin, A.; Pais, M.; Monneret, C. *J. Chem. Soc. Chem. Comm.* **1983**, 306-311; ibid. *Tetrahedron Lett.* **1986**, 27, 575; Thiem, J.; Klaffke, W. *J. Org. Chem.* **1989**, 54, 2006-2012.
32. Gerken, M.; Krause, M.; Blank, S.; Kolar, C.; Hermentin, P. Abstr. Pap. IV. Europ. Carbohydr. Symposium, Darmstadt **1987**, Abst. A143.
33. Gerken, M.; Blank, S.; Kolar, C.; Hermentin, P. *J. Carbohydr. Chem.* **1989**, 8, 247-254.
34. Kolar, C.; Gerken, M.; Kraemer, H. P.; Krohn, K.; Linoh, H. *J. Carbohydr. Chem.* **1990**, 9, 233-234.
35. Broadhurst, M. J.; Hassall, C. H.; Thomas, G. J. *J. Chem. Soc., Perkin Tr. 1* **1982**, 2249-2251.
36. Kimura, Y.; Suzuki, M.; Matsumoto, T.; Abe, R.; Terashima, S. *Chem. Lett.* **1984**, 4, 501-504.
37. Kolar, C.; Dehmel, K.; Knödler, U.; Paal, M.; Hermentin, P.; Gerken, M. *J. Carbohydr. Chem.*, **1989**, 8, 295-305.
38. Kraemer, H. P.; Sedlacek, H. H. *Behring Inst. Mitt.* **1984**, 74, 301-328.
39. Bagshawe, K. D. *Br. J. Cancer* **1987**, 56, 531-535.
40. Bagshawe, K. D.; Springer, C. J.; Searle, F.; Antoniv, P.; Sharma, S. K.; Melton, R. G.; Sherwood, R. G. *Br. J. Cancer* **1988**, 58, 700-703.
41. Philpott, G. W.; Bower, R. J.; Parker, C. W. *Surgery*, **1973**, 74, 51-58.
42. Philpott, G. W.; Shearer, W. T.; Bower, R. J.; Parker, C. W. *J. Immunol.* **1973**, 111, 921-929; ibid. *Cancer Res.* **1974**, 34, 2159-2163; ibid. **1979**, 39, 2084-2088.
43. Knowles, D. M.; Sullivan. T. J.; Parker, C. W.; Williams; R. C. jr. *J. Clin. Invest.* **1973**, 52 1443-1449.
44. Bode, C.; Matsueda, G. R.; Hui, K. Y.; Haber, E. *Science* **1985**, 229, 765-769.
45. Senter, P. D.; Schreiber, G. I.; Hirschberg, D. L.; Ashe, S. A.; Hellström, K. E.; Hellström, I. *Cancer Res.* **1989**, 49, 5789-5795.

46. Sahin, U.; Hartmann, F.; Senter, P.; Pohl, C.; Engert, A.; Diehl, V.; Freudenschuh, M. *Cancer Res.* **1990**, 50, 6955-6959.
47. Poljak, R. J.; Amzel, L. M.; Avey, H. P.; Chen, B. I.; Phizackerley, R. P.; Saul, F. *Proc. Natl. Acad. Sci. USA* **1973**, 70, 3305-3310.
48. Segal, D. M.; Padlan, E. A.; Cohen, G. H.; Rudikoff, S.; Potter, M.; Davies, D. R. *Proc. Natl. Acad. Sci. USA* **1974**, 71, 4298-4302.
49. Oshima, A.; Kyle, J. W.; Miller R. D. *Proc. Natl. Acad. Sci. USA* **1987**, 84, 685-691.
50. Wissenschaftliche Tabellen Geigy, 5. Edition, Ciba-Geigy AG, Basel, **1979**.
51. Lorenz, P.; Schuermann, M.; Seemann, G. *J. Cancer Res. Clin. Oncol.* **1991**, 16, 9-14.
52. Chakravarty, P. K.; Carl, P. L.; Weber, M. J.; Katzenellenbogen, J. A. *J. Med. Chem.* **1983**, 26, 633-638; ibid. **1983**, 26, 638-644.
53. Connors, T. A.; Whisson, M. E. *Nature*, **1966**, 210, 866-871.
54. Connors, T. A.; Farmer, P. B. *Biochem. Pharmakol.* **1980**, 22, 1071-1079.
55. Bundgaard, H. *Drug of the Future* **1991**, 16, 443-458.
56. Kolar, C.; Czech, J.; Bosslet, K.; Seemann, G.; Sedlacek, H. H. DE 3935016 A1 **1991**.
57. Gerken, M.; Bosslet, K.; Seemann, G.; Hoffmann, D.; Sedlacek, H. H. EP 0441218 A2 **1991**.
58. Carl, P. L.; Chakravarty, P. K.; Katzenellenbogen, J. A. *J. Med. Chem.* **1981**, 24, 479-480.
59. Kolar, C.; Dehmel, K.; Hoffmann, D.; Kraemer, H.P.; Gronau, T.; Ger. Offenl. DE 3924655. **1991**.
60. Chakravarty, P. K.; Carl, P. L.; Weber, M. J.; Katzenellenbogen, J. A. *J. Med. Chem.*, **1983**, 26, 633-638; ibid. *J. Med. Chem.* **1983**, 26, 638-644.
61. Jacquesy, J. C.; Gesson, J. P.; Monneret, C.; Mondon, M.; Renoux, J. B.; Florent, J. C.; Koch, M.; Tillequin, F.; Sedlacek, H.H.; Gerken, M.; Kolar, C. EP 511917, WO 92 19639
62. Florent, J. C.; Dong, X.; Monneret, C. *16. Intern. Carbohydr. Symposium* (Paris), Jun. 5-10, **1992**, Abst. A262; Gesson, J, P.; Jacquesy, J. C.; Mondon, M. *16. Intern. Carbohydr. Symposium* (Paris), Jun. 5-10, **1992**, Abst. A263; Andrianomenjanahary, S.; Koch, M.; Tillequin, F. *16. Intern. Carbohydr. Symposium* (Paris), Jun. 5-10, **1992**, Abst. A264
63. Florent, J. C.; Dong, X.; Monneret, C. *12th Int. Symp. Med. Chem.* (Sept. 13-17, 1992, Basel) **1992**, Abst. OC-024
64. Bosslet, K.; Czech, J.; Lorenz, P.; Sedlacek, H. H.; Schuermann, M.; Seemann, G. *Brit. J. Cancer*, **1992**, 65, 234-243.
65. Bosslet, K.; Czech, J.; Seemann, G.; Hoffmann, D. Proceedings, 10th *Internat. Hammersmith Conference in Clinical Oncology*, Paphos, Zypern, 3-5. May **1993**.

RECEIVED July 26, 1994

Chapter 5

Synthetic Options for Reversal of Anthracycline Resistance and Cardiotoxicity

Claude Monneret[1], Jean-Claude Florent[1], Jean-Pierre Gesson[2], Jean-Claude Jacquesy[2], François Tillequin[3], and Michel Koch[3]

[1]Laboratoire de Chimie, Institut Curie-Biologie, 26 rue d'Ulm, 75005 Paris, France
[2]Laboratoire de Chimie XII, 40 rue du recteur Pineau, 86022 Poitiers cedex, France
[3]Laboratoire de Pharmacognosie, Université Paris V, 4 avenue de l'Observatoire, 75006 Paris, France

Chiral pool syntheses of enantiomerically pure anthracyclinones from leucoquinizarin, α-D-isosaccharino-lactone and α-D-glucosaccharino-lactone are reported. Subsequently, a more versatile approach was developed from diacetone glucose to give access to 9-alkyl anthracyclinones, precursors of 3'-morpholino-9-alkylanthracyclines claimed to hold great promise for the circumvention of multidrug resistance. With respect to the antibody-directed enzyme prodrug therapy (ADEPT) concept, a series of new prodrugs of anthracyclines was synthesized to deliver anthracyclines specifically to tumors cells and therefore avoid or minimize side effects such as cardiotoxicity. The low cytotoxicity of these prodrugs, their high stability in plasma, and their easy conversion into active drugs in the presence of the antibody-enzyme conjugate fulfill the required conditions for *in* vivo experimentation.

The clinical usefulness of anthracyclines is mainly limited by two major problems which are cumulative cardiotoxicity, and the appearance of an acquired resistance. Several options can be taken to overcome these problems. In order to contribute, as organic chemists, to the search for solutions to these problems, we have essentially proceeded with two separate paths:

-   The first is the development of general methods for the preparation of 9-alkyl aglycones, which can be subsequently combined with 3'-morpholino-sugars.

-   The second is a new prodrug strategy for the delivery of anthracyclines specifically to tumor cells.

Total Syntheses of Anthracyclinones

When we initiated our anthracycline synthesis program in 1981, numerous syntheses of daunorubicin, doxorubicin, and analogs had already been reported.(*1*) However approaches permitting the synthesis of the aglycone component in enantiomerically pure form, were rare, and those that did exist were mainly based upon the use of chiral auxiliary reagents (2), chiral catalysts (3), or even classical resolution methods (4).

The "chiron" approach to the anthracyclinones had been practically unexplored when we became interested in this area. Our initial idea was to prepare fully functionalized ring A possessing the correct C-9 absolute configuration and to use it to control the C-7 configuration during an annelation process.

In this regard, we were looking for a suitable monochiral synthon containing a tertiary alcohol and an alkyl side chain in the right configuration which would eventually become the C-9 center. Starting materials containing this system were not readily available from natural sources. Nevertheless, we recognized that the well-known α–D-isosaccharino-lactone 1 and α-D-glucosaccharino-lactone 2 could serve as excellent starting materials for the construction of the A-ring in our anthracycline targets. Both lactones were readily obtained by alkaline treatment of lactose (5) or fructose (6) with calcium hydroxide followed by neutralization of the calcium salts of the corresponding acids.

Moreover, by far the most important reaction in anthracyclinone synthesis-using anthraquinones as starting materials-was the electrophilic addition of aldehydes. Reaction of such an aldehyde with the reduced form of quinizarin (the "leuco form") under Marschalk conditions(7) was simultaneously investigated by Sih et al. (8), Krohn et al.(9), and Morris and Brown (10).

As regards the chiron approach, 4-demethoxy anthracyclinones were considered interesting targets because 4-demethoxydaunorubicin (or idarubicin) and 4-demethoxydoxorubicin are known (11) to be five to ten times more potent than the parent compound. We thus decided to prepare analogs of 4-demethoxydaunomycinone according to the retrosynthetic pathway shown below, in which leucoquinizarin A was used as the component of rings B, C, and D and ring A was elaborated from the chiron B readily obtained from iso- and glucosaccharino-lactones (Figure 1).

Initially we synthesized the 4-demethoxy-9-deacetyl-9-hydroxymethyl-daunomycinone 8, with the hydroxymethyl side chain coming from the isopropylidene derivative 1b of α-D-isosaccharino-lactone. After several improvements (12,13) the best synthesis was achieved as depicted in scheme 1.

The main features of the synthesis were the very high stereocontrol observed in the final cyclization reaction of 7 under Marschalk conditions giving exclusively the 7S isomer and the fact that only one chromatography had to be done at the penultimate step.

The scale-up of this synthesis allowed us to prepare several glycosides (14,15) by coupling 8 with deoxy or amino-deoxy-sugars. Among them, anthracyclines 9, 10 and 11 displayed high antitumor activity, but the most interesting compound was moflomycin 11. The latter showed (16) a high antiproliferative activity against two human and one murine leukemia cell lines (HL-60, REH, and L1210) (10 times more potent than doxorubicin and daunorubicin), and a high antitumor activity against L1210 leukemia in mice (T/C 341 at 2.2 mg/kg with 4/6 LTS *versus* 350 at 3.75 mg/kg for doxorubicin). More complete evaluation of this compound is ongoing and will be discussed elsewhere (Figure 2).

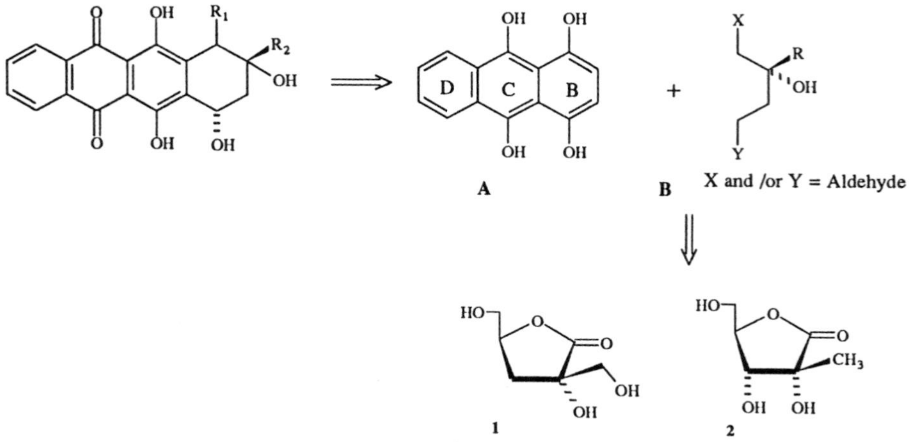

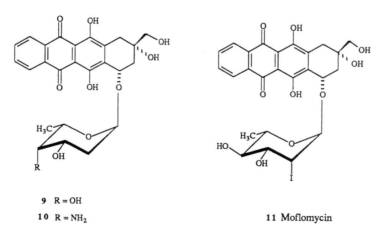

Figure 1. Retrosynthetic pathway for 4-demethoxydaunomycinone analogues.

9   R = OH

10   R = NH$_2$

11 Moflomycin

Figure 2. Anthracyclines with high antitumor activity.

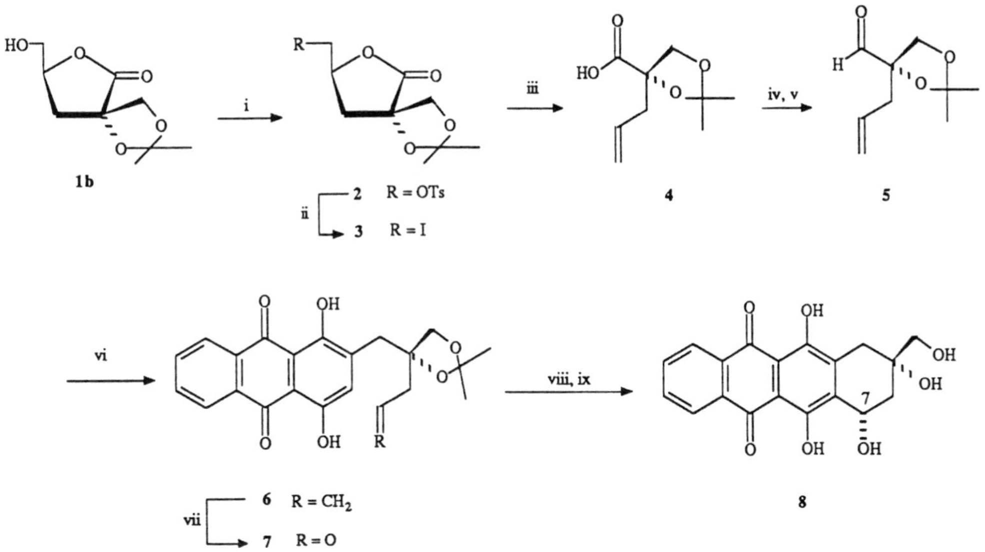

Scheme 1

Reagents and conditions: i: TsCl, pyridine, r.t., 72 h, 75%; ii: NaI, 2-butanone, reflux 15 h, 82%; iii: Zn, THF, AcOH-H_2O, 80%; iv: LAH, THF, reflux, 18 h, 80%; v: PDC CH_2Cl_2, 90%; vi: leucoquinizarin, piperidine, AcOH, iPrOH, reflux, 80%; vii: O_3, CH_2Cl_2, -78°C, then Me_2S, 90%; viii: $Na_2S_2O_4$, KOH, THF-10°C, 10 min, 75%; ix: H_3O^+

In a subsequent synthesis (13) the hydroxymethyl side chain was converted into an ethyl group as indicated in scheme 2 for preparing 4-deoxy-γ-rhodomycinone 19. This transformation involved a reaction of the mesylate 13 with Me₂CuLi in ether as the key step.

The best way to achieve the synthesis of 19 required the temporary protection of the diol adjacent to the ethyl chain in order to get 15 and subsequently an open-chain aldehyde as 17. After condensation of 17 with leucoquinizarin under Lewis conditions (17), acidic hydrolysis and oxidation, conversion of the α-hydroxy-aldehyde 18 led to 19 along with a small amount of the cis-epimer, under conditions previously reported by Krohn (18).

More fully ring-A hydroxylated anthracyclinones 20 and 21 were further obtained (19) as depicted in scheme 3 from unsaturated aldehyde 5 (three steps).

Using the glucosaccharinolactone 2, a first attempt as shown in scheme 4 was made to synthesize the anthracyclinone 29 (20). Transformation of such a lactone into the pivotal unsaturated alcohol 22 was conveniently achieved in six steps and 70% overall yield. Unfortunately, condensation of the aldehyde 23, resulting from pyridinium dichromate oxidation of 22, with leucoquinizarin afforded a rather low yield (33%) of a mixture of diastereoisomeric alkylanthraquinones 24 that could not be separated. Since all attempts to deoxygenate or oxidize the benzylic OH group were unsuccessful, the terminal double bond was oxidized to give 25. However, since the cyclization reaction provided a complex mixture that could not be characterized, we turned our attention towards the synthesis of the corresponding 7-deoxy-aglycone 30.

Thus, aldehyde 27, fixed in the open-chain form, was easily obtained from 22 by benzylation and oxidation. Condensation of 27 with leucoquinizarin, benzylic deoxygenation, and alcohol deprotection were carried out in a one-pot procedure and 60% overall yield, giving alkylanthraquinone 28. Finally, the synthesis of 30 was cleanly achieved via an intramolecular aldolization reaction, after oxidation, and acidic hydrolysis (Figure 3).

The addition-cyclization techniques of Kraus (21) and Hauser (22) for anthraquinone synthesis have been widely adapted by Li (23), Hauser (24), and Swenton (25) to achieve regiospecific anthracyclinone syntheses by the condensation of phtalide anions with either 1(4H)-naphthalenone derivatives or tetralin-type quinone monoketals. Concurrently with the first route, we explored the last methodology, which has been shown to be compatible with a fully functionalized ring A, to be adaptable on a large scale, and to occurr without loss of chiral integrity. This allowed us to prepare several aglycones as anthracyclinone 8 or 1-O-methyl-γ-rhodomycinone (26), and as anthracyclinone 29 (20) via the tetralin derivatives X, Y, and Z, which were prepared from α-D-isosaccharino-lactone 1 and α-D-glucosaccharino-lactone 2, respectively.

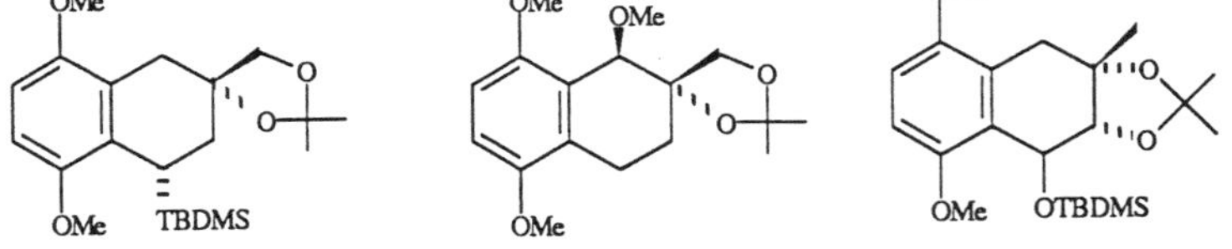

X   Y   Z

(TBDMS= tertbutyldimethylsilyl)

Figure 3. Tetralin derivatives for preparation of anthracyclinone 29.

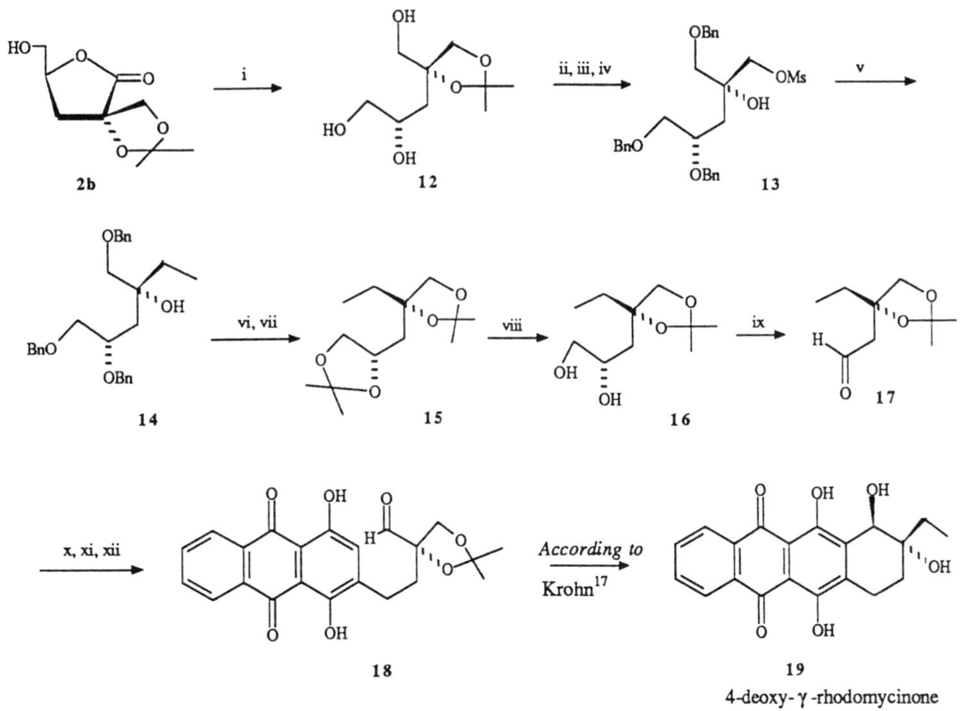

Scheme 2

Reagents and conditions: i: LAH, THF, 90%; ii: NaH, BnBr, Bu₄NI, r.t. 72 h, 55% ; iii:MeOH HCl; iv: MsCl, pyrid., 94%; v: CuI, MeLi, ether, 65%; vi: Pd/C, 10%, MeOH, 98%; vii: α,α dimethoxypropane, camphorsulfonic acid, DMF, 73%; viii: MeOH, AcOH-H₂O, r.t. 10 h: ix: NaIO₄, MeOH, r.t.,15 min, 92%; x: leucoquinizarin, piperidine, AcOH, iPrOH, reflux, 18 h; 80%; xi: H₃O⁺; xii: Moffatt oxid. 45%.

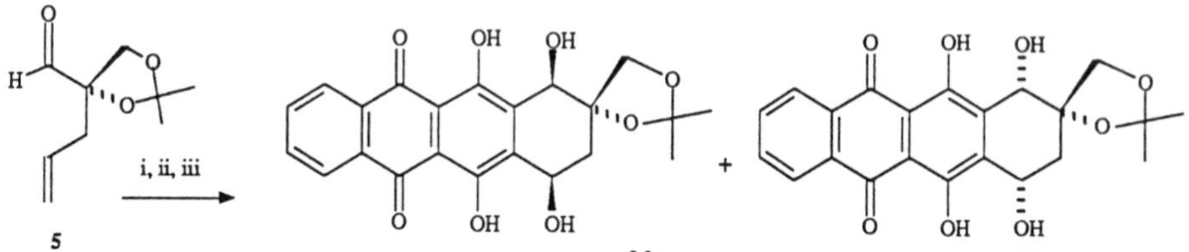

Reagents and conditions: i: leucoquinizarin, DBU, THF, 55% overall yield; ii: OsO₄ -NaIO₄, 98%; iii: Na₂S₂O₄, NaOH, THF-MeOH (1:1), -20°C, 5 min, 70-75%.

Scheme 3

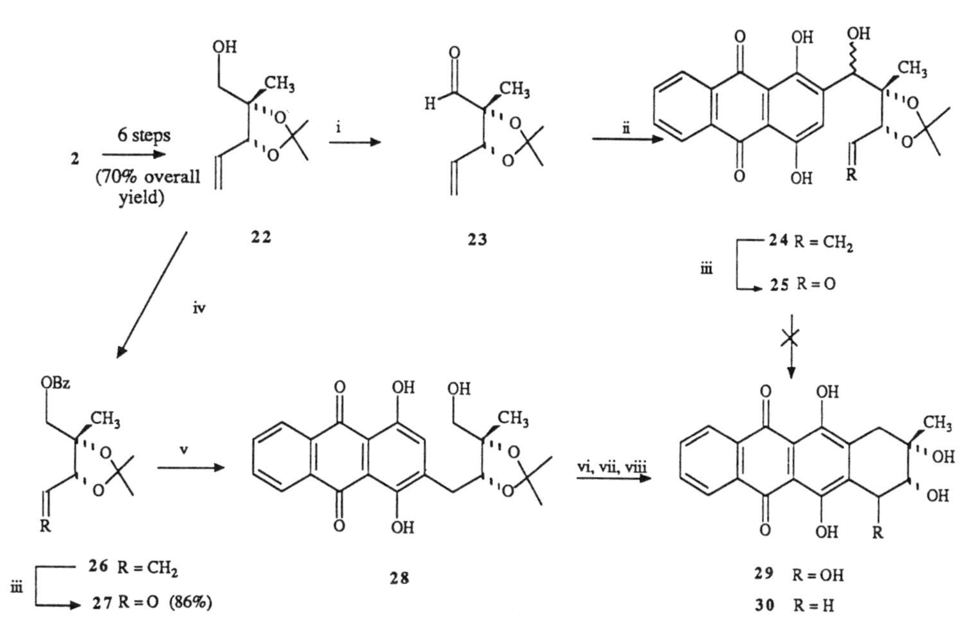

Scheme 4

Reagents and conditions i: PDC, Ac$_2$O, molecular sieves, CH$_2$Cl$_2$ (93%); ii: leucoquinizarin, DBU, THF, r.t., 0.5 h, 33%; iii: OsO$_4$- NaIO$_4$, ether-H$_2$O; iv: BzCl, pyridine, r.t.,1 h (95%); 95%; v: cf ii then Na$_2$S$_2$O$_4$, 80°C, 0.5 h, 60%; vi N-chlorosuccinimide then Et$_3$N, > 90%; vii: Na$_2$S$_2$O$_4$, KOH, H$_2$O, MeOH-THF (1:1), 0°C, 0.5 h (60%); viii AcOH-H$_2$O (8:2), 90°C,

Total   Syntheses   of   9-Alkyl-Anthracyclinones

Very recently, it was reported that 9-alkyl anthracyclines (27) and, in particular, 9-alkylanthracyclines combined with a 3'-morpholino-3'-deamino-daunosaminyl (or 2',3',6'-trideoxy-3'-morpholino-L-*lyxo*-hexopyranosyl) moiety exhibited high antitumor activity against a number of doxorubicin-resistant cell lines (28). It was postulated that the lipophilic character at the 9-alkyl position and probably in the sugar moiety could increase the rate of cellular uptake while at the same time reducing the affinity for efflux pumps such as P-glycoprotein (29). Undoubtedly, as deduced from structure-activity relationship studies noted by the same authors, there is a clear trend for decreasing resistance factors with increasing 9-alkyl side-chain length, by comparing 9-methyl with 9-ethyl and 9-isopropyl-containing anthracyclines.

Therefore, in order to contribute to the circumvention of multidrug resistance as found for example with MX2 or KRN8602 developed in Japan (30), the synthesis of new 9-alkyl anthracyclines was our next objective. We started our first experiments (31) with α-D-isosaccharino-lactone 1 and selected 4-demethoxy-feudomycinone C 39 as target molecule. The synthesis was readily achieved as indicated in scheme 5 and was essentially based upon the initial conversion of 1 into the methyl lactone 31 according to Bock and co-workers (32). Final intramolecular cyclization of 36 was carried out under Marschalk conditions at 0°C to diastereoselectively afford the desired compound 39 along (ratio 6:4) with its *trans* -isomer 38.

Although this approach seemed quite satisfactory for preparing 9-methyl anthracyclinones, all subsequent efforts to convert the hydroxymethyl side chain in an ethyl or propyl side chain as in 41 were unsuccessful. This was illustrated by the fact that the tosyl derivative 40 or the anhydro derivative 42 were completely unreactive when treated with RLi/CuI in THF or with mixed cuprate, even in the presence of Lewis acids (scheme 6) (33).

Then, our feeling was to focus on other suitable chirons; among them, the aldehydo-derivative 44 prepared from the corresponding 3,4-unsaturated analog (34) and the cyano-derivative 45 (35), both easily obtained from diacetone glucose 43, appeared to be the most convenient according to Figure 4.

The first approach would involve electrophilic addition of 44 to leucoquinizarin A and alkylation of the intermediate B, whereas a reverse sequence should be used in the second approach with first alkylation of the cyano-derivative 45, followed by the subsequent electrophilic addition of compound C to leucoquinizarin A.

Following the first route, electrophilic addition of 44 afforded the alkylanthraquinone 46, which was methylated (scheme 7) and oxidized to give 47. Unfortunately, alkylation of 47 with acetaldehyde under enolizable conditions gave the desired product 48 with less than 20% yield in the best case.

Therefore, without wasting time, we zealously took advantage of the second route (36). Stereoselective alkylations (Figure 5) of the cyano-sugar template 45, easily prepared from diacetone-D-glucose 43, which can be considered as a chiral protected cyanohydrine, afforded β–alkyl derivatives as 49 with electrophiles as MeI, EtBr, or 2-(iodomethyl)-3,5-dioxa-hex-1-ene in the presence of KHMDS as base. The *cis*-relationships between the cyano group and the acetal ring was unambiguously demonstrated by X-ray analysis.

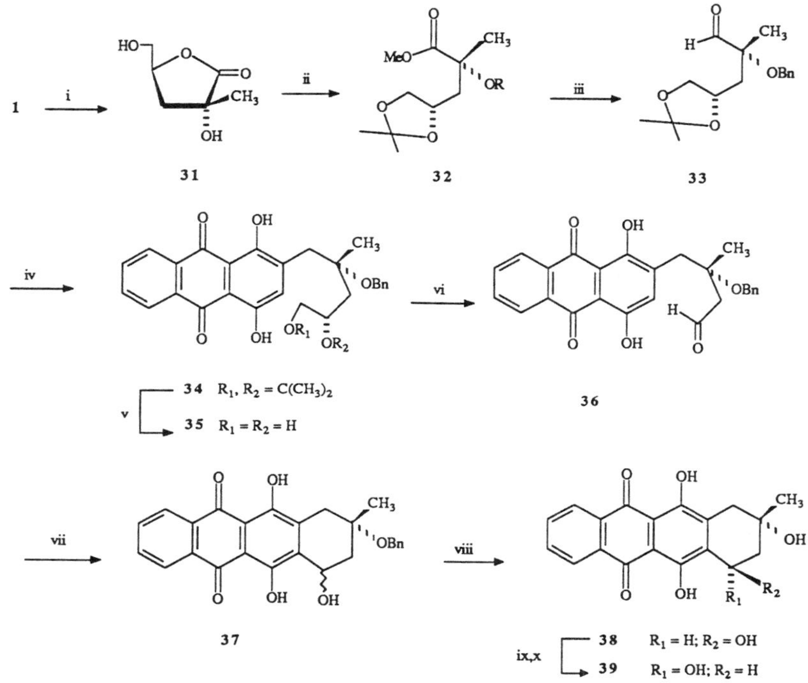

Reagents and conditions: i: Ref. [31]; ii: dimethoxypropane, Amberlyst 15, r.t., 48 h, 62% then BnBr, NaH, DMF, -20°C, 6 h, 82%; iii: DiBAL-H, toluene, -78°C, 3h, 75%; iv: leucoquinizarin, i-PrOH,piperidine, r.t. 24 h, 75%; v: AcOH - H_2O, r.t., 20 h; vi: $NaIO_4$, H_2O, CH_2Cl_2, r.t. 20 h, 78% from 34; vii: $Na_2S_2O_4$, NaOH, THF, MeOH, 20°C, 15 min., 67%; viii: BBr_3, CH_2Cl_2, - 78°C, 5 h, 62%; ix: $PhB(OH)_2$, TsOH; x: 2-methyl-2,4 pentane-diol, H^+

Scheme 5

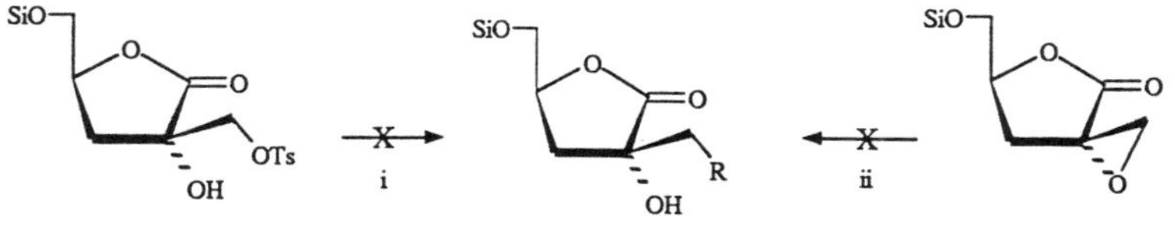

40   41 R= $(CH_2)_nCH_3$ n = 1 or 2   42

Reagents and conditions: i: RLi, CuI, THF; ii: $Me_2Cu(CN)Li_2$, BF_3-Et_2O

Scheme 6

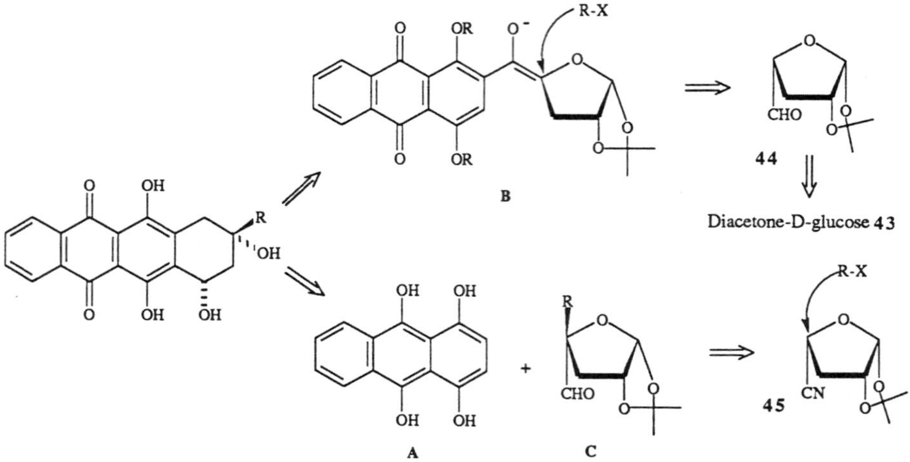

Figure 4. Retrosynthetic pathway for chirons obtained from diacetone glucose **43**.

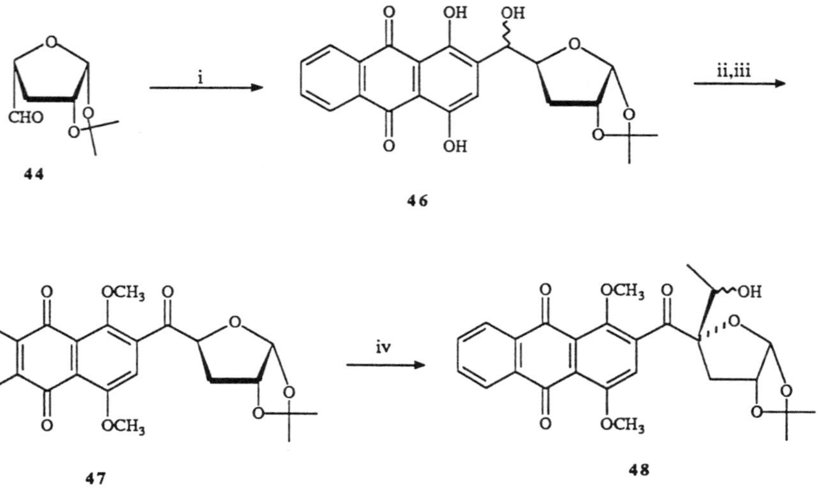

Reagents and conditions: i: leucoquinizarin, DBU, DMF, 75%; ii: acetone, Na$_2$CO$_3$, MeI; iii: PCC, CH$_2$Cl$_2$; iv: MeCHO, NaOH, MeOH, < 20% or LDA, -78°C in THF, then MeCHO, 0 -5%.

Scheme 7

The synthetic utility of these 4-branched-chain cyano-sugars **49a** was next demonstrated (scheme 8) by the synthesis of 4-demethoxyfeudomycinone **39** in six steps and 42% overall yield. Higher stereoselectivity in favor of the cis isomer **39** was observed during the annelation process than during the corresponding cyclization with a benzyl ether at C-9 (see **36** --> **37**). We have so far demonstrated that the cyano compound **49a** is a useful chiral synthon for the preparation of anthracyclinone analogs with C-7 and C-9 configurations as in natural products.

Furthermore, as regards this original strategy, we are now preparing completely new anthracyclines with increased 9-alkyl side-chain lipophilicity in order to test whether this feature may be useful in circumventing multidrug resistance.

Prodrugs of Anthracyclines for use in Antibody-Directed Enzyme Prodrug Therapy (or ADEPT Concept).

Anthracycline antibiotics are widely used in chemotherapy but their clinical efficacy is limited by a severe dose-cumulative cardiotoxicity and by the appearance of an acquired resistance. Rational design to avoid these side effects can be represented by drug targeting or prodrug synthesis. Ideally, the activation of a prodrug should be restricted to its required site of action.

In this regard the combination of prodrugs, and tumor specific enzymes, for use as therapeutic agents was reported in 1978 by Connors (*37*). This approach has been extended in the case of antitumor drugs, by Bagshawe (*38*) and then by Senter (*39*) and is known as antibody-directed enzyme prodrug therapy (ADEPT) (Figure 6).

The ADEPT approach entails the use of *monoclonal antibodies* (MAbs), which are directed against a particular tumor and are covalently bonded to a prodrug cleaving enzyme. Therefore the antibody enzyme conjugate is first injected and localized at the tumor cell surface antigen. Later the non cytotoxic prodrug is applied and must be converted into the cytotoxic species at the tumor cell surface.

For the ADEPT approach to be successful at least two things are required: i) the enzymes must be able to convert relatively non cytotoxic drug precursors (prodrugs) into active drugs; ii) it must be ensured that the enzymes selected can be conjugated to antibodies and remain stable and active in the extracellular fluid compartment of tumors for relatively long periods. Moreover there are advantages in selecting enzymes with broad substrate specificities that are mammalian in origin (i.e., reduced immunogenicity) but preferably not present extracellularly (e.g., in human blood or in the human gastrointestinal tract to avoid extradelivery of the drug). The active product is likely to be highly diffusible through the tumor. A short half-life (measured in seconds rather than minutes) would help to avoid toxic effects on normal cell renewal tissues.

In this context, a number of prodrugs have recently been developed that can be transformed into active anticancer drugs by enzymes of both mammalian and non-mammalian origin. For example, among the former approaches (*40,41*), the targeted enzyme alkaline phosphatase (AP, calf intestine) was used to convert etoposide phosphate (EP) into the clinically approved anticancer drug etoposide but other phosphorylated drug derivatives including mitomycin phosphate, or doxorubicin phosphate (*42*), and mustard phosphate (*43*) were subsequently prepared as potential prodrugs that could be activated by AP. All the phosphorylated drug derivatives were found to be much less cytotoxic than their hydrolysis products, and it has clearly been demonstrated that the MAb-AP conjugates are able to convert the prodrugs into cytotoxic anticancer drugs. Results also indicate that the MAb-AP conjugates were present in the tumors and that the localized enzyme was still active.

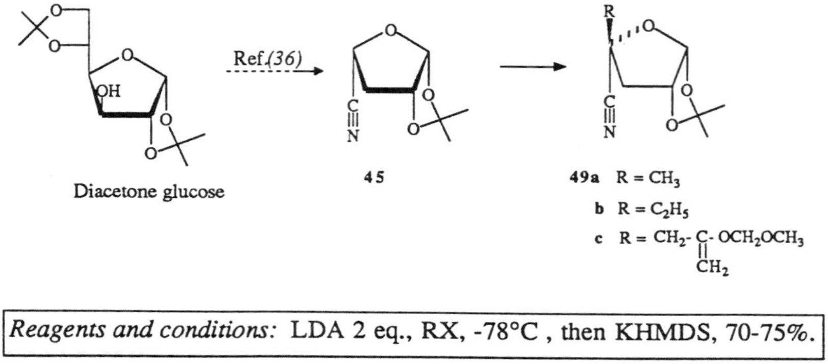

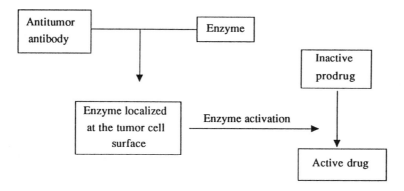

Figure 5. Stereoselective alkylations of cyano–sugar template **45.**

Figure 6. The antibody-directed enzyme prodrug therapy approach.

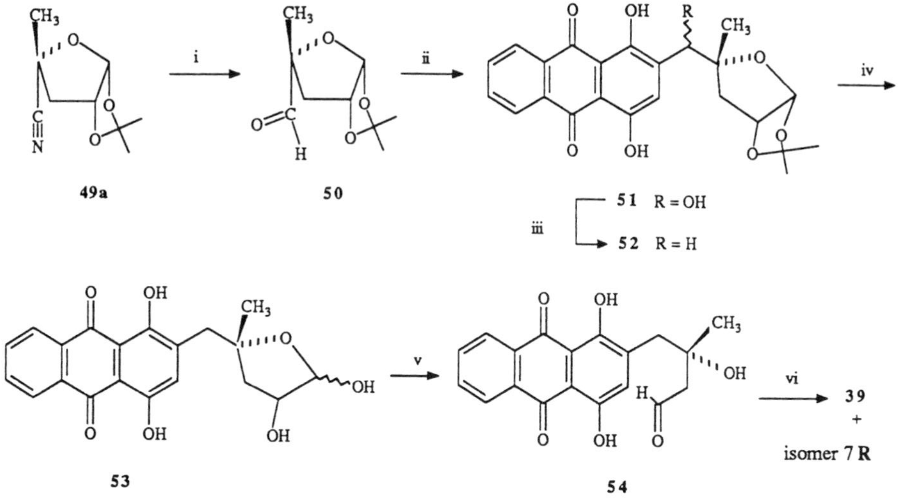

Scheme 8

Reagens and conditions i: Ni Raney, H₂O, Pyr., AcOH (1:1:1), 80%; ii: leucoq., DBU DMF, r.t., 90%; iii: Na₂S₂O₄, DMF - H₂O, 90%; iv: AcOH - H₂O(4:1) reflux, 90%; v: NaIO₄, AcOH-H₂O, 90%; vi: Na₂S₂O₄, NaOH, MeOH-THF, 0°C, 0.5 h, 80%, ratio 7(**S**) / 7(**R**) = 4 : 1

A great number of other enzymes have also been used, including carboxypeptidase A2 (44-46) and G2 (38,47,48), penicillin V-amidase (49), penicillin G-amidase (50), β-lactamase (51-55), cytosine deaminase (41,42), o-nitroreductase (56), and β-glucuronidase (57). Moreover, a first clinical trial has been realized (58) with a bacterial enzyme carboxypeptidase G2 conjugated to anti-HCG and anti-CEA antibodies in patients with advanced colorectal cancer.

For their part, Bosslet and co-workers (Behring Institute ,Marburg, Germany), with whom we are collaborating, prepared a fusion protein (59) consisting of the humanized Fab fragment of the anti CEA MAb BW 431, and human lysosomal β-glucuronidase. The human enzyme was selected to minimize the antibody response in patients.

Specific activation of glucuronide prodrugs at tumor cells with β-glucuronidase-antibody conjugates may possess advantages over other prodrug-enzyme combinations. The absence of high serum β-glucuronidase activities in humans should minimize premature activation of prodrug.[60] β-Glucuronidase is highly specific for the glucuronyl residue, but has little specificity for the conjugated aglycone, suggesting that a wide variety of glucuronide prodrugs can be synthesized. Moreover, glucuronide prodrugs also appear to be less toxic than sulfate or phosphate prodrugs (61).

In a preliminary investigation Gerken and co-workers (62) elaborated a model including a direct linkage of a glucuronyl residue with the 14-OH of doxorubicin, whereas Haisma and co-workers (63) have studied the glucuronide of epirubicin isolated from the urine of patients treated with this drug (Figure 7). The doxorubicin prodrug was found to be approximately 100 times less toxic than the corresponding doxorubicin, but the enzyme gives only a slow release of doxorubicin from the prodrug so that the cytotoxic concentration of the drug was not achieved. The decrease in enzymatic activity which was also observed in the case of epirubicin prodrug, may have resulted from the proximity of the enzyme site to the bulky aglycon. This led us to imagine new prodrugs, including a self-immolative spacer between the anthracycline part and the glucuronyl residue, and to attach this spacer to the 3′-NH$_2$ of the daunosaminyl moiety of the anthracycline. The conception of this new kind of prodrug was essentially based upon two considerations:

On the one hand in 1981 Katzenellenbogen (64) reported a chemical linkage he considered useful for solving certain problems in prodrug design. Thus, he demonstrated that it is possible to prepare a model of prodrug in which a connector between a specifier moiety and a drug is released by a sequence of hydrolytic steps: the first involving an enzymatic cleavage, the second a solvolysis that proceeds spontaneously but only after the first step occurs.

On the other hand, 1,4 or 1,6 eliminations from hydroxy- or amino-substituted benzyl systemes have long been known . Their high reactivity is due to the strong electrodonor ability of the O or NH$_2$ group in these positions, and the ease of elimination depends on the nature of the leaving group. Studies by Wakselman (65) have shown that this elimination is a fast reaction under mild conditions when the group in *para* position is a good leaving group such as a halide, ester, or carbamate.

Therefore, elaborations of new prodrugs of the general formula in Figure 8 were undertaken.

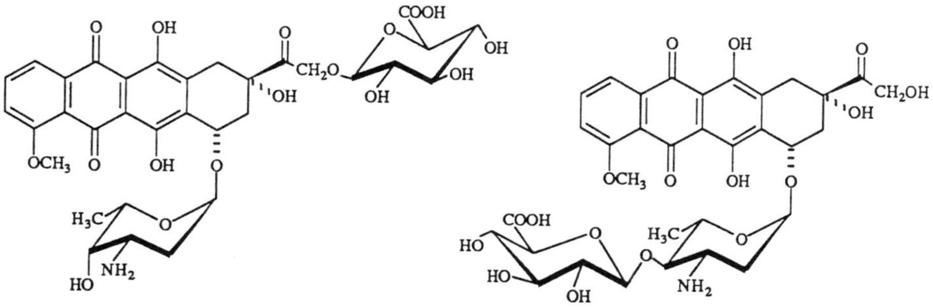

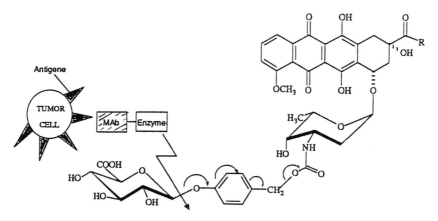

Gerken et al.*(62)*

Haisma et al.*(63)*

Figure 7.   Direct linkage of glucuronyl residue to doxorubicin (**62**) and the glucuronide of epirubicin isolated from human urine (**63**).

Figure 8.   New prodrugs with a self-immolating spacer between the anthracycline and the glucuronyl residue.

After a preliminary approach using galactose as the sugar-residue, (66) the prodrug of daunorubicin **61** (scheme 9) was synthesized by coupling the methyl peracetylglucuronate with *p*-cresol. Conversion of the methyl group, as present in **55** into an hydroxymethyl group, as in **58**, was achieved in three steps *via* bromination, solvolysis, and reduction. Activation of **58** with *N,N'*-disuccinimidyl carbonate afforded **59**, which was condensed with daunorubicin in the presence of triethylamine to give **60**. Complete deprotection of the sugar residue led to **61**.

Although in vitro enzymatic cleavage of **61** proceeded normally, the insufficient acidity of the phenolic function (pK_a =10) prevented self-elimination. Thus, **61** gave rise to **62**, but not to daunorubicin itself. Therefore in order to enhance the acidity of this phenol, synthesis of analogs bearing an *o*-chloro or an *o*- nitro electron withdrawing substituents (pKa = 7-8) were undertaken.

The synthesis of the *o*-nitro derivative F910754 which is depicted in scheme 10, started from the methyl peracetyl glucuronate bromide **63** and commercially available *o*-nitro phenol **64**.

Preliminary *in* vitro experiments carried out at Behring Institute have shown that the prodrug F910754 which was easily converted into daunorubicin in the presence of β-glucuronidase or of the fusion protein[59], could be suitable for use following the A.D.E.P.T. concept. In both cases, this enzymatic cleavage proceeded easily and the spacer was rapidly eliminated (t_1 = 132 min. and t_2< 5 min.). Moreover F910754 was found to be 100 times less cytotoxic (L1210 leukemia cells) than daunorubicin and stable in plasma, since 91% was recovered after 24 h at 37°C (Figure 9).

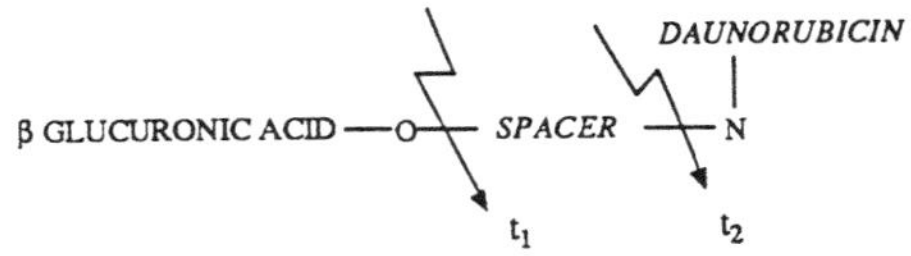

β GLUCURONIC ACID —O— SPACER —N— DAUNORUBICIN

t_1 t_2

Products	Spacers	IC_{50}* μg/mL	t_1	t_2
F910754	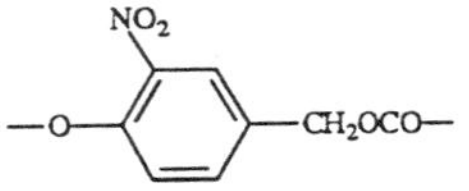 —O—⟨NO₂⟩—CH₂OCO—	2	132 min.	< 5 min.
F910722	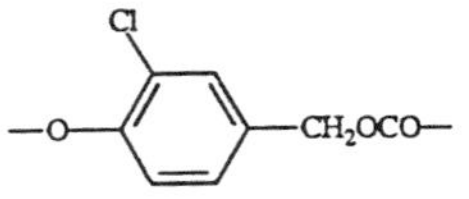 —O—⟨Cl⟩—CH₂OCO—	2.5	28 min.	< 5 min.
F910733	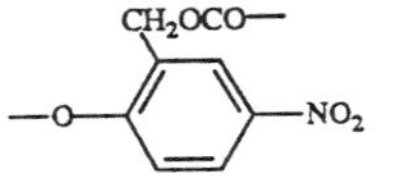 —O—⟨CH₂OCO—⟩—NO₂	3.1	130 min.	320 min.
F910765	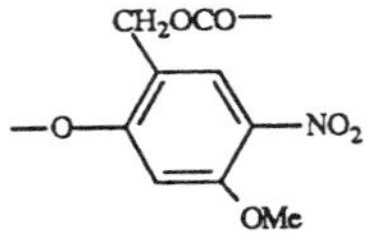 —O—⟨CH₂OCO—, OMe⟩—NO₂	> 10	342 min.	

* For doxorubicin: IC_{50} = 0.015 μg/mL

IC_{50} = drug concentration that inhibited cell growth (L1210 leukemia cells) by 50% compared to untreated control cultures.

Figure 9. New prodrugs that are less cytotoxic than daunorubicin when tested against L1210 leukemia cells.

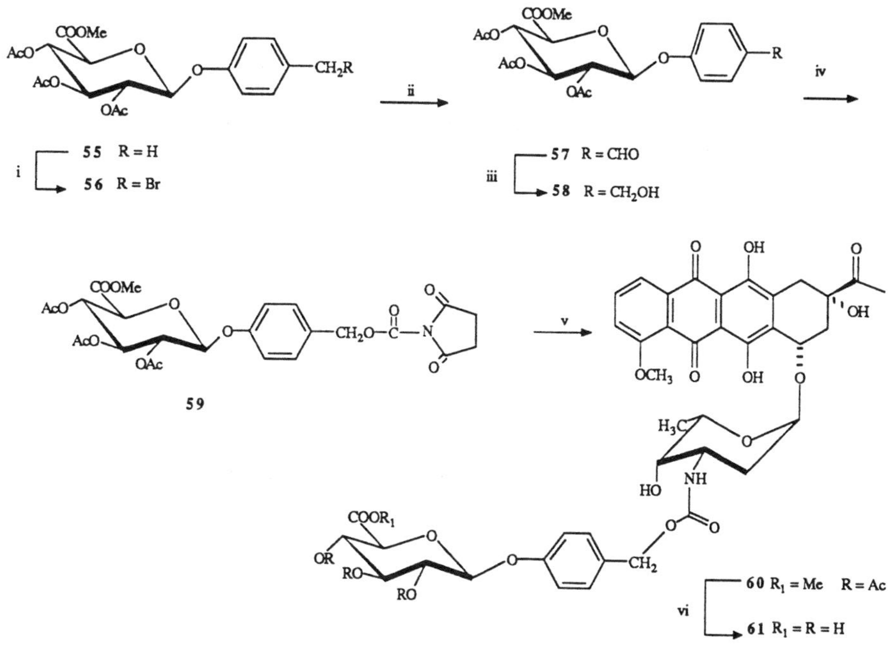

Reagents and conditions i: N-bromosuccinimide, CCl₄, 80°C, 79%; ii: acetone-H₂O, AgNO₃, r.t. 2 h; iii: NaBH₄, iPrOH, silice; iv: N,N'-disuccinimidyl carbonate (DSC), CH₃CN, CH₂Cl₂, Et₃N, 52%; v: daunorubicin, DMF, Et₃N, 70%; vi: MeONa-MeOH, 0°C, 15 min, then BaO, MeOH, r.t., 3 h, 80%

Scheme 9

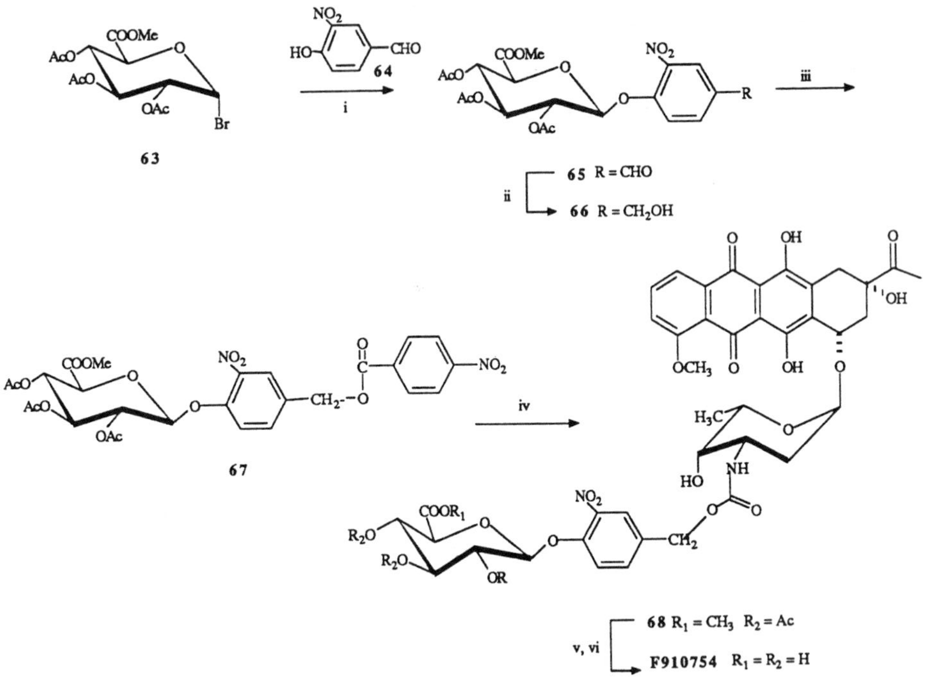

Reagents and conditions: i: Ag₂O, CH₃CN;, r.t., 4 h, 80%; ii: NaBH₄, iPrOH, 72%; iii: 4-nitro-benzylchloroformiate, pyrid., CHCl₃, Et₃N, 92%; *v:* daunorubicin, Et₃N, 46%; *v:* MeONa-MeOH, 0°C, 30 min.; *vi:* NaOH, 2.5 eq., THF, r.t., 98%.

Scheme 10

Starting from the adequate *o*- or *p*-susbtituted phenols, prodrugs F910722 was similarly prepared as well as the regioisomer F910733 or the *p*-nitro-*m*-methoxy analog F910765. As depicted in the next table, all these prodrugs were considerably less cytotoxic than daunorubicin when tested against L1210 leukemia cells. Cleavage of these prodrugs was studied with the fusion protein and the relative rate of hydrolysis and self-elimination differed dramatically between the two series *o* or *p*-substituted phenols. Plasma stability was controlled by addition of the prodrugs to freshly prepared plasma, incubation at 37°C, and analysis by HPLC after 24 h.

Studies concerning the in vivo activities of the fusion protein[59] in combination with these prodrugs are now under investigation by Bosslet et al. at Behring Institute in Marburg.

Acknowledgments : We would like to thank Drs. F. Bennani, B. Deguin, and E. Bertounesque for their contribution to the first part of this account. We gratefully acknowledge Drs Bosslet, Szech and Hoffmann from Behring Institute for their collaboration and for the biological evaluation of the prodrugs. Thanks are also due to the Centre National de la Recherche Scientifique (Unités associées 1387, 1468 and 1310) and to "Laboratoires Hoechst, France" for financial support.

Literature cited

1. Arcamone, F. *Doxorubicin*, Academic Press: New York, 1981.
2. Terashima, S.; Jew, S.-S.; Koga, K. *Tetrahedron Lett.* **1978**, 19, 4937; Terashima, S.; Tanno, K.; Koga, K. *Tetrahedron Lett.* **1980**, 21, 2749.
3. Terashima, S.; Tanno, N.; Koga, K. *Tetrahedron Lett.* **1980**, *21*, 2753; Rama Rao, A.V.; Yadav, J.S.; Bal Reddy, K.; Mehendale, A.R. *J. Chem. Soc., Chem. Commun.* **1984**, 453; *Tetrahedron* **1984**,*40* , 4643.
4. Arcamone F.; Bernardi, L.; Patelli, B.; Di Marco, A. Ger. Patent 2,601,785 (July 29, **1976**).
5. Whistler, R.L.; BeMiller, J.N. *J. Org. Chem.* **1961**, 2886; In *"Methods in Carbohydr. Chem."*, Academic Press: New York, 1963; vol II p. 477.
6. Whistler, R.L.; BeMiller, J.N. In *"Methods in Carbohydr. Chem."*, Academic Press: New York, 1963; vol. II p. 484.
7. Marschalk, C.; Koenig, J.; Ouroussoff, N. *Bull. Soc. Chim. Fr.* **1936**, 1545; Marschalk C., *Bull. Soc. Chim. Fr.* **1939**, 655.
8. Suzuki, F.; Gleim, R.D.; Trenbeath, S.; Sih, C.J. *Tetrahedron Lett.* **1977**, 2303; Suzuki, F.; Trenbeath, S.; Gleim, R.D.; Sih, C.J. *J. Am. Chem. Soc.* **1978**, *100* 2272.
9. Krohn, K.; Radeloff, M. *Chem. Ber.* **111** (1978) 3823; Krohn, K.; Hemme, C. *Liebigs Ann. Chem.* **1979**, 19.
10. Morris, M.J.; Brown, J.R. *Tetrahedron Lett.* **1978**, 2937.
11. Penco, S.; Casazza, A.M.; Franchi, G.; Barbieri, B.; Bellini, O.; Podesta, A.; Savi, G.; Pratesi, G.; Geroni, C.; Di Marco, A.; Arcamone, F. *Cancer Treatment Reports* **1983**,*67*, 665.
12. Bennani, F.; Florent J.-C. ; Koch, M.; Monneret, C. *Tetrahedron* **1984**,*40* , 4669.
13. Florent, J.-C.; Ughetto-Monfrin, J.; Monneret, C. *J. Org. Chem.* **1987**,*52* , 1051.
14. Monneret, C.; Florent, J.-C.; Bennani, F.; Sedlacek, H.H.; Kraemer, H.P. *Actual. Chim. Thér.* 15th series **1988**, p 253; Florent, J.-C.; Genot, A.; Monneret, C. *Journal of Antibiot* . **1989**,*42*, 1823.
15. Florent, J.C.; Gaudel G.; Monneret, C.; Hoffmann, D.; Kraemer, H. P. *J. Med. Chem.*. **1993**,*36*, 1364.

16. Nafziger, J.; Auclair, C.; Florent, J.-C.; Guillosson, J.-J.; Monneret, C. *Leukemia Res.* **1992**, 15, 709.
17. Lewis, C.E. *J. Org. Chem.* **1970**, *35*, 2938.
18. Krohn, K.; Behnke, B. *Liebigs Ann. Chem.* **1983**, 1818.
19. Bertounesque, E.; Florent, J.-C.; Monneret, C. *Tetrahedron Lett.* **1990**, *31*, 7153.
20. Deguin, B.; Florent, J.-C.; Monneret, C. *J. Org. Chem.* **1991**, *56*, 405.
21. Kraus, G.A.; Sugimoto, H. *Tetrahedron Lett.* **1978**, 2263.
22. Hauser, F.M.; Rhee, R.P. *J. Am. Chem. Soc.* **1977**, *99* , 4553; *J. Org. Chem.* **1978**, *43* , 178.
23. Li, T.T.; Walsgrove, T.C. *Tetrahedron Lett.* **1981**, *22* , 3741; Li, TT.; Wu, Y.-L.; Walsgrove, T.C. *Tetrahedron* **1984**,*40*, 4701.
24. Hauser, F.M.; Baghdanov, V.M. *Tetrahedron* **1984**, *40* , 4719 and references cited therein.
25. Chenard, B.L.; Dolson, M.G.; Sercel, A.D.; Swenton, J.S. *J. Org. Chem.* **1984**, 49, 318 and references cited therein.
26. Génot, A.; Florent, J.-C.; Monneret, C. *Tetrahedron Lett.* **1989**, *39*, 711.
27. Scott, C.A.; Wesmacott, D.; Broadhurst, M.J.; Thomas, J.; Hall, M.J. *Br. J. Cancer* **1986**, *53*, 595; De Vries, E.G.E.; Zijstra, J.G. *Eur. J. Cancer* **1986**, *26*, 659.
28. Coley, H.M.; Twentyman, P.R.; Workman, P. *Eur. J. Cancer* **1986**, *26*, 655.
29. Coley, H.M.; Twentyman, P.R.; Workman, P. *Anticancer Drug Design* **1992**, *7*, 471.
30. Watanabe, M.; Komeshima, N.; Naito, M.; Isoe, T.; Otake, N.; Tsuruo, T. *Cancer Research* **1991**, *51*, 157 and references cited therein.
31. Bertounesque, E.; Florent, J.-C.; Monneret, C. *Synthesis* (1991) 270.
32. Bock, K.; Maya Castilla, I.; Lundt, I.; Pedersen, C. *Acta Chemica Scand.* **1987**, *B41*, 13.
33. Millal, F.; Bertounesque, E.; Florent, J.-C.; Monneret, C., unpublished results.
34. Sun, K.M.; Fraser-Reid, B. *Synthesis* **1982**, 28.
35. Weidman, H.; Schwarz, H., *Monatsh. Chem.* **1972**, *103*, 218.
36. Cousson, A.; Florent, J.-C.; Le Gouadec, G.; Monneret, C. *J. Chem. Soc., Chem. Commun* . **1993**, 388.
37. Connors, T.A. *Biochemie* **1978**, *60*, 979.
38. Bagshawe, K.D. *Br. J. Càncer* **1987**, *56*, 531; Bagshawe, K.D.; Springer, C.J.; Searle, F.; Antoniw, P.; Sharma, S.K.; Melton, R.G.; Sherwood, R.F. *Br. J. Cancer* **1988**, *58*, 700.
39. Senter, P.D.; Saulnier, M.G.; Schreiber, G.J.; Hirschberg, D.L.; Brown, J.P.; Hellstrom, I.; Hellstrom, K.E. *Proc. Natl. Acad. Sci. USA* **1988**, *85*, 4842.
40. Senter, P.D.; Schreiber, G.J.; Hirschberg, D.L.; Ashe, S.A.; Hellstrom, K.E.; Hellstrom, I. *Cancer Res.* **1989**, *49*, 5789.
41. Senter, P.D. The *FASEB Journal* **1990**, 188.
42. Senter, P.D.; Wallace, P.M.; Svensson, H.P.; Kerr, D.E.; Hellstrom, I.; Hellstrom, K.E. *Adv. Exp. Med. Biol.* **1991**, *303*, 97.
43. Wallace, P.M.; Senter, P.D. *Bioorg. Chem.* **1991**, *2*, 349.
44. Haenseler, E.; Esswein, A.; Vitols, K.S.; Montejano, Y.; Mueller, B.-M.; Reisfeld, R.A.; Huennekens, F.M. *Biochem.* **1992**, *31*, 891.
45. Esswein, A.; Haenseler, E.; Montejano, Y.; Vitols, K.S.; Huennekens, F.M. *Adv. Enzyme Regul.* **1991**, *31*, 3.
46. Mann, J.; Haase-Held, M.; Springer, C.J.; Bagshawe, K.D. *Tetrahedron* **1990**, *46*, 5377.

47. Searle, F.; Buckley, R.G. *Brit. J. Cancer* **1986**, *53*, 277.
48. Springer, C.J.; Bagshawe, K.D.; Sharma, S.K.; Searle, F.; Boden, J.A.; Antoniw, P.; Burke, P.J.; Rogers, G.T.; Sherwood, R.F.; Melton, R.G. *Eur. J. Cancer* **1991**, *27*, 1361.
49. Kerr, D.E.; Senter, P.D.; Burnett, W.V.; Hirschberg, D.L.; Hellstrom, I.; Hellstrom, K.E. *Cancer Immunol. Immunother.* **1990**, *31*, 202.
50. Vrudhula, V.M.; Senter, P.D.; Fischer, K.J.; Wallace, P.M. *J. Med. Chem.* **1993**, *36*, 919.
51. Shepherd, T.A.; Jungheim, L.N.; Meyer, D.L.; Starling, J.J. *Bioorgan. Med. Chem. Lett.* **1991**, *1*, 21.
52. Alexander, R.P.; Beeley, N.R.A.; O'Driscoll, M.; O'Neill, F.P.; Millican, T.A.; Pratt, A.J.; Willenbrock, F.W. *Tetrahedron Lett.* **1991**, *32*, 3239.
53. Svensson, H.P.; Kadow, J.F.; Vrudhula, V.M.; Wallace, P.M.; Senter, P.D. *Bioconjugate Chem.* **1992**, *3*, 176.
54. Meyer, D.L.; Jungheim, L.N.; Mikolajczyk, S.D.; Shepherd, T.A.; Starling, J.J.; Ahlem, C.N. *Bioconjugate Chem.* **1992**, *3*, 42.
55. Jungheim, L.N.; Shepherd, T.A.; Meyer, D.L. *J. Org. Chem.* **1992**, *57*, 2334.
56. Sunters, A.; Baer, J.; Bagshawe, K.D. *Biochem. Pharmacol.* **1991**, *41*, 1293.
57. Roffler, S.R.; Wang, S.-M.; Chern, J.-W.; Yeh, M.-Y.; Tung, E. *Biochem. Pharmacol.* **1991**, *42*, 2062; Wang, S.-M.; Chern, J.-W.; Yeh, M.-Y.; Co Ng J.; Tung, E.; Roffler, S.R. *Cancer Res.* **1992**, *52*, 4484.
58. Bagshawe, K.D.; Sharma, S.K.; Springer, C.J.; Antoniw, P.; Rogers, G.T.; Burke, P.J.; Melton, R.; Sherwood, R. *Immun. and radiopharmaceuticals* **1991**, *4*, 915.
59 Bosslet, K.; Czech, J.; Lorenz, P.; Sedlacek, H.H.; Schuermann, M.; Seemann, G.*Br. J. Cancer* **1992**, *65*, 234.
60. Stahl, P.D.; Fishman, W.H. In *"Methods of Enzymatic Analysis "* 3rd ed.; Bergmeyer HU ed.; Verlag Chemie: Weinheim, 1984; Vol.4, pp 246-256.
61. Bukhari, M.A.; Everett J.L.; Ross, W.C.J. *Biochem. Pharmacol.* **1972**, *21*, 963.
62. Gerken, M.; Bosslet, K.; Seeman, G.; Hoffmann,D.; Sedlack, H.-H. *Eur. Patent EP* 0441 218 , 28.01.1991.
63. Haisma, H.J.; Boven, E.; Van Muijen, M.; De Jong, J.; Van der Viggh, W.J.F.; Pinedo, H.M. *Br. J. Cancer* **1992**, *66*, 474.
64. Carl, P.L.; Chakravarty, P.K.; Katzenellenbogen, J.A. *J. Med. Chem.* **24** (1981) 47
65. Wakselman, M. *Nouveau Journal de Chimie* **1987**, *7*, 439.
66. Andrianomenjanahary, S.; Dong, X.; Florent, J.-C.; Gaudel, G.; Gesson, J.-P.; Jacquesy, J.-C.; Koch, M.; Michel, S.; Mondon, M.; Monneret, C.; Petit, P.; Renoux, B.; Tillequin, F. *Bioorgan. Med. Chem. Letters* **1992**, *2*, 1093.

RECEIVED June 3, 1994

Chapter 6

Synthesis and Biological Activities of Fluorinated Daunorubicin and Doxorubicin Analogues

Tsutomu Tsuchiya and Yasushi Takagi

Institute of Bioorganic Chemistry, 1614 Ida, Nakahara-ku, Kawasaki 211, Japan

Preparation and antitumor activities of a variety of the anthracycline antibiotics having a fluorine at C-2' are discussed. Suitably protected 2-deoxy-2-fluoroglycosyl halides have been coupled with daunomycinone, and derived into doxorubicin analogues. The anthracycline glycosides introduced a fluorine at the α-side of C-2' of doxorubicin and replaced the 3'-amino group with a hydroxyl group showed high antitumor activity with decreased toxicity. A $2'R$-configuration was necessary to display activity; $(2'S)$-2'-fluoro and 2',2'-difluoro analogues were devoid of activity. While modifications at C-3' of the 2'-fluoroanthracycline antibiotics diminished their activity, most of the modifications at C-4' involving inversion of the configuration of the substituent did not significantly change activity.

The anthracycline glycosides of which daunorubicin and doxorubicin are representative examples, are clinically important antitumor antibiotics. However, cumulative cardiotoxicity and other undesirable side-effects have restricted their extensive use. To overcome these drawbacks and expand their use, many efforts have been made over the past two decades. Most of the anthracycline glycosides possess a 2-deoxy sugar linked to the anthracyclinone moiety and are susceptible to acid hydrolysis. Since neat aglycones do not show antitumor activity, strengthening the glycosidic bond will give highly effective analogues. We therefore undertook to prepare anthracycline glycosides having a fluorine at C-2', that atom being the most electron-withdrawing of all atoms stabilizes the glycosidic bond against hydrolysis. The relatively small van der Waals radius of fluorine (135 pm; cf. O, 140 pm; H, 120 pm) will also give only a small steric effect. Generally, these characteristics of fluorine, namely, the high electron density on the fluorine atom polarizing the C-F bond, the small atomic size, and the strong C-F bond-energy, make fluorine a unique element in comparison with the other halogens, and give the fluorinated analogues a special position among the anthracycline glycosides.

In this account, we want to describe the preparation of 2'-fluoroanthracycline glycosides and discuss their structure-antitumor activity relationships (*1*).

2'-Fluoroanthracyclines with Neutral Sugars

Most of the anthracycline antibiotics have a 3'-amino or 3'-dimethylamino group, which was once believed to be essential for antitumor activity. Horton et al. (*2 – 4*), however, found that replacement of the 3'-amino group of daunorubicin and doxorubicin by a hydroxyl group gave derivatives of fairly good antitumor activity and weak toxicity. This led us to prepare daunorubicin and doxorubicin analogues having a neutral sugar with a fluorine at C-2. During our synthesis, Horton et al. (*5 – 7*) reported a stimulating study, along the same lines as ours, to synthesize analogues having 3'-deamino-3'-hydroxy-2'-iodo, -chloro, and -bromo groups and found that all derivatives with 2'*R* and 2'*S* configurations are active and inactive, respectively (*8, 9*). Based on their results, we set out to prepare, as the first targets, 7-*O*-(2,6-dideoxy-2-fluoro-α-L-talopyranosyl)daunomycinone (FTDM) (*10, 11*), 7-*O*-(2,6-dideoxy-2-fluoro-α-L-talopyranosyl)adriamycinone (FTADM) (*10, 11*), and their 4-demethoxy analogues **1** and **2** (*12*).

			R¹	R²
Daunorubicin	R = H	FTDM	H	OMe
Doxorubicin	R = OH	FTADM	OH	OMe
		1	H	H
		2	OH	H

2,6-Dideoxy-2-fluoro-α-L-talopyranose was prepared from L-fucose. Introduction of fluorine at C-2 of sugars sometimes presents difficulties. Thus, when methyl 6-deoxy-3,4-*O*-isopropylidene-α-L-galactopyranoside **3**, readily prepared from L-fucose, was treated with diethylaminosulfur trifluoride (DAST), the desired 2-fluoro-L-*talo* derivative was not obtained, and instead a five-membered compound was produced, possibly by ring contraction. Our attempt to substitute the triflate of **3** with fluoride also gave the same result. A similar reaction was reported recently by Baer et al.(*13*). Thus, we decided to introduce C(2*R*)-fluorine through a 2,3-epoxy-ring opening as shown in Scheme I. Compound **3** was converted into the L-*gulo*-2,3-epoxide **4** *via* a reaction sequence involving 2-*O*-acetylation, deacetonation, 3-*O*-tosylation, 4-*O*-benzylation, and sodium methoxide treatment. Reaction of **4** with potassium hydrogenfluoride in ethylene glycol at 180°C gave the 2-deoxy-2-fluoro-L-idopyranoside **5** in a moderate yield (44%). Inversion of HO-3 in **5** was performed by an oxidation-reduction sequence, and the resulting L-talopyranoside **6** was converted into the glycosyl bromide **7** *via* three steps. Coupling of **7** with daunomycinone or 4-demethoxydaunomycinone was performed by a Koenigs–Knorr type of reaction (yellow mercury(II) oxide and mercury(II) bromide with molecular sieves in dichloromethane) to give the α-L-glycosides in good yields (82 and 71%, respectively). Subsequent deacetylation gave FTDM and **1**. These compounds were further transformed into the corresponding 14-hydroxy compounds (FTADM and **2**) substantially according to Arcamone's procedure (*14*). Bromination at C-14 by bromine in the presence of methyl orthoformate followed by acetonation gave the 14-

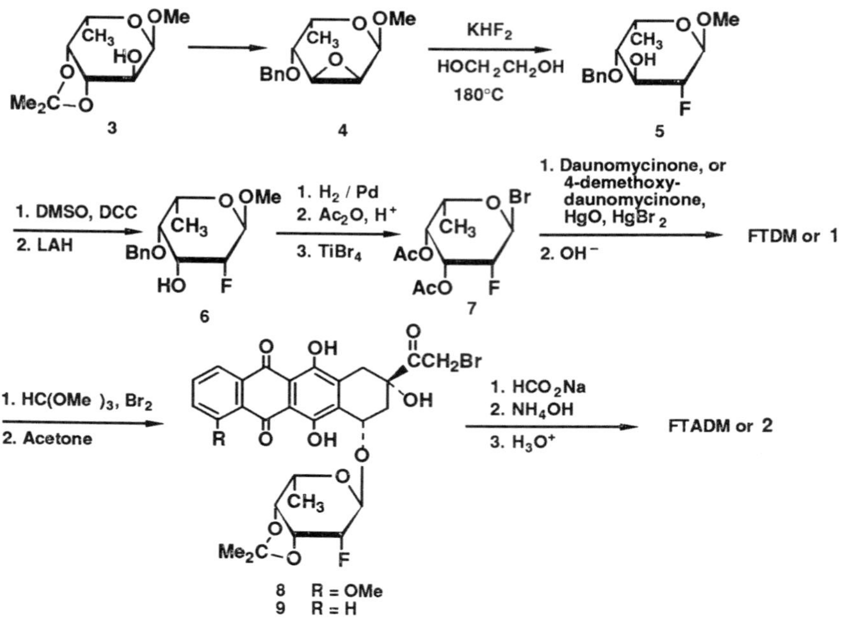

Scheme I

bromo-3',4'-*O*-isopropylidene derivatives **8** and **9**, which were converted into FTADM and **2** by treatment with sodium formate and deprotection. The glycosidic bond of the 2'-fluoroglycosides was very stable, as expected, against acid hydrolysis. As shown in Table I, FTADM displayed significant antitumor activity in vivo against murine leukemia L1210 and showed decreased toxicity in comparison with doxorubicin. The 4-demethoxy analogue **2** also showed strong activity but was more toxic than FTADM. FTADM thus prepared was a candidate for clinical use but being scarcely soluble in water due to its lack of an amino group, its use for intravenous administration was limited.

To increase the solubility of FTADM the corresponding 14-hemiesters were prepared (*15*); the mono sodium salts of some dicarboxylic acids were condensed with the 14-bromo derivatives **8** and **9** according to the method of Israel et al. (*16*) giving, after deacetonation, 14-hemisuccinate **10**, 14-hemiglutarate **11**, 14-hemiadipates **12** and **15**, 14-hemipimelate **13**, and 14-hemisuberate **14**. All of these compounds showed better

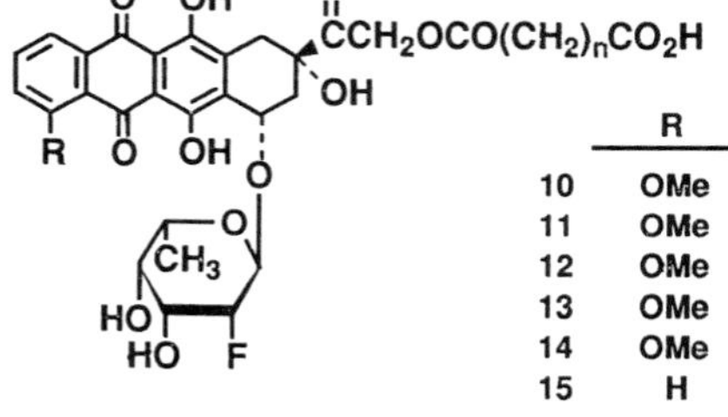

	R	n
10	OMe	2
11	OMe	3
12	OMe	4
13	OMe	5
14	OMe	6
15	H	4

Table I.   Antitumor Activities (T/C %) of 2'-Fluoro and 2'-Methoxy Anthracyclines in Comparison with Daunorubicin (DNR) and Doxorubicin (DOX) against Murine Leukemia L1210[a]

Compound	Dose (μg/mouse/day)							
	100	50	25	12.5	6.25	3.13	1.56	0.50
FTDM	184	217	171	125	105	105		
FTADM	>740	>352	275	185	182	127		
2			104*	>676*	>484	300	144	
11	>674	>329	>407	211	169	112		
12	>674	>340	>419	219	112	107		
13	>674	>556	>486	>458	124	140		
15			>296*	242*	166	130	99	
28	160*	280	>309	189	123	109		
29	211*	231	>426	>363	>440	157		
30	>463	>686	>346	329	274	123		
43	174*	200*	222	140	133	117		
44	97*	187*	168*	203	135	123		
45		22*	25*	33*	33*	49*	53*	144
64	>314*	260	146	124	117	111		
65	124*	>761*	>501	279	152			
66	212	132	102	105	99	95		
67	151*	>484	171	122	113	110		
76	>452	121	117	108	102	95		
82	>389	137	109	94				
83	153*	>477	>494	172	230	141	107	
DNR	117*	151*	193	166	133	130		
DOX	177*	273*	330	208	132	140		

[a] Leukemia L1210 cells (10^5) were inoculated ip into CDF_1 mice (20 ± 1 g). Drugs were administered ip daily, starting 24 h after inoculation, from day1 to 9. Survival studies were continued up to 60 days.
* Toxic

solubility in water and higher activity than that of parent FTADM. 4-Demethoxy analogue **15** again showed higher toxicity than **12**. 14-Hemipimelate **13** is now under trial for clinical use.

As the sugar component of **13**, 2,6-dideoxy-2-fluoro-L-talopyranose was prepared from L-fucose, which is expensive and therefore unsuitable as a starting material for scale-up preparation. Thus, we tried to prepare the sugar part (*17*) from D-fructose. The synthesis was performed based on head-to-tail inversion of D-fructose, in that C-2 is reduced and C-6 oxidized as shown in Scheme II. Treatment of 3-*O*-benzyl-4-*O*-*tert*-

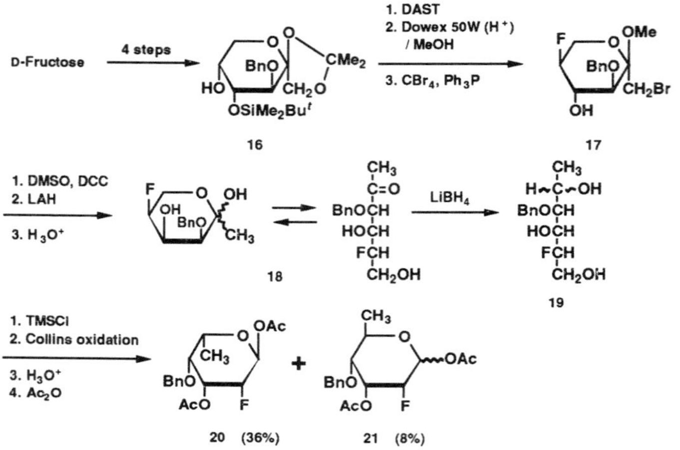

Scheme II

butyldimethylsilyl-1,2-*O*-isopropylidene-β-D-fructopyranose **16**, prepared from D-fructose in four steps, with DAST gave the 5-deoxy-5-fluoro-α-L-sorbopyranose, which was methanolyzed and selectively brominated to give the 1-bromo derivative **17**. After oxidation, the 4-oxo derivative was treated with lithium aluminum hydride to inverse the configuration at C-4 (in **17**) and reductively debrominate the bromine simultaneously. Final hydrolysis gave **18** as an equilibrium mixture of two pyranoids and an acyclic keto form. Reduction of **18** with lithium borohydride gave a diastereoisomeric mixture of **19** in a ratio of 2S:2R = 3.5:1. Selective oxidation (*18*) of the primary hydroxyl group of **19** followed by acetylation gave, after chromatographic separation, the desired 2,6-dideoxy-2-fluoro-L-talopyranose **20** in 36% yield, together with the 5-epimer (D-*allo*, 8%). As shown in Scheme III, another route to the 2-deoxy-2-fluoro sugar with fewer steps was

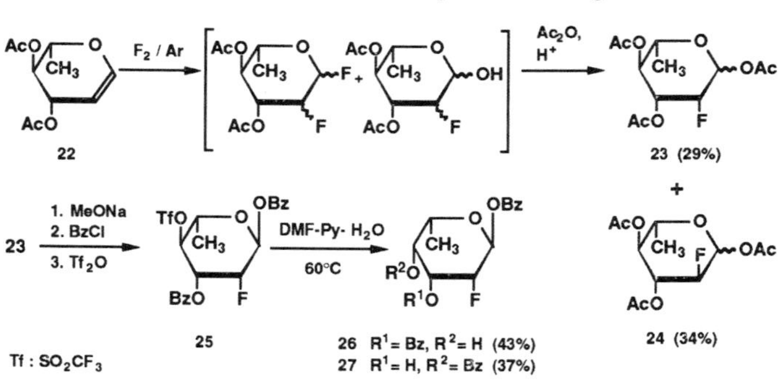

Scheme III

developed by using L-rhamnose. Addition of fluorine to L-rhamnal diacetate **22** followed by acetolysis gave a mixture of 1,3,4-tri-*O*-acetyl-2,6-dideoxy-2-fluoro-L-manno **23** and L-glucopyranose **24** in a ratio of 1:1.2. Deacetylation of the former compound followed by selective benzoylation gave the 1,3-dibenzoate. After trifluoromethanesulfonylation, the 4-triflate **25** was treated with pyridine in DMF to give 2,6-dideoxy-2-fluoro-L-talopyranose dibenzoates **26** and **27** *via* a benzoxonium ion intermediate.

4'-Epidaunorubicin and 4'-epidoxorubicin (*19*) are known to be more potent than their parent anthracyclines.   Thus, the above described 2,6-dideoxy-2-fluoro-L-mannopyranose triacetate **23** was utilized to prepare the 4'-epi analogues (*20*) of FTDM and FTADM. Coupling of the glycosyl bromide derived from **23** with daunomycinone and followed by usual reactions gave **28**. The 14-hydroxy analogue **29** and the 14-hemipimelate **30** were also prepared. As shown in Table I, compounds **28** and **29** were

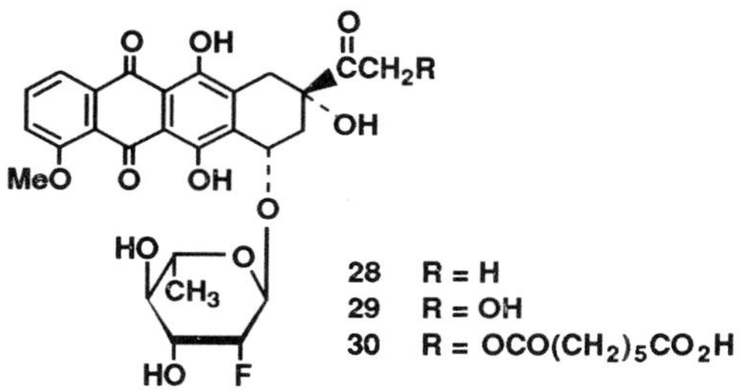

28   R = H
29   R = OH
30   R = OCO(CH$_2$)$_5$CO$_2$H

slightly more toxic than the L-*talo* analogues but more effective at lower dose-ranges, whereas **30** showed similar activity to the corresponding 14-hemipimelate **13**. These results suggest that the configuration at C-4' of 2'-fluoroanthracyclines affects little on the activity.

The significant antitumor activity and decreased toxicity of FTADM and its 4'-epimer prompted us to clarify the relation between biological activities and the configuration at C-2'; thus, the 2'-epifluoro derivative **33** with a 2'*S* structure was prepared. Treatment of methyl 6-deoxy-3,4-*O*-isopropylidene-β-L-talopyranoside **31** (which was prepared by inversion of the HO-2 of methyl β-L-fucoside) with DAST readily gave the 2-deoxy-2-fluoro-L-galactopyranoside without ring contraction. This is in sharp contrast with the reaction of **3** with DAST. The glycosyl bromide **32** was then coupled with daunomycinone. The reaction proceeded smoothly as for **7**, but gave a mixture of α-L- and β-L-glycosides in a ratio of 1:1.3, which (together with the deprotected products) could not be separated because of their similar mobility. However, the corresponding 3',4'-*O*-isopropylidene derivatives were separated by chromatography. Deacetonation gave the desired 2'-epifluoro isomer **33** (*12*) of FTDM (see Scheme IV). Compound **33** as well as the β-L-glycoside **34** were inactive against L1210 at doses up to 100 μg/mouse/

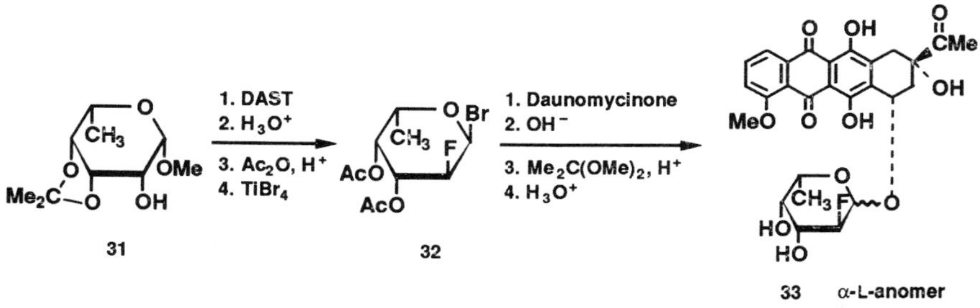

33   α-L-anomer
34   β-L-anomer

Scheme IV

day, indicating that the 2'R but not the 2'S configuration is essential for antitumor activity. This was also true for the other 2'-halo, namely, 2'-iodo (6), 2'-bromo (7), and 2'-chloro (7) derivatives.

To examine the effect of the 2'-fluoro substituent further, the 2',2'-difluoro derivative 40 (12) was prepared (see Scheme V). Benzyl 6-deoxy-3,4-O-isopropylidene-β-L-

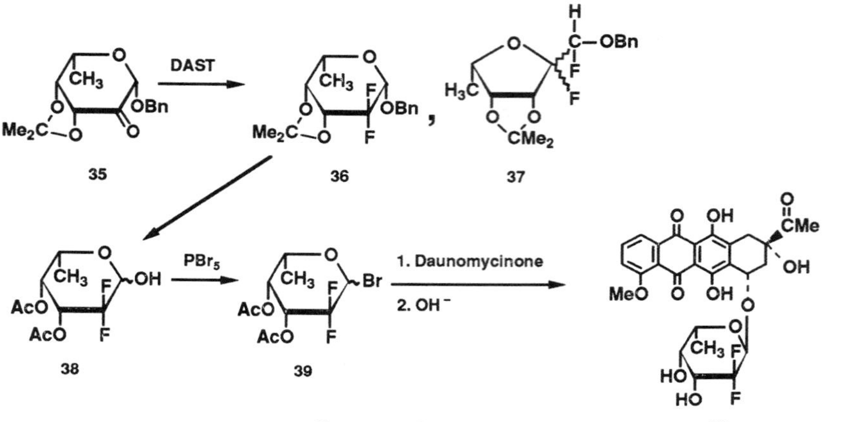

Scheme V

lyxo-hexopyranosid-2-ulose 35 prepared from L-fucose in six steps was treated with DAST to give the 2-deoxy-2,2-difluoro-β-L-*lyxo* derivative 36 and a ring-contraction product 37 in equal amounts. Similar treatment of the α-L isomer of 35 gave 37 almost exclusively. Deprotection of 36 followed by acetylation gave 38 or its 1-acetate, but both of them failed to give the 1-bromide by treatment with thionyl bromide or titanium tetrabromide. Bromination was successfully performed only when 38 was treated with phosphorous pentabromide in refluxing dichloroethane, giving 39 in 97% yield. Coupling of 39 with daunomycinone by a Koenigs–Knorr type reaction gave the α-L-glycoside in a moderate yield, and subsequent deprotection gave 40. Compound 40 was also inactive against L1210 at doses up to 100 μg/mouse/day, suggesting that the presence of an equatorial fluoro group at C-2' causes loss of activity.

The reason why the 2'S isomer 33 is inactive is not clear, but this may be due to poor fitness of the compound to the DNA helix. Inspection of the computer graphic model of the drug–oligonucleotide complex prepared in reference to the daunomycin–d(CGATCG) complex taken from the Protein Data Bank, 1D10 (21) suggests that the equatorial fluorine atom of 33 comes close to the O-2 of the cytidine residue. This may create electrostatic repulsion between the groups in question, thus, destabilizing the DNA–drug complex. On the other hand, the axial fluorine in FTDM and one of the oxygen atoms of phosphoryl group in DNA dispose in such a direction that a water molecule can be inserted by forming hydrogen bondings between them stabilizing the complex.

2'-Fluoroanthracyclines with Amino Sugars

Since the compounds having a (2R)-2-deoxy-2-fluoro-3-hydroxy sugar exhibited strong antitumor activity and decreased toxicity, preparation of analogues having (2R)-3-amino-2,3-dideoxy-2-fluoro sugar, which resembles natural anthracycline antibiotics in structure, was undertaken. Displacement of the 3-triflate of 5 with an azide followed by catalytic reduction gave methyl 3-amino-2,3,6-trideoxy-2-fluoro-α-L-talopyranoside 41. After trifluoroacetylation of the amino group followed by acetolysis, the resulting 1,4-diacetate 42 was coupled with daunomycinone according to the method reported (22), and the product was deprotected by alkaline hydrolysis to give the α-L-glycoside 43

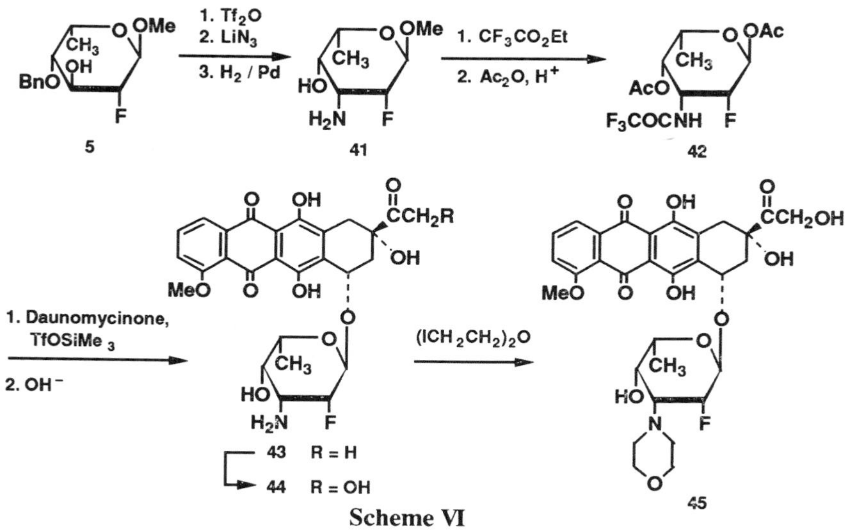

Scheme VI

(*23*) in a moderate yield (Scheme VI). Conversion to the doxorubicin analogue **44** was performed in the usual manner, and **44** was further transformed into the 3'-deamino-3'-morpholino derivative **45** by treatment with bis(2-iodoethyl) ether in the presence of triethylamine. Although **43** and **44** showed activities similar to those for daunorubicin and doxorubicin against L1210, their toxicities increased compared with those for FTDM and FTADM (Table I). This indicates that a 3'-hydroxyl group is superior to a 3'-amino group. Recently, Castillon et al. (*24*) and Baer et al. (*25*) reported the synthesis of (2'S)-2'-fluorodaunorubicin, that is, the 2'-epimer of **43**, which proved to have weaker activity against L1210 (T/C, 129, at 64 mg/kg) than did **43**. Baer et al. (*26*) also reported the synthesis of (2'R)-2'-fluorocarminomycin and found that the compound had activity similar to its parent, carminomycin. These results indicate that, as the cases for 3'-hydroxy analogues, (2'R)-configuration is necessary for the 3'-amino analogues to be able to display antitumor activity. 3'-Deamino-3'-morpholino derivative **45**, like 3'-deamino-3'-morpholinoadriamycin (*27*), showed strong toxicity, and its effective dose-range was very narrow. It is noteworthy that **45** was not cross-resistant with doxorubicin in the P388/ADM subline (IC_{50}, 1.8 ng/ml against P388/S and 2.2 ng/ml against P388/ADM; resistance factor of 1.2).

Analogues Modified at C-3' and C-4'

Multiple drug resistance is a big obstacle in cancer chemotherapy. Although many efforts have been made to overcome this problem, no satisfactory outcome has been obtained. The 14-hemipimelate **13** was shown to be partially active (*28*) against resistant tumor cells in vivo but its activity was not strong enough. While the morpholino derivative **45** exhibited significant cytotoxicity against resistant cells in vitro, strong toxicity in vivo decreased its usefulness.   Recently, numerous studies have been undertaken to understand the mechanism of multiple drug resistance, and some characteristic features have been disclosed. For instance, it was reported (*29*) that in multidrug-resistant cells, the accumulation of a variety of chemically unrelated drugs in the cells was inhibited by an efflux function of the cells, presumably by the action of P-glycoprotein (*30*), with an increase in the Ca^{++} content on the plasma membrane. Administration of calcium antagonist such as verapamil with antitumor agents resulted in effective accumulation of the drugs in the resistant cells, thus enhancing the cytotoxicity of the drugs against the resistant cells (31).

As FTADM and its analogues have vicinal hydroxyl groups at C-3' and C-4', Ca^{++} may chelate between them. If such a chelation promotes the efflux of the drugs, inhibition of the chelation by replacing one of the hydroxyl groups with a substituent should cause increased accumulation in resistant cells. Under this assumption, we undertook to prepare FTDM derivatives modified at C-3' and C-4'.

Synthesis of Derivatives of FTDM Modified at C-3'. As the first target of this series, we synthesized 3'-O-methyl-FTDM **56** (12) starting from **6**. The suitably protected glycosyl bromide **47** prepared by a reaction sequence as shown in Scheme VII was coupled with daunomycinone. The compound **56** thus prepared showed a moderate activity against L1210 even at a high dosage (T/C, 174, 200 μg/mouse/day), but showed only small cross-resistance with doxorubicin in vitro. It is suggested that the modification at C-3' might afford compounds effective against resistant tumor cells. This is in accord with the activity of the 3'-deamino-3'-morpholino derivative **45**. The relatively weak activity of **56** might be attributable to the bulkiness of the methoxyl group. Therefore, 3'-deoxy **57** (12), 3'-deoxy-3'-fluoro **58** (12), and 3'-deoxy-3'-iodo analogues **59** (12) were prepared by coupling of the corresponding glycosyl bromides (Scheme VII) with

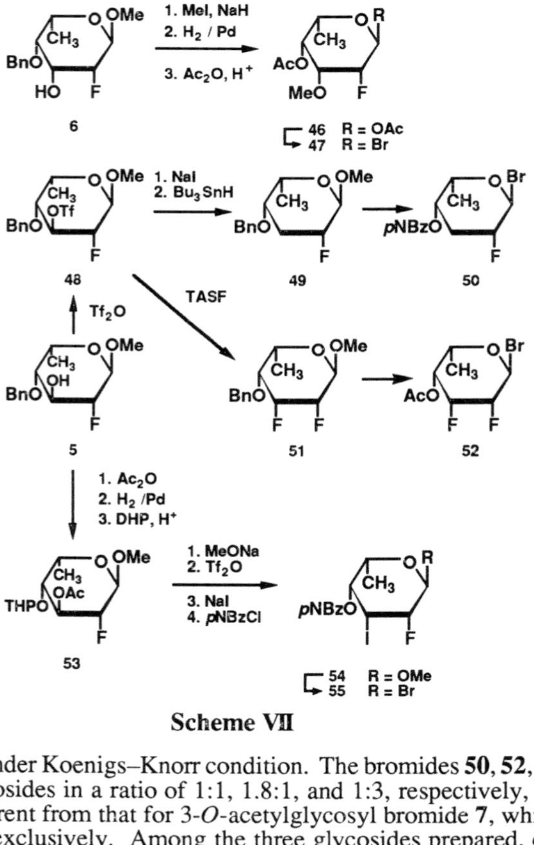

Scheme VII

daunomycinone under Koenigs–Knorr condition. The bromides **50**, **52**, and **55** afforded α-L- and β-L-glycosides in a ratio of 1:1, 1.8:1, and 1:3, respectively, the ratios being considerably different from that for 3-O-acetylglycosyl bromide **7**, which yielded α-L-glycoside almost exclusively. Among the three glycosides prepared, only **58** showed moderate activity against L1210 (T/C, 148, 100 μg/mouse/day). Thus, it seems that

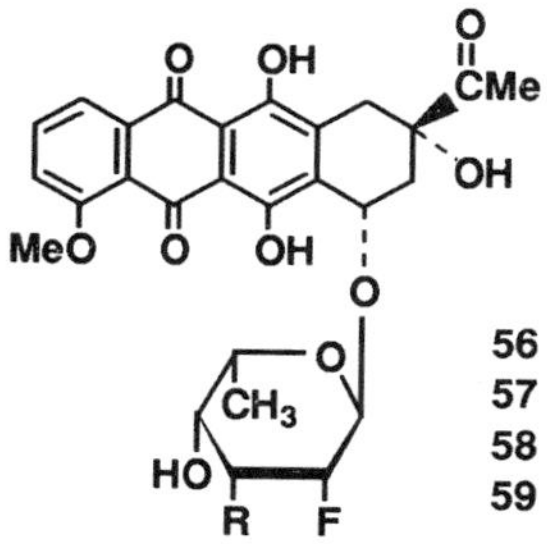

56	R = OMe
57	R = H
58	R = F
59	R = I

modification at C-3' reduces potency. Nevertheless, this position is still worthy of future consideration in relation to the resistance mechanism.

Synthesis of Derivatives of FTDM Modified at C-4'. Synthesis of 4'-O-methyl (*12*) and 4'-deoxy derivatives (*12*) was undertaken next. Glycosyl bromides **61** and **63** were prepared from **6** and **26**, respectively, as shown in Scheme VIII, and then coupled with

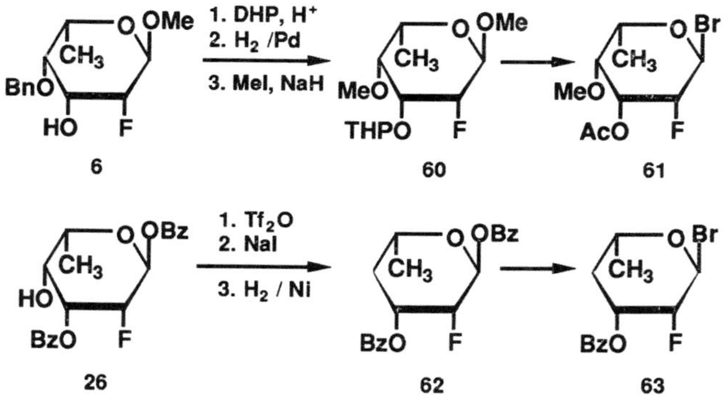

Scheme VIII

daunomycinone. Deprotection of the products gave the α-L-glycosides **64** and **66**. Both of them were further transformed into the 14-hydroxy derivatives **65** and **67** as described

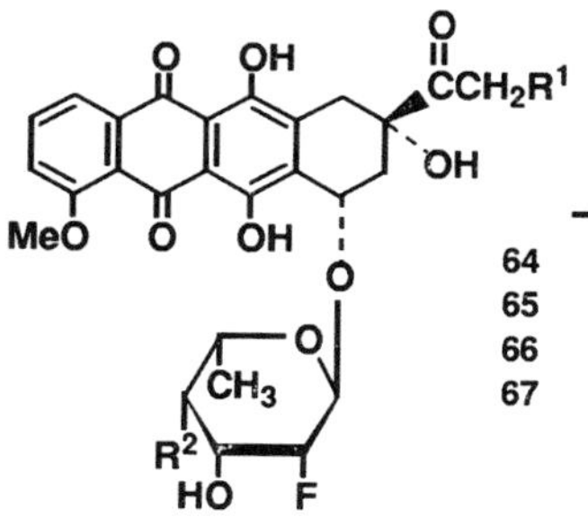

	R¹	R²
64	H	OMe
65	OH	OMe
66	H	H
67	OH	H

previously (*11*). All of these compounds displayed strong antitumor activity (see Table I). This suggests that modification at C-4' does not influence activity as much as does modification at C-3'. Compounds **64** and **67** showed higher activity against resistant tumor cells than FTDM, whereas **66** showed comparable activity.

Several natural and semisynthetic 4'-O-glycosyl anthracycline antibiotics such as aclacinomycins (*32*) and (2"*R*)-4'-*O*-tetrahydropyranyladriamycin (*33*) are known. These compounds exhibit lower toxicity and higher antitumor activity than doxorubicin. Furthermore, the latter compound was reported (*34*) to be readily taken up by tumor cells. With these facts in mind, we have undertaken to prepare several 4'-*O*-glycosyl derivatives of FTDM. Our first synthetic compounds along this line were 4'-*O*-(3-fluorotetrahydropyran-2-yl) derivatives **70–72** (Scheme IX). As these compounds have

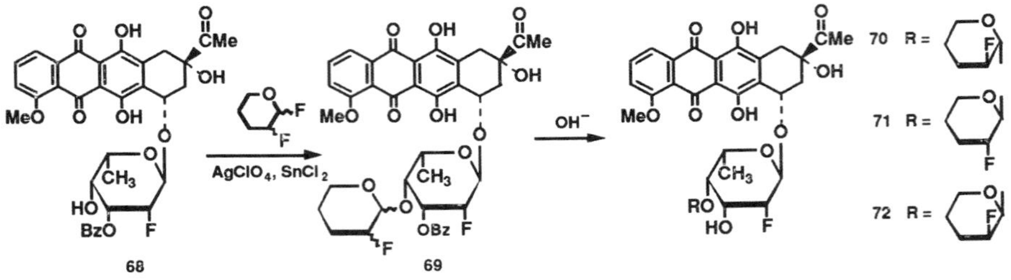

Scheme IX

a tetrahydropyranyl moiety bearing a fluorine at the position vicinal to the glycosyl bond, they are expected to resist acid-catalyzed hydrolysis in contrast to aclacinomycins or 4'-*O*-tetrahydropyranyl derivatives. FTDM was selectively benzoylated to give the 3'-*O*-benzoyl derivative **68**, which was coupled with 2,3-difluorotetrahydropyran, prepared from 3,4-dihydro-2*H*-pyran by reaction with fluorine, in the presence of silver perchlorate, stannous chloride, and molecular sieves in dichloromethane. A mixture of 4'-*O*-(3-fluorotetrahydropyran-2-yl) derivatives **69** was obtained, which, after deprotection, was resolved into three compounds, **70, 71,** and **72**. The structures of these compounds were determined from ^{1}H and ^{19}F NMR spectra data and by comparison of their optical rotations. The three compounds had similar activity (*35*) to that of FTDM without exhibiting toxicity at doses up to 100 μg/mouse/day.

As an extension of this line, another four 4'-*O*-glycosyl derivatives of FTDM, namely, **73, 75, 76,** and **80**, were prepared. The sugars attached at *O*-4' are L-daunosamine (for **73**), 2,6-dideoxy-2-iodo-L-talopyranose (for **75**), 2,6-dideoxy-L-*lyxo*-hexopyranose (for **76**) and 2,6-dideoxy-2-fluoro-L-talopyranose (for **80**). Compounds **73, 75,** and **76** were prepared by coupling of the protected glycals with 3'-*O*-benzoyl FTDM **68**. Thus, **68** and 1,5-anhydro-2,3,6-trideoxy-4-*O*-*p*-nitrobenzoyl-3-trifluoroacetamido-L-*lyxo*-hex-1-enitol (*36*) were coupled in the presence of trimethylsilyl trifluoromethanesulfonate, triethylamine, and molecular sieves (in dichloromethane at -60°C) to give the desired α-L-glycoside in 68% yield along with the 9,4'-di-*O*-glycosylated derivative (20%), while coupling catalyzed by *p*-toluenesulfonic acid afforded the 3'-*O*-glycoside in low yield as a result of acyl migration. Subsequent deprotection gave **73**. Coupling of 3,4-di-*O*-acetyl-L-fucal (*37*) with **68** in the presence of *N*-iodosuccinimide (NIS) under alkoxyhalogenation gave the α-L-glycoside **74** in 85% yield, which was deacetylated to give **75**. Treatment of **74** with tributyltin hydride in the presence of 2,2'-azobis(isobutyronitrile) (AIBN) followed by deprotection gave **76** (Scheme X). To synthesize **80**, however, an alternative method was necessary. Since all efforts to couple **68** with suitably protected 2,6-dideoxy-2-fluoro-L-talopyranose derivatives failed, the disaccharide portion was prepared first and then it was coupled with daunomycinone. Thus, benzyl 3-*O*-benzoyl-2,6-dideoxy-2-fluoro-β-L-talopyranoside **77**, prepared from the corresponding glycosyl bromide **7** in three steps, was coupled with **7** by the Koenigs–Knorr reaction to give the 4-*O*-α-L-glycosyl derivative **78**, which was then converted to the 1-bromide **79**. Coupling of **79** with daunomycinone followed by deprotection of the product gave **80** (Scheme XI). Among the compounds prepared, **76** displayed the strongest activity against L1210 (Table I) as a daunorubicin analogue prepared in our laboratory to date; the other three compounds were moderately or weakly active; all

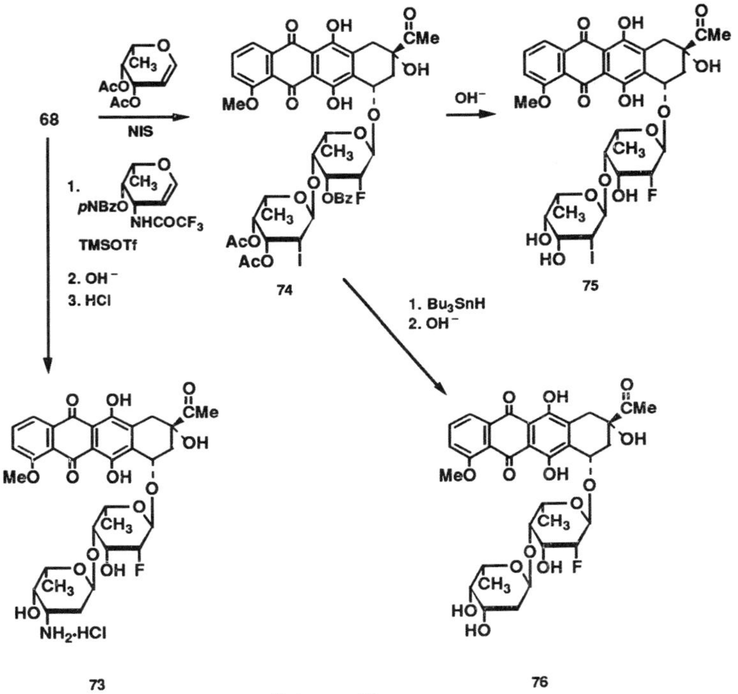

Scheme X

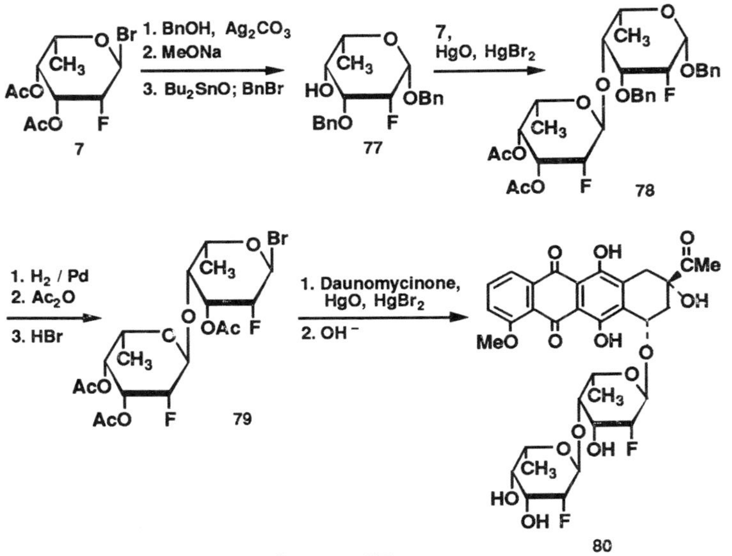

Scheme XI

compounds were nontoxic at doses up to 100 μg/mouse/day. Together, these results suggest that glycosylation at HO-4' of FTDM and FTADM gives derivatives less toxic than the parent anthracycline antibiotics.

Methoxyanthracyclines

After our efforts to synthesize a variety of compounds having fluorine at C-2', we turned to preparing compounds having other electron-withdrawing substituents at C-2'. As the first approach along this line, we prepared anthracycline glycosides having an α-side methoxyl group; methoxyl group is expected to be slightly weaker than fluorine in electron-withdrawal ability. Glycosyl bromide 81 was prepared from 31 as shown in Scheme XII and it was coupled with daunomycinone or 14-O-tert-butyldimethylsilyl-adriamycinone (4). Deprotection of the products gave the desired 2'-O-methyl derivative 82 and its doxorubicin analogue 83 (38). Both compounds had stronger activity against L1210 and weaker toxicity than daunorubicin and doxorubicin, respectively (see Table I). Although 83 exhibited toxicity at a dose of 100 μg/mouse/day, the activity was comparable with that of FTADM.

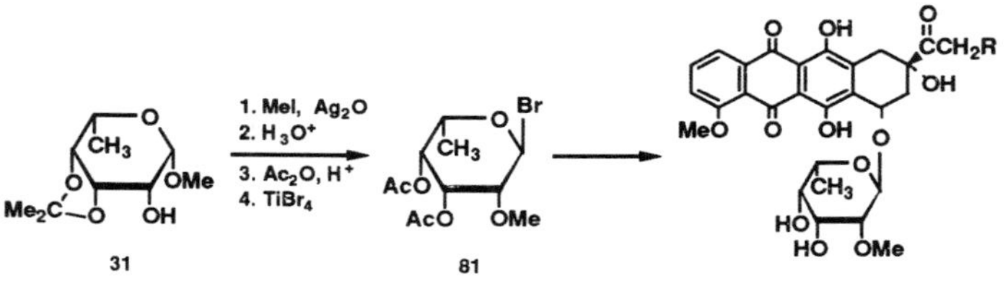

Scheme XII

Conclusions

We have discussed the preparation and the structure-antitumor relationships of the anthracycline glycosides having a fluorine at C-2' and shown a 2'R configuration is requisite for exhibiting activity. Although modification at C-3' of the 2'-fluoroanthracycline glycosides seems to reduce the potency, this position is worth of future consideration in relation to the multiple drug resistance mechanism. On the other hand, modification at C-4' did not significantly change activity and further modification at this point is expected to give rise to more potent antibiotics.

Acknowledgments

The authors are grateful to Dr. Tomio Takeuchi, the chief director of Institute of Microbial Chemistry (IMC) for conducting the in vivo assays, evaluating biological properties, and valuable discussions, and to Professor Sumio Umezawa of this Institute for supporting this project. We also wish to express our appreciation to Dr. Yoji Umezawa of IMC for carrying out the calculations for the oligonucleotide–anthracycline complex and to Meiji Seika Kaisha Ltd. for conducting the in vitro assays. We also thank Kwang-dae Ok, Hae-il Park, Min Sun Chang, and Guen-Jho Lim of Dong-A Pharm. Co. Ltd. Korea; Drs. Shunzo Fukatsu and Toshio Yoneta of Meiji Seika Kaisha Ltd.; to Dr. Toshiaki Miyake, Hiromi Sohtome and Kazuko Takimura of our Institute; to Naoki Kobayashi, Tetsuro

Kondo, Hirohito Kaminaga, Hiroaki Akane, and Ken Nakai of Keio University, and to Yuji Tokuoka of Kitasato University for their earnest cooperation in synthesizing the many compounds. The authors also gratefully acknowledge partial support of this research by grant-in-aid for Cancer Research No. 62015093 from the Ministry of Education, Science and Culture in Japan.

Literature Cited

1. Brief summary: Takagi,Y.; Tsuchiya, T.; Miyake, T; Takeuchi, T.; Umezawa, S. *Yuki Gosei Kagaku Kyokai Shi* **1992**, *50*, 131.
2. Fuchs, E.; Horton, D;. Weckerle, W. *Carbohydr. Res.* **1977**, *57*, C36.
3. Fuchs, E.; Horton, D;. Weckerle, W.; Winter-Mihaly, E. *J. Med. Chem.* **1979**, *22*, 406.
4   Horton, D.; Priebe, W.; Varela, O. *J. Antibiot.* **1984**, *37*, 853.
5. Horton, D.; Priebe, W.; Varela, O. *Carbohydr. Res.* **1984**, *130*, C1.
6. Horton, D.; Priebe, W. *Carbohydr. Res.* **1985** *136*, 391.
7. Horton, D.; Priebe, W.; Varela, O. *Carbohydr. Res.* **1985**, *144*, 305.
8. Naff, M. B.; Plowman, J.; Narayanan, V. L. In *Anthracycline Antibiotics*; El Khadem, H. S., Ed.; Academic Press: New York, NY, 1982, pp 1 – 57.
9. Horton, D.; Priebe, W.; In *Anthracycline Antibiotics*; El Khadem, H. S., Ed., Academic Press: New York, NY, 1982, pp 197 – 224.
10. Tsuchiya,T.; Takagi, Y.; Ok, K.; Umezawa, S.; Takeuchi, T.; Wako, N.; Umezawa, H. *J. Antibioti.* **1986**, *39*, 731.
11. Ok, K.; Takagi, Y.; Tsuchiya, T.; Umezawa, S.; Umezawa, H. *Carbohydr. Res.* **1987**, *169*, 69 .
12. Umezawa, S.; Tsuchiya, T.; Takagi, Y.; Sohtome, H.; Chang, M. S.; Kobayashi, N.; Tokuoka, Y.; Takeuchi, T. *Abstr. Pap. Int. Carbohydr. Symp. XV.* **1990**, D037.
13. Baer, H. H.; Mateo, F. H.; Siemsen, L. *Carbohydr. Res.* **1989**, *187*, 67.
14. Arcamone, F.; Bernardi, L.; Giardino, P.; Di Marco, A. *Ger. Offen.* 2, 652, 391, **1977**, May 26 .
15. Tsuchiya, T.; Takagi, Y.; Umezawa, S.; Takeuchi, T.; Komuro, K.; Nosaka, C.; Umezawa, H.; Fukatsu, S.; Yoneta, T. *J. Antibiot.* **1988**, *41*, 988.
16. Israel, M.; Potti, P. G.; Seshadri, R. *J. Med. Chem.* **1985**, *28*, 1223.
17. Takagi, Y.; Lim, G.; Tsuchiya, T.; Umezawa, S. *J. Chem. Soc. Perkin trans. 1.* **1992**, 657.
18. Mahrwald, R.; Theil, F.; Schick, H.; Schwarz, S.; Palme, H-J.; Weber, G. *J. Prakt. Chem.*, **1986**, *328*, 777.
19. Arcamone, F.; Penco, S.; Vigevani, A.; Redaelli, S.; Franchi, G.; Di Marco, A.; Casazza, A. M.; Dasdia, T.; Formelli, F.; Necco, A.; Soranzo, C. *J. Med. Chem.* **1975**, *18*, 703.
20. Miyake, T.; Takimura, K.; Takagi, Y.; Tsuchiya T.; Umezawa, S.; Takeuchi, T. *Abstr. Pap. Chem. Soc. Jpn. Meet.* 58. **1989**, II-4IL16.
21. Frederick, C. A.; Williams, L. D.; Ughetto, G.; van der Marel, G. A.; van Boom, J. H.; Rich, A.; Wang, A. H.-J. *Biochemistry.* **1990**, *29*, 2538.
22 Kimura, Y.; Suzuki, M.; Matsumoto, T.; Abe, R.; Terashima, S. *Bull. Chem. Soc. Jpn.* **1986**, *59*, 423.
23. Takagi, Y.; Park, H.; Tsuchiya, T.; Umezawa, S.; Takeuchi, T.; Komuro, K.; Nosaka, C. *J. Antibiot.* **1989**, *42*, 1315.
24. Castillon, S.; Dessinges, A.; Faghih, R.; Lukacs, G.; Olesker, A.; Thang, T. T. *J. Org. Chem.*, **1985**, *50*, 4913.
25. Baer, H. H.; Siemsen, L. *Can. J. Chem.* **1988**, *66*, 187.
26. Baer H. H.; Mateo, F. H. *Can. J. Chem.* **1990**, *68*, 2055.
27. Takahashi, Y.; Kinoshita, M.; Masuda, T.; Tatsuta, K.; Takeuchi, T.; Umezawa, H. *J. Antibiot.* **1982**, *35*, 117.

28.   Tsuruo, T.; Yusa, K.; Sudo, Y.; Takamori, R.; Sugimoto, Y. *Cancer Res.* **1989**, *49*, 5537
29.   Tsuruo, T.; Iida, H.; Kawabata, H.; Tsukagoshi, S.; Sakurai, Y. *Cancer Res.* **1984**, *44*, 5095.
30.   Riordan, J. R.; Deuchars, K.; Kartner, N.; Alon, N.; Trent, J.; Ling, V. *Nature* **1985**, *316*, 817.
31.   Tsuruo, T.; Ida, H.; Tsukagoshi, S.; Sakurai, Y. *Cancer Res.* **1981**, *41*, 1967.
32.   Oki, T.; Matsuzawa, Y.; Yoshimoto, A.;   Numata, K.; Kitamura, I.; Hori, S.; Takamatsu, A.; Umezawa, H.; Ishizuka, M.; Naganawa, H., Suda, H.; Hamada, M.; Takeuchi, T.; *J. Antibiot.* **1975**, *28*, 830.
33.   Umezawa, H.; Takahashi, Y.; Kinoshita, M.; Naganawa, H., Masuda, T.; Ishizuka, M.; Tatsuta, K.; Takeuchi, T. *J. Antibiot.* **1979**, *32*, 1082.
34.   Kunimoto, S.; Miura, K.; Takahashi, Y.; Takeuchi, T.; Umezawa, H. *J. Antibiot.* **1983**, *36*, 312.
35.   Takagi, Y.; Sohtome, H.; Tsuchiya, T.; Umezawa, S.; Takeuchi, T. *J. Antibiot.* **1992**, *45*, 355.
36.   Umezawa, H.; Takahashi, Y.; Kinoshita, M.; Naganawa, H.; Tatsuta, K.; Takeuchi, T. *J. Antibiot.* **1980**, *33*, 1581.
37.   El Khadem, H. S.; Swartz, D. L.; Nelson, J. K.; Berry, L. A.; *Carbohydr. Res.* **1977**, *58*, 230.
38.   Takagi, Y.; Kobayashi, N.; Tsuchiya, T.; Umezawa, S.; Takeuchi, T.; Komuro, K.; Nosaka, C. *J. Antibiot.* **1989**, *42*, 1318.

RECEIVED August 16, 1994

Chapter 7

Redox Chemistry of Anthracyclines and Use of Oxomorpholinyl Radicals

Tad H. Koch[1] and Giorgio Gaudiano[2]

[1]Department of Chemistry and Biochemistry, University of Colorado, Boulder, CO 80309–0215
[2]Consiglio Nazionale delle Ricerche, Istituto di Medicina Sperimentale, Viale Marx 15–43, 00137 Rome, Italy

Recent experiments with oxomorpholinyl radicals and anthracycline redox chemistry are summarized. Analysis of oxomorpholinyl radical formation via bond homolysis of radical dimers and the reduction potential of oxomorpholinyl radical dimers are presented. The saponification of daunomycin hydroquinone heptaacetate as it relates to the state responsible for glycosidic cleavage is discussed. The effect of water/dimethylsulfoxide (DMSO) medium on transient lifetimes and reactivity is presented. Synthesis and reactivity of 5-deoxydaunomycin are reported. Of particular significance are further evidence for glycosidic cleavage at the hydroquinone state, long semiquinone and quinone methide lifetimes in water/DMSO, and resistance of reduced 5-deoxydaunomycin to air oxidation and glycosidic cleavage relative to reduced daunomycin.

The biomedically important anthracycline drugs, daunomycin (**1**), adriamycin (**2**), aclacinomycin A (**3**), and menogaril (**4**), likely derive a portion of their biological activity through in vivo reductive activation. The concept of bioreductive activation of quinone antitumor drugs was first proposed by Sartorelli and co-workers (*1*) and then promoted by Moore (*2*). We have been studying the in vitro redox chemistry of the anthracyclines for the past 15 years. Most of our earlier results were reviewed 2 years ago (*3*) and will be only briefly summarized here.

   Our investigations were prompted in part by our discovery of a low-toxicity reducing agent for quinones, bi(3,5,5-trimethyl-2-oxomorpholin-3-yl) (TM-3 dimer) (*4*). It operates through facile bond homolysis of the 3,3'-bond to give 3,5,5-trimethyl-2-oxomorpholin-3-yl (TM-3), which functions as a one-electron reducing agent. The product of oxidation of TM-3 is 5,6-dihydro-3,5,5-trimethyl-1,4-oxazin-2-one (**5**). The bond homolysis and probably also the electron transfer are assisted by solute-solvent interactions, especially with protic solvent. TM-3 dimer is useful both in vitro and in vivo (*5*). We and others have also employed dithionite as an in vitro reducing agent for the anthracyclines. It operates in a similar manner through homolysis of the sulfur-sulfur bond to form a sulfur dioxide radical anion which is also a one-electron reducing agent.

0097–6156/95/0574–0115$08.00/0

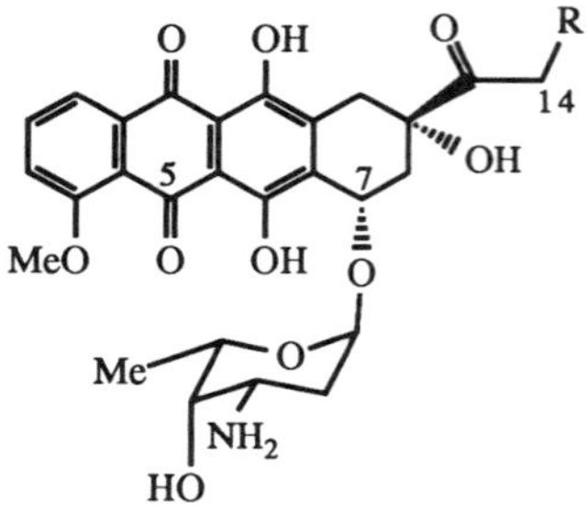

R = H, daunomycin (1)
R = OH, adriamycin (2)

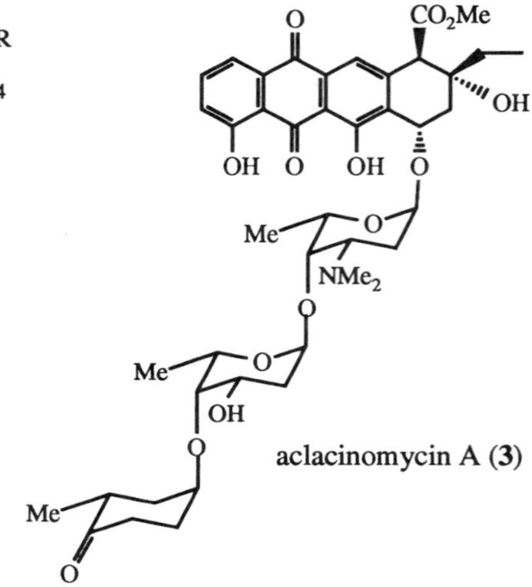

aclacinomycin A (3)

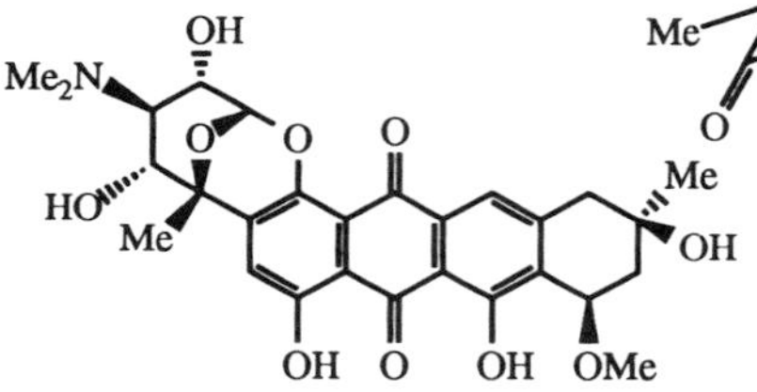

menogaril (4)

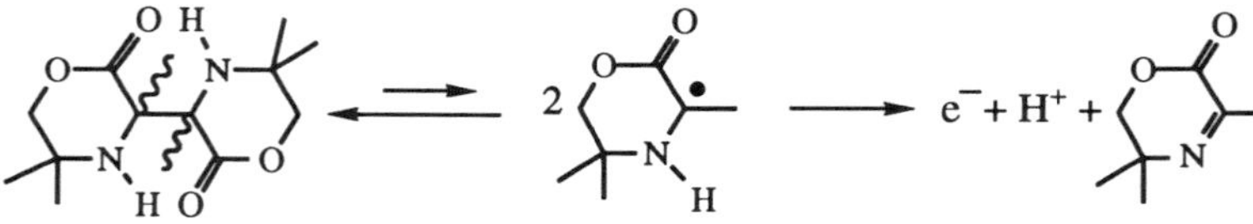

TM-3 Dimer   TM-3   5

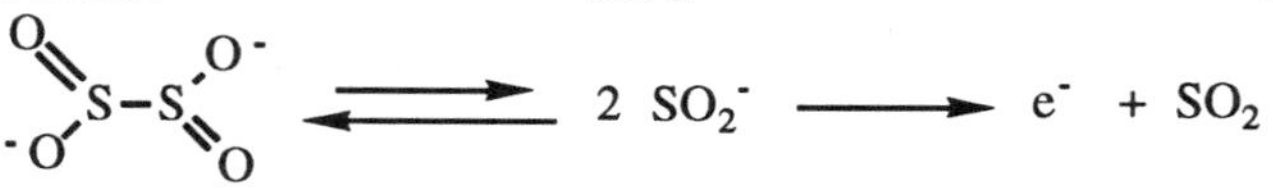

dithionite

The primary redox processes for the anthracyclines as illustrated with daunomycin in Scheme 1 are sequential one-electron reductions to the semiquinone 6 and the hydroquinone 7. Both the semiquinone and hydroquinone react rapidly with molecular oxygen to generate superoxide with formation of quinone and semiquinone, respectively. The process of reduction followed by molecular oxygen oxidation has come to be known as redox recycling and is an important aspect of drug cytotoxicity (6). In an anaerobic medium, anthracyclines in a reduced state undergo cleavage of the substituent at the 7-position. The specific state in which anthracyclines undergo cleavage remains under debate; however, most investigations conclude that it is the hydroquinone state (7). Recent evidence in support of this will be presented below. The product of the cleavage process then is the quinone methide 8, which shows both electrophilic and nucleophilic reactivity at the 7-position. The principle of bioreductive

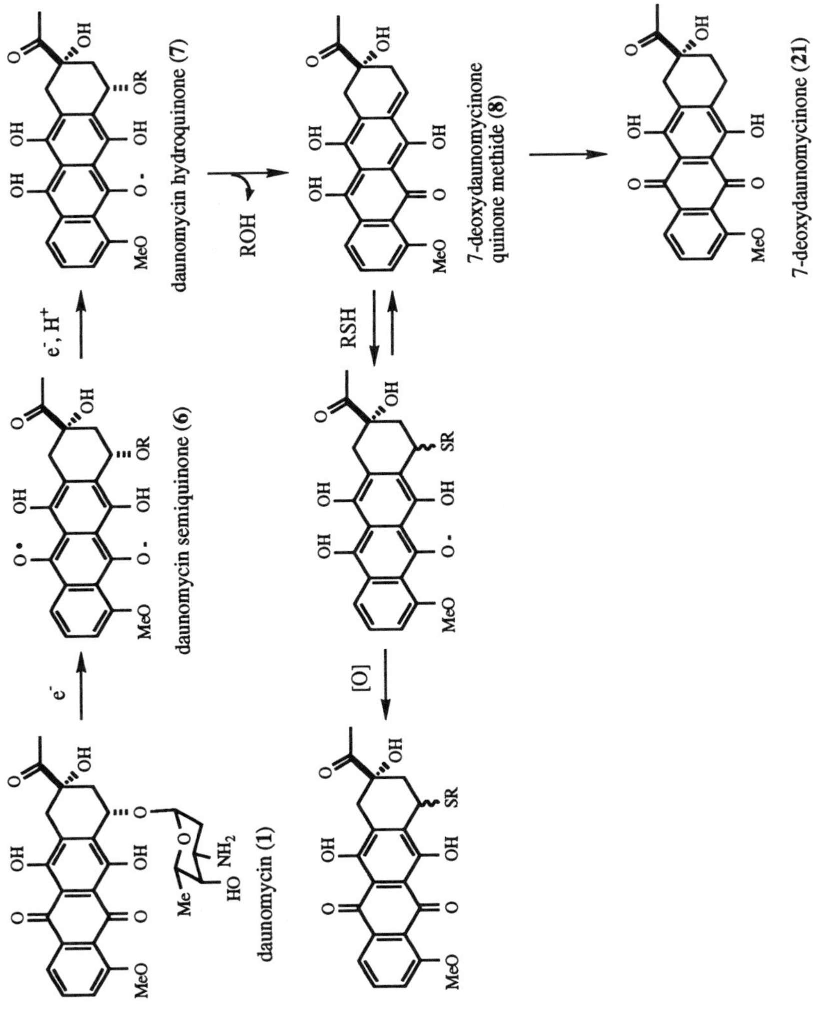

Scheme 1

activation requires that the quinone methide alkylate a nucleophilic site in a critical biological macromolecule. However, the biological importance of this process for anthracycline-derived quinone methides still awaits definitive experimental evidence. Model reactions have demonstrated reactivity with divalent-sulfur (8), nitrogen (9), and oxygen nucleophiles (10). The quinone methide from reduction of menogaril has been shown to couple to the 2-amino substituent of guanosine (11). A significant problem associated with covalent bond formation via nucleophilic addition to a quinone methide is formation of the adduct in its hydroquinone redox state. In competition with oxidation of the adduct hydroquinone is elimination of the nucleophile to restore the quinone methide. Also in competition with nucleophilic addition is irreversible reaction with the ever-present electrophile, the proton. Protonation of the quinone methide yields the 7-deoxyaglycon.

The biological activity of the 7-deoxyaglycon when formed inside a cell is unknown; as a drug, it is inactive. In vitro, elaborate redox chemistry has been demonstrated (12). A major difference between daunomycin and its 7-deoxyaglycon is solubility in water; the 7-deoxyaglycon shows very low water solubility. The change in solubility with reductive cleavage is much less with menogaril where only methanol is lost.

The quinone methide from reduction of daunomycin is also very reactive with molecular oxygen. The products daunomycinone (9), 7-epidaunomycinone (10), 7-deoxy-7,13-epidioxydaunomycinol (11), 7-deoxy-7-ketodaunomycinone (12), and bi(7-deoxydaunomycinon-7-yl) (13) suggest initial oxidation of quinone methide to semiquinone methide 14 with formation of superoxide as shown in Scheme 2 (13).

Successful in vitro reductive activation of the anthracyclines with TM-3 dimer prompted subsequent investigation of in vivo activation to therapeutic advantage. For these studies a more water soluble derivative, bi(3,5-dimethyl-5-hydroxymethyl-2-oxomorpholin-3-yl) (DHM-3 dimer) was synthesized (14). This material also shows low animal toxicity and is highly effective for the treatment of extravasation necrosis incident to therapy with anthracyclines and mitomycin C in a swine model (15). Low toxicity may result from the product of oxidation of DHM-3, 5,6-dihydro-3,5-dimethyl-5-hydroxymethyl-1,4-oxazin-2-one (15), hydrolyzing to pyruvate and 2-amino-2-methyl-1,3-propanediol. The basis for the treatment appears to be extracellular reduction of the quinone antitumor drug to inactive forms such as 7-deoxy-adriamycinone. Also, dramatic results were obtained for the intraperitoneal treatment of L-1210 leukemia in mice with the combination of high doses of adriamycin followed by DHM-3 dimer; the treatment cohorts showed 70% long-term survivors (16). An attractive scenario for the response is rapid uptake of adriamycin by tumor cells followed by intracellular activation by DHM-3 dimer and less rapid uptake by normal cells coupled with extracellular deactivation by DHM-3 dimer to 7-deoxyadriamycinone. ^{14}C-Labeled DHM-3 dimer has been shown to pass through cell membranes (17). Further progress, however, in the use of oxomorpholinyl radical dimers awaits the design of protocols and/or structures that minimize the consumption of the radical dimer through reduction of molecular oxygen to form hydrogen peroxide before reduction of the quinone antitumor drug.

DHM-3 Dimer   DHM-3   15

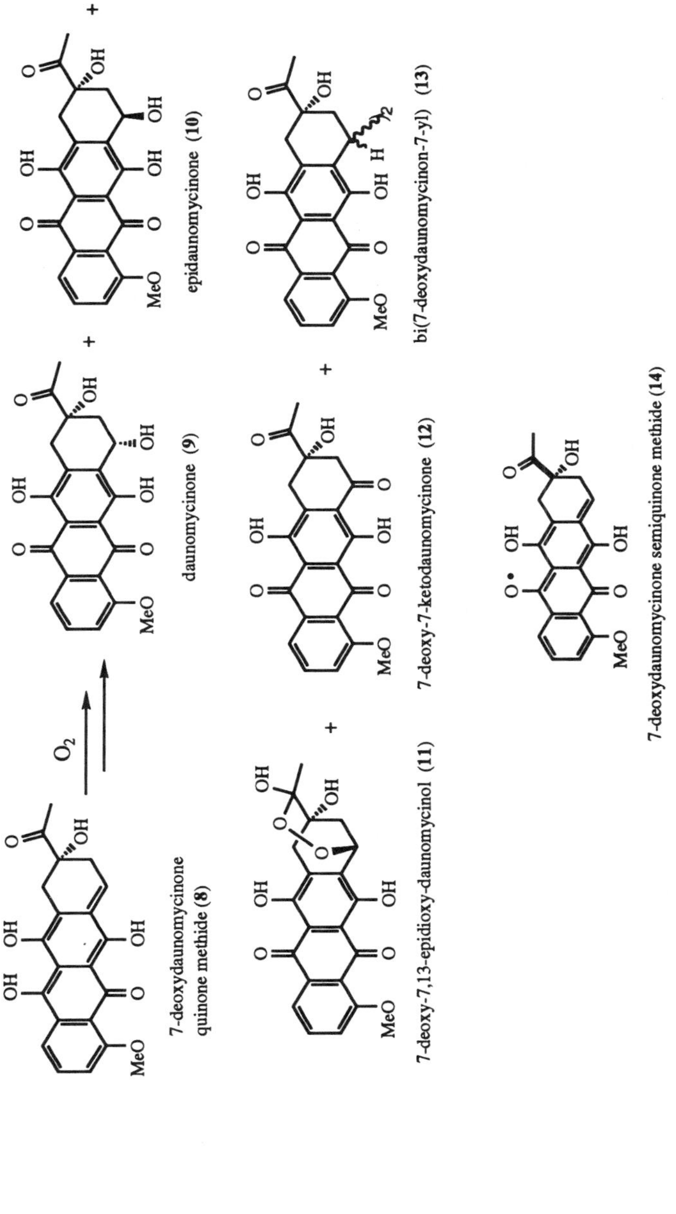

epidaunomycinone (10)

bi(7-deoxydaunomycinon-7-yl) (13)

daunomycinone (9)

7-deoxy-7-ketodaunomycinone (12)

7-deoxydaunomycinone semiquinone methide (14)

7-deoxydaunomycinone quinone methide (8)

7-deoxy-7,13-epidioxy-daunomycinol (11)

Scheme 2

Recent Results and Discussion.

Captodative Radical Reducing Agents. TM-3 radical is one of the most interesting examples of the class of radicals known as captodatively stabilized (*18*) or merostabilized (*19*). These are free radicals which are proposed to derive some stability from the synergetic interaction of the radical center with an electron withdrawing substituent and an electron donating substituent (*20*). The ability of TM-3 and DHM-3 dimers to function as effective reducing agents for the anthracyclines is based upon the free energy of bond homolysis and the oxidation potential of the radical or, more commonly, the reduction potential of the product, dihydrooxazinone **5** or **15**, respectively. Recent photoacoustic calorimetric measurements of the reaction of *t*-butoxyl radical with 3,5,5-trimethyl-2-oxomorpholine (**16**) to form TM-3, in collaboration with Clark and Wayner at the National Research Council of Canada, indicate a captodative stabilization energy of approximately 5 kcal/mol for TM-3 (*21*). The measurements also indicate significant relief of strain upon radical formation; relief of strain is anticipated to be even higher for homolysis of TM-3 dimer. Hence bond homolysis is facile because of both electronic and steric effects.

16   TM-3

The reduction potential of TM-3 dimer in methanol solvent at apparent pH 8 is -0.56 V versus the normal hydrogen electrode (NHE) (*22*). This reduction potential includes the free energy of bond homolysis. The measurement also provided an estimate of the reduction potential of oxazinone **5** to TM-3 equal to -0.85 V versus NHE. The reduction potential for daunomycin is not perfectly established; however, several investigations place it in the range of -0.31 to -0.46 V versus NHE (*23*). Hence, reduction with TM-3 dimer is exergonic by at least 2.3 kcal/mol. For comparison, the reduction potential for dithionite in water at pH 7 is in the range of -0.5 V versus NHE (*24*).

The Redox State Responsible for Glycosidic Cleavage. Whether glycosidic cleavage occurs after one-electron reduction or two-electron reduction is still debated. Clearly, the state of reduction will determine the subsequent chemistry. Glycosidic cleavage at the semiquinone state will yield the semiquinone methide, and glycosidic cleavage at the hydroquinone state will yield the quinone methide. A major problem associated with establishing the reactive redox state is the facile redox equilibrium of semiquinones with quinones and hydroquinones (*25*). Recently, Danishefsky and coworkers proposed a non-redox pathway to hydroquinone as a method for observing the reactivity of hydroquinone in the absence of semiquinone (*26*). They synthesized daunomycin hydroquinone heptaacetate (**17**) and then, during a 14-h period, saponified the heptaacetate in tetrahydrofuran/water with lithium hydroxide under an inert gas atmosphere, presumably to N-acetyldaunomycin hydroquinone (**18**). Upon aerobic work-up of the products, they did not observe products characteristic of glycosidic cleavage but isolated 43% *N*-acetyldaunomycin (**19**). They then raised the possibility that glycosidic cleavage does not occur at the hydroquinone state.

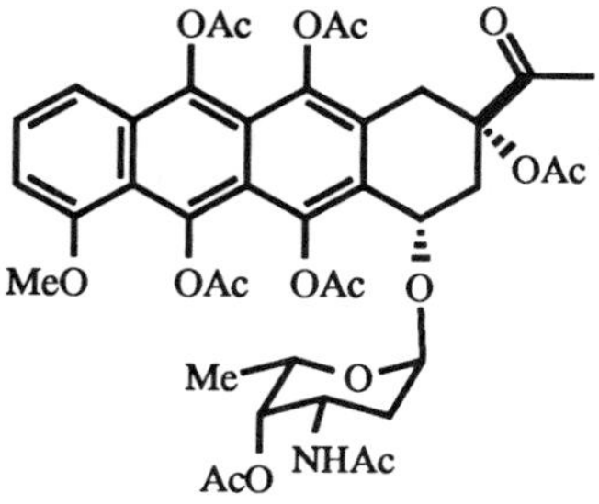

daunomycin hydroquinone
heptaacetate (**17**)

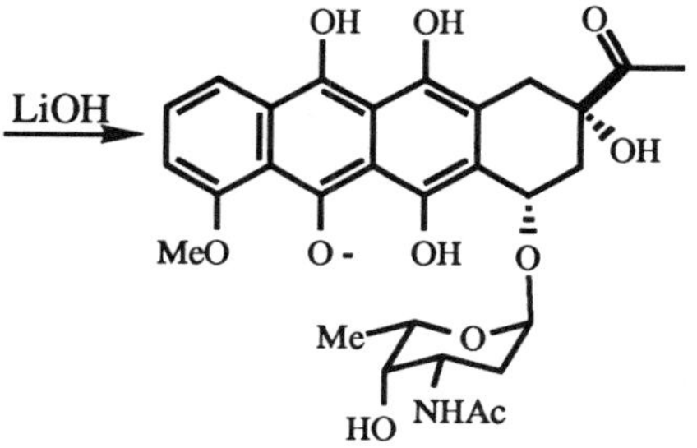

N-acetyldaunomycin
hydroquinone anion (**18⁻**)

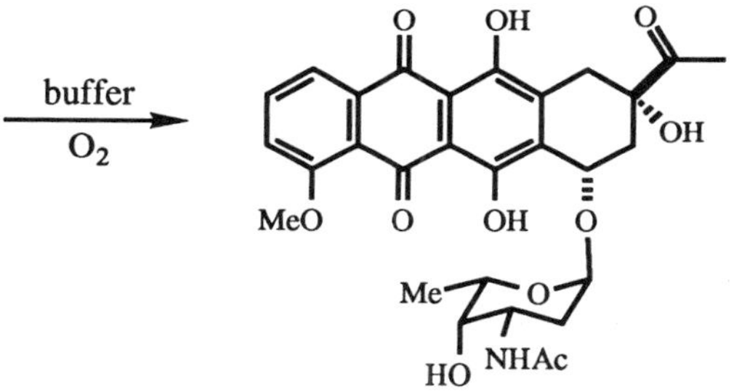

N-acetyldaunomycin (**19**)

Earlier we performed a similar type of experiment. Reduction of daunomycin with dithionite to its hydroquinone immediately followed by lowering the pH to 3 yielded a mixture of air-stable, diastereomeric, daunomycin hydroquinone tautomers which we named leucodaunomycins **20** because of their light yellow color (*27*). When leucodaunomycins were redissolved in anaerobic water at pH 7, they yielded

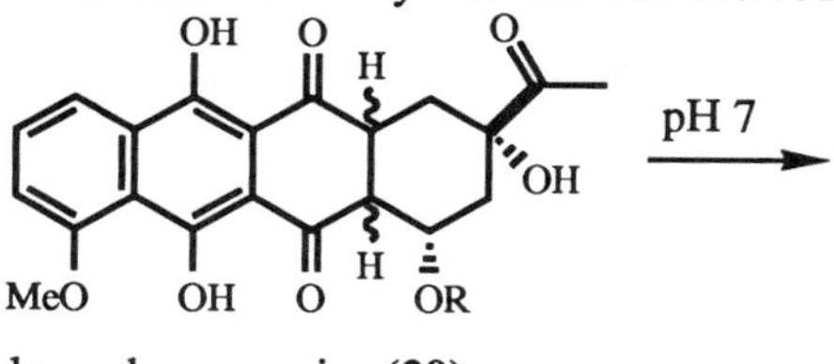

leucodaunomycins (**20**)

daunomycin
hydroquinone (**7**)

7-deoxydaunomycinone
quinone methide (**8**)

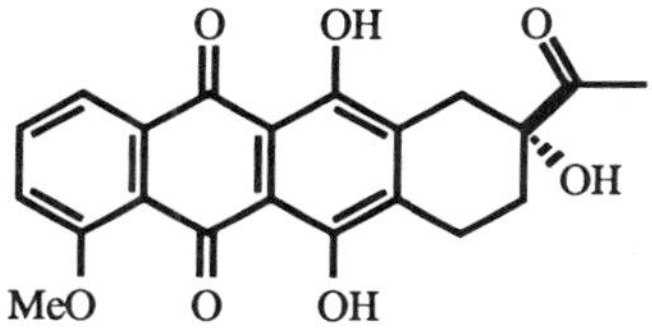

7-deoxydaunomycinone (**21**)

primarily 7-deoxydaunomycinone (21) from tautomerization of spectroscopically observed 7-deoxydaunomycinone quinone methide (8). Most likely, the reaction proceeded via tautomerization of leucodaunomycin back to daunomycin hydroquinone (7) followed by glycosidic cleavage.

The inconsistency of these two experiments clouded our understanding of the redox state responsible for glycosidic cleavage. Initially, we questioned two differences between our experiments and those of Danishefsky and coworkers: the acetylation of the amino group of the sugar and the unusual medium (28). However, reduction of N-acetyldaunomycin in protic medium with dithionite yielded hydroquinone followed by 7-deoxydaunomycinone quinone methide (8) and then 7-deoxydaunomycinone (21) with kinetics similar to that observed upon dithionite reduction of daunomycin. A similar result was observed upon reduction of daunomycin in tetrahydrofuran/water at high pH.

We next questioned the degree of anaerobicity maintained during the saponification of daunomycin hydroquinone heptaacetate over a 14-h period. Molecular oxygen oxidation of daunomycin hydroquinone competes very effectively with glycosidic cleavage. Slow formation of hydroquinone together with a slow leak of molecular oxygen into the reaction vessel would explain the lack of glycosidic cleavage and isolation of N-acetyldaunomycin in the experiment by Danishefsky and coworkers. We performed the saponification after freeze-pump-thaw degassing of the tetrahydrofuran/water reaction medium employing a vacuum of 10^{-6} torr and sealing the reaction mixture with a torch. With this rigorous technique for creating and maintaining anaerobicity, the saponification of daunomycin hydroquinone heptaacetate (17) with lithium hydroxide for 19 h yielded upon aerobic work-up 3% daunomycinone (9), 9% 7-epidaunomycinone (10), 15% N-acetyldaunomycin (19), 13% 7-deoxydaunomycinone (21), 29% N-acetyl-5-deoxydaunomycin (22), and 9% 2-acetyl-11-hydroxy-7-methoxy-5,12-naphthacenedione (23) plus some unidentified products. Epidaunomycinone and daunomycinone are proposed to result from nucleophilic addition of hydroxide to the 7-deoxydaunomycinone quinone methide 8. Hence, these two products together with the 13% 7-deoxydaunomycinone indicate 25% formation of quinone methide. Formation of 19 possibly occurred through oxidation of a partially deacetylated hydroquinone by molecular oxygen during work-up or by 21 during the reaction. Monitoring of the reaction by UV-vis absorption spectroscopy revealed the presence of two long-lived transients in equilibrium (24, 25) bearing acetoxy groups at the 5- and 6-positions. As proposed in Scheme 3, subsequent saponification of the acetoxy group at the 6-position leads to eventual formation of 22 and 23, and saponification of the acetoxy group at the 5-position leads to glycosidic cleavage with formation of a quinone methide. UV-vis spectroscopy indicated that disappearance of 24 and 25 actually required 5 days. Now, the anaerobic saponification of 17 no longer appears to be inconsistent with glycosidic cleavage at the hydroquinone redox state.

Medium Effects. Recently, we have explored the redox chemistry of daunomycin in 5% water/95% dimethylsulfoxide (DMSO) medium (29). Interest in less aqueous media stems from anthracycline biological activity in cellular, mitochondrial, and nuclear membranes (30). DMSO dramatically affects the equilibrium of daunomycin semiquinone with daunomycin quinone and hydroquinone and the rate at which it is established. In protic media the equilibrium strongly favors quinone and hydroquinone and equilibration is rapid, whereas in DMSO a significant concentration of semiquinone is observed by UV-vis and EPR spectroscopy and equilibration is slow. We have summarized in Table I UV-vis absorption data for the anthracyclines, redox transients, and reduction products as a function of structure and medium collected over a period of 15 years.

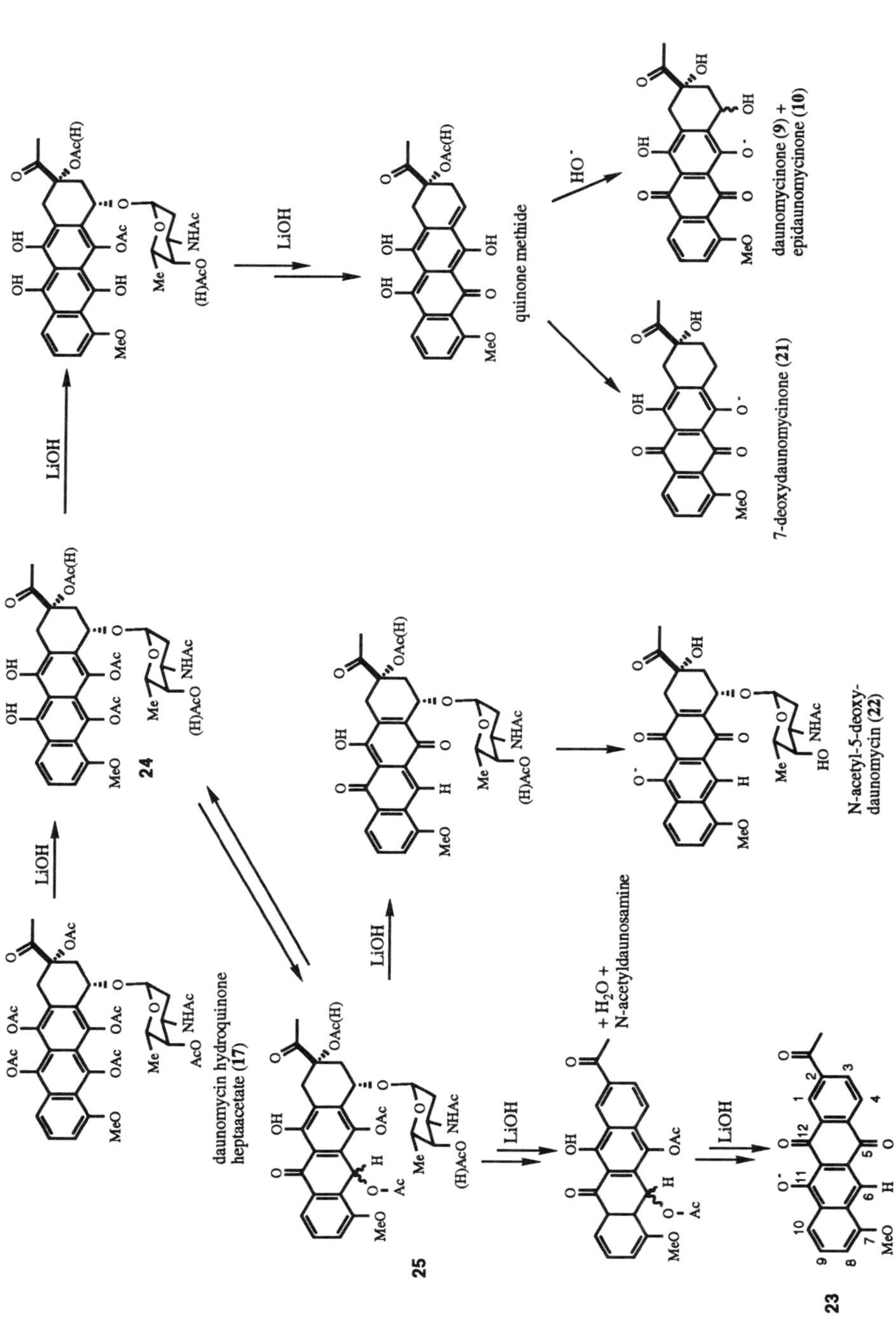

Scheme 3

Table I. UV-vis Data Above 300 nm for Anthracyclines and Their Redox Transients and Products

Compound	Solvent	pH	λ_{max}	Reference
daunomycin (1)[a]	MeOH	8	480	7
	H_2O	7	490	27
	DMSO		480, 494	29
daunomycin semiquinone (6)	DMSO		510	29
daunomycin hydroquinone (7)	MeOH	8	420	7
	DMSO		440	29
7-deoxydaunomycinone quinone methide (8)	MeOH	8	380, 610	7
	DMSO		380, 420, 680	29
7-deoxydaunomycinone (19)	MeOH	8	480	7
	DMSO		480, 494	29
leucodaunomycin (18)	H_2O	7	420, 440	27
7-deoxy-7-(N-acetyl-S-cysteinyl)-daunomycinone[b] bi(7-deoxy-daunomycin-7-yl) (13)	MeOH	8	490	27
	DMSO		480, 494	31
aclacinomycin A (3)	MeOH	8	430	b
7-deoxyaklavinone	MeOH	8	430	b
bi(7-deoxyaklavin-7-yl	MeOH	8	430	b
menogaril (4)	MeOH	8	475	c
menogaril hydroquinone	MeOH	8	420	c
7-deoxynogarol	MeOH	8	475	c
bi(7-deoxynogarol-7-yl)	MeOH	8	475	c
4-demethoxydaunomycin	MeOH	8	480	d
4-demethoxy-7-deoxydauno-mycinone quinone methide	MeOH	8	375, 600	d
11-deoxydaunomycin	MeOH	8	420	d
7,11-dideoxydaunomycinone quinone methide	MeOH	8	340, 530	d
7,11-dideoxydaunomycinone hydroquinone	MeOH	8	395	d
5-iminodaunomycinone (27)	MeOH	8	550, 592	e
7-deoxy-5-iminodaunomycinone	MeOH	8	540, 580	e
naphthacenedione 23	MeOH	8	480	e, 33
8-acetyl-1-methoxy-7,9,10,12-tetrahydro-6,8,11-trihydroxy 5(8H)-naphthacenone	MeOH	8	397	e
5-deoxydaunomycin (29)	MeOH	8	490	33
naphthacenone 33	MeOH	8	398	33

[a]Daunomycin and adriamycin, and their redox transients and products have essentially the same absorption spectra. [b]Kleyer, D. L. ; Gaudiano, G.; Koch, T. H. *J. Am. Chem. Soc.*, **1984**, *106*, 1105. [c]Bolt, M.; Gaudiano, G.; Haddadin, M. J.; Koch, T. H. *J. Am. Chem. Soc.* **1989**, *111*, 2283. [d]Boldt, M.; Gaudiano, G.; Koch, T. H. *J. Org. Chem.* **1987**, *52*, 2146. [e]Bird, D. M.; Boldt, M.; Koch, T. H. *J. Am. Chem. Soc.* **1987**, *109*, 4046.

A consequence of the equilibrium and kinetics of reduction in 5% water/95% DMSO with dithionite and TM-3 dimer is a clear picture of glycosidic cleavage occurring at the hydroquinone state. UV-vis spectroscopic monitoring of the anaerobic reduction of daunomycin with 1 mol equiv of dithionite showed rapid disappearance of the quinone during the first 6 s of the reaction with appearance and disappearance of bands characteristic of semiquinone and hydroquinone during the next 60 s. As the semiquinone and hydroquinone bands disappeared, bands for quinone methide appeared and maximized at 60 s. Similar reduction with TM-3 dimer led to the semiquinone band maximizing at 35 min and the quinone methide band maximizing at 54 min; a strong hydroquinone band was not observed. Slower formation of semiquinone resulted from the significantly slower rate of bond homolysis of TM-3 dimer relative to the rate of bond homolysis of dithionite. If glycosidic cleavage had occurred at the semiquinone state, semiquinone would not have maximized at 35 min because with dithionite reduction semiquinone disappears in approximately 60 s. It maximizes at 35 min because it must wait for a second reduction by TM-3, which is slowly formed from TM-3 dimer, or for its own disproportionation, which is also slow in DMSO.

The first half-life for disappearance of the quinone methide in water/DMSO is about 2000 s. The products are 7-deoxydaunomycinone (**21**) and bi(7-deoxy-daunomycinon-7-yl) (**13**). Under these conditions **13** is formed via one quinone methide serving as a nucleophile and the second as an electrophile as shown in Scheme 4 in contrast with formation of **13** from molecular oxygen oxidation of quinone methide as shown in Scheme 2. The first half-life for decay of the quinone methide in water/DMSO is approximately 100 times longer than the half-life for decay of the quinone methide in water. Furthermore, in water no **13** is formed. The longer lifetime for the quinone methide and the formation of **13** suggest that a less hydrophilic medium may foster coupling of the quinone methide to biological nucleophiles.

When quinone methide was formed by reduction in 5% D_2O/95% DMSO, deuterium incorporation occurred at both the 7- and 14-positions of 7-deoxy-daunomycinone. A control experiment showed no deuterium incorporation at the 14-position upon reduction of 7-deoxydaunomycinone in the same medium. Deuterium incorporation at the 14-position is rationalized via intramolecular protonation at C-7 of the quinone methide by the methyl at C-14 to yield the 14-enolate **26**. Subsequent reaction of the enolate with D_2O gives 7-deoxy-14-deuteriodaunomycinone (**27**) as shown in Scheme 5.

Both dimerization and enolate formation are a consequence of the longer lifetime of quinone methide in water/DMSO medium. Another consequence of the long quinone methide lifetime, when quinine methide is formed with excess dithionite, is subsequent reaction with dithionite as a nucleophile (*31*). Upon aerobic work-up, diastereomeric 7-deoxydaunomycinon-7-yl sulfonates **28** were isolated together with 7-deoxydaunomycinone (**21**) and the products of reaction of quinone methide with molecular oxygen **9-13**. Addition of dithionite to quinone methide was reversible as indicated by the ratio of **28** to **21** as a function of time. The reactive sulfur nucleophile was proposed to be dithionite as shown in Scheme 6 rather than hydrogen sulfite, a byproduct of reduction with dithionite, because the presence of a large initial excess of bisulfite did not affect the rate of disappearance of the quinone methide.

Because of the long lifetime of quinone methide in DMSO, we were able to observe and characterize an anthracycline-derived quinone methide by ^1H NMR spectroscopy. The quinone methide selected was that from reductive cleavage of 11-deoxydaunomycin because high transient quinone methide yields could be achieved with 1 mol equiv of dithionite with formation of only minimal traces of paramagnetic semiquinones. Of particular note was the chemical shift of 7.27 ppm for the vinyl proton at position 7. This suggested that the carbonyl of the quinone methide may reside in the B ring as shown in Scheme 7 rather than in the traditionally assigned C ring. Group equivalent calculations were also consistent with a B ring carbonyl.

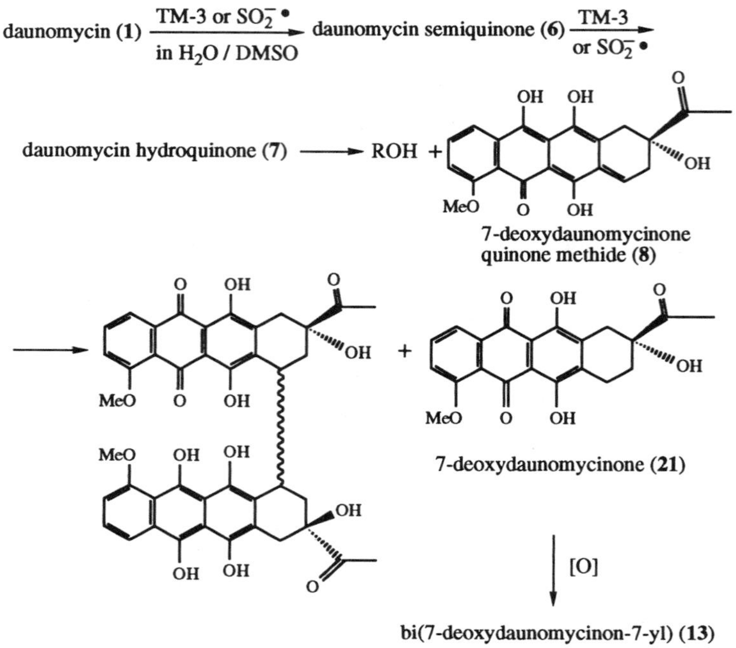

Scheme 4

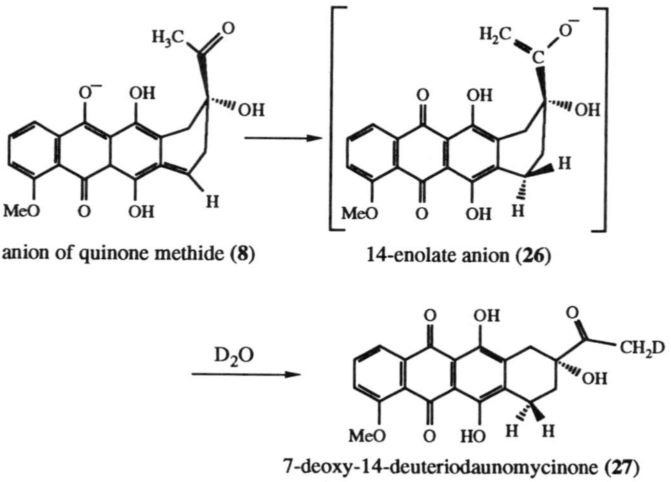

Scheme 5

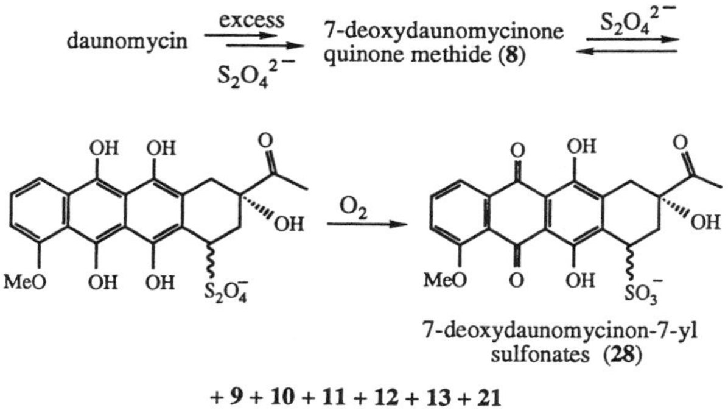

daunomycin $\xrightarrow[S_2O_4^{2-}]{\text{excess}}$ 7-deoxydaunomycinone quinone methide (**8**) $\xrightarrow{S_2O_4^{2-}}$

7-deoxydaunomycinon-7-yl sulfonates (**28**)

+ 9 + 10 + 11 + 12 + 13 + 21

Scheme 6

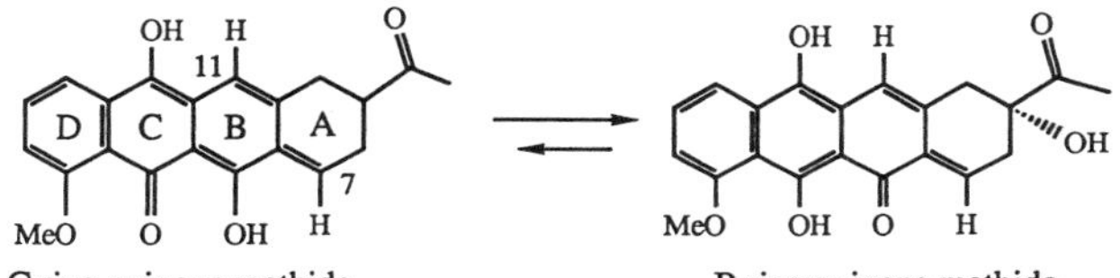

C ring quinone methide   B ring quinone methide

Scheme 7

Synthesis and Reduction of 5-Deoxydaunomycin (29). Cameron and coworkers have reported that catalytic hydrogenation of daunomycin on platinum in acidic methanol yields 5-deoxydaunomycin (**29**) and a mixture of stereoisomeric 5-deoxydaunomycinols (*32*). Similar results were obtained upon reduction of adriamycin. The authors further reported that the 5-deoxyanthracyclines showed significant anticancer activity, comparable with that of the parent anthracyclines. We have observed that reduction of 5-iminodaunomycin (**30**) with dithionite followed by lowering the pH to 3 results in high yields of 5-deoxydaunomycin (*33*). UV-vis spectroscopic monitoring of the reaction showed the intermediacy of a transient assigned tetrahydronaphthacenone structure **31** shown in Scheme 8 based upon absorption at 396 nm. Protonation of the amino substituent of 5-iminodaunomycin hydroquinone results in isomerization of the C ring followed by elimination of ammonia. In contrast, at pH 8, 5-iminodaunomycin hydroquinone undergoes rapid glycosidic cleavage to form the respective quinone methide.

Anaerobic reduction of 5-deoxydaunomycin with TM-3 dimer in methanol at apparent pH 8 yielded 26% recovered 5-deoxydaunomycin, 56% 5,7-dideoxydaunomycinone (**32**), and 18% naphthacenedione **23** upon aerobic work-up. UV-vis monitoring of the reaction showed the intermediacy of a long-lived transient assigned tetrahydronaphthacenone structure **33** as shown in Scheme 9 based upon absorption at 398 nm. When the reduction reaction mixture was saturated with air at the point of maximum transient concentration, HPLC analysis showed 88% recovered **29** and 12% **23**. Kinetic analysis of the anaerobic and aerobic reactions indicated that the half-life of **33** with respect to glycosidic cleavage was 17 h and with respect to air oxidation, 18 min at 25 °C and 630 mm atmospheric pressure. These half-lives are substantially longer than the comparable half-lives for disappearance of transients upon reduction of daunomycin, for example, 8000 times longer for glycosidic cleavage.

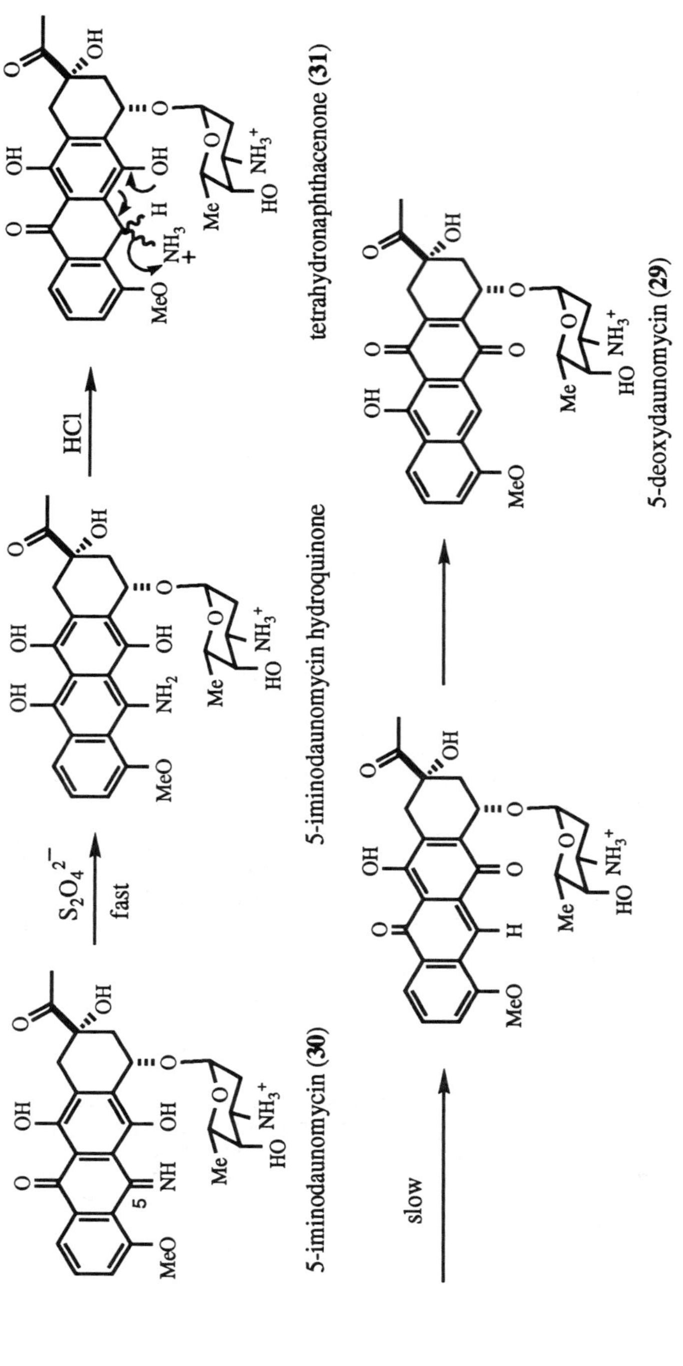

tetrahydronaphthacenone (**31**)

5-iminodaunomycin hydroquinone

5-deoxydaunomycin (**29**)

5-iminodaunomycin (**30**)

HCl

$S_2O_4^{2-}$
fast

slow

Scheme 8

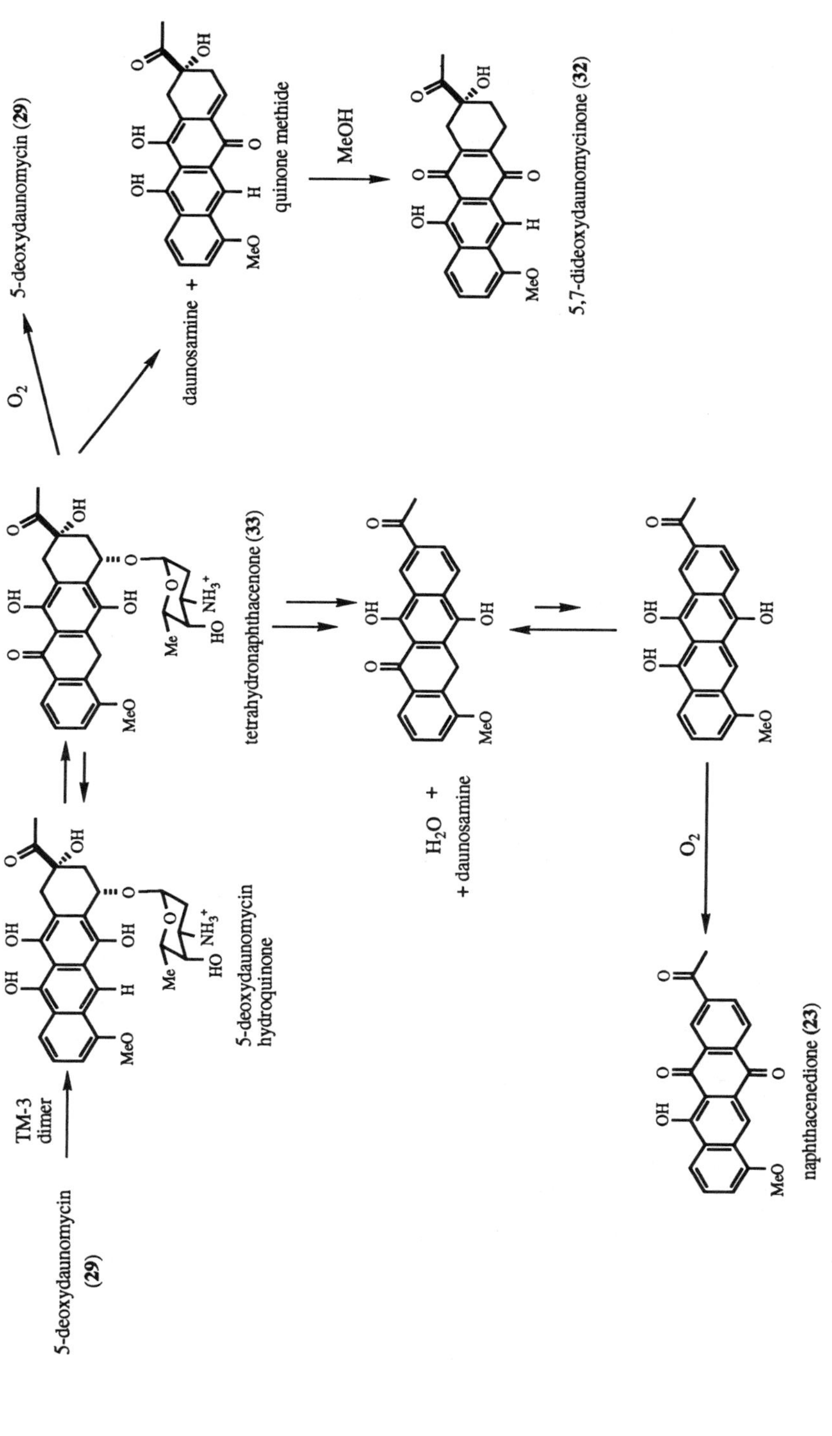

Scheme 9

The lower rate of air oxidation of redox transients predicts less in vivo redox cycling to produce reactive oxygen species and possibly lower cardiotoxicity. The similarity in tumor response to daunomycin and 5-deoxydaunomycin brings into question the role of glycosidic cleavage and the quinone methide state. Possibly, the transient tetrahydronaphthacenone **33** protects the reduced 5-deoxydaunomycin until it finds a suitable target.

Closing Remarks.

We conclude by noting that tetrahydronaphthacenones are relatively long lived intermediates in the saponification of the heptaacetate of daunomycin hydroquinone, in the deamination of protonated 5-iminodaunomycin hydroquinone, and in the reactions of 5-deoxydaunomycin hydroquinone. Hence, the presence of an acetoxy group or a positively charged ammonium substituent at the 5-position or the absence of a hydroxy group at the 5-position fosters tautomerization of the anthracycline hydroquinone functionality to the naphthacenone functionality in competition with glycosidic cleavage. The competition in part probably results from a slower rate of glycosidic cleavage.

The future of in vivo anthracycline therapy with chemical reductive activation may reside in the development of delivery techniques that protect the reducing agent from molecular oxygen and the anthracycline from the reducing agent as they both travel towards their target. The reducing agent can be protected from molecular oxygen by a hydrophobic medium. The rate constant for the bond homolysis of the TM-3 dimer decreases by three orders of magnitude in going from hydrophilic to hydrophobic medium (*34*). The slow step in the reaction with molecular oxygen is the radical dimer bond homolysis. Possible intravenous hydrophobic delivery tools include liposomes and carrier proteins.

Acknowledgments.

The authors thank the U.S. National Institutes of Health and National Science Foundation for financial support and Dr. Sergio Penco of Farmitalia Carlo-Erba, Milan, Italy, for a generous gift of daunomycin. GG also thanks the CNR, Progetto Finalizzato ACRO, for financial support.

Literature Cited.

1. Lin, A. J.; Cosby, L. A.; Shansky, C. W.; Sartorelli, A. C. *J. Med. Chem.* **1972**, *15*, 1247. Lin, A. J.; Sartorelli, A. C. *J. Med. Chem.*, **1976**, *19*, 1336.
2. Moore, H. W.; Czerniak, R. *Med. Res. Rev.* **1981**, *1*, 249. Moore, H. W.; Czerniak, R.; Hamdan, A. *Drugs Exptl. Clin. Res.* **1986**, *12*, 475.
3. Gaudiano, G.; Koch, T. H. *Chem.Res. Toxicol.* **1991**, *4*, 2.
4. Burns, J. M.; Wharry, D. L.; Koch, T. H. *J. Am. Chem. Soc.* **1981**, *103*, 849.
5. Banks, A. R.; Jones, T.; Koch, T. H.; Friedman, R. D.; Bachur, N. R. *Cancer Chemother. Pharmacol.* **1983**, *11*, 91.
6. Doroshow, J. H. *Proc. Natl. Acad. Sci. U.S.A.* **1986**, *83*, 4514. Doroshow, J. H.; *Biochem. Biophys. Res. Commun.* **1986**, *135*, 330.
7. Kleyer, D. L.; Koch, T. H. *J. Am. Chem. Soc.* **1984**, *106*, 2380. Fisher, J.; Abdella, B. R. J.; McLane, K. E.; *Biochemistry* **1985**, *24*, 3562. Anne, A.; Moiroux, J. *Nouv. J. Chim.* **1985**, *9*, 83. Land, E. J.; Mukherjee, T.; Swallow, A. J.; Bruce, J. M. *Br. J. Cancer* **1985**, *51*, 515.
8. Ramakrishnan, K.; Fisher, J. *J. Med. Chem.* **1986**, *29*, 1215.

9. Gaudiano, G.; Egholm, M.; Haddadin, M. J.; Koch, T. H. *J. Org. Chem.* **1989**, *54*, 5090.
10. Gaudiano, G.; Frigerio, M.; Bravo, P.; Koch, T. H. *J. Am. Chem. Soc.* **1990**, *112*, 6704.
11. Egholm, M.; Koch, T. H. *J. Am. Chem. Soc.* **1989**, *111*, 8291.
12. Brand, D. J.; Fisher, J. *J. Am. Chem. Soc.* **1986**, *108*, 3088.
13. Gaudiano, G.; Koch, T. H. *J. Am. Chem. Soc.* **1990**, *112*, 9423.
14. Gaudiano, G.; Koch, T. H. *J. Org. Chem.* **1987**, *52*, 3073.
15. Averbuch, S. D.; Boldt, M.; Gaudiano, G.; Stern, J. B.; Koch, T. H.; Bachur, N. R. *J. Clin. Invest.* **1988**, *81*, 142.
16. Averbuch, S. D.; Gaudiano, G.; Koch, T. H.; Bachur, N. R. *Cancer Res.* **1985**, *45*, 6200.
17. Averbuch, S. D.; Sankar, I. U.; Gaudiano, G.; Koch, T. H., unpublished data.
18. Viehe, H. G.; Janousek, Z.; Merényi, R.; Stella, L. *Acc. Chem. Res.* **1985**, *12*, 148.
19. Baldock, W. R.; Hudson, P.; Katritsky, A. R. *J. Chem. Soc., Perkin Trans. I* **1974**, 1422.
20. Sustmann, R.; Korth, H. *Adv. Free Radical Chem.* **1990**, *26*, 131.
21. Clark, W. B.; Wayner, D. D. M.; Demirdji, S. H.; Koch, T. H. *J. Am. Chem. Soc.* **1993**, *115*, 2447.
22. Mahoney, R. P.; Fretwell, P. A.; Demirdji, S. H.; Mauldin, R. L., III; Benson, O., Jr.; Koch, T. H. *J. Am. Chem. Soc.* **1992**, *114*, 186.
23. Berg, H.; Horn, G.; Jacob, H.-E.; Fiedler, U.; Luthardt, U.; Tresselt, D. *Bioelectrochem. Bioenerg.* **1986**, *16*, 135. Svingen, B. A.; Powis, G. *Arch. Biochem. Biophys.* **1981**, *209*, 119. Rao, G. M.; Lown, J. W.; Plambeck, J. A. *Electrochem. Soc.* **1978**, *125*, 534.
24. Mayhew, S. G. *Eur. J. Biochem.* **1978**, *85*, 535.
25. Mukherjee, T.; Land, E. J.; Swallow, A. J.; Bruce, J. M. *Arch. Biochem. Biophys.* **1989**, *272*, 450.
26. Sulikowski, G. A.; Turos, E.; Danishefsky, S. J.; Shulte, G. M. *J. Am. Chem. Soc.* **1991**, *113*, 1373.
27. Bird, D. M.; Gaudiano, G.; Koch, T. H. *J. Am. Chem. Soc.* **1991**, *113*, 308.
28. Schweitzer, B. A.; Koch, T. H. *J. Am. Chem. Soc.* **1993**, *115*, 5446.
29. Gaudiano, G.; Frigerio, M.; Bravo, P.; Koch, T. H. *J. Am. Chem. Soc.* **1992**, *114*, 3107.
30. Pedersen, J. Z.; Marcocci, L.; Rossi, L.; Mavelli, I.; Rotilio, G. *Biochem. Biophys. Res. Commun.* **1990**, *168*, 240. Alegria, A. E.; Rodrguez, M. S.; Hernandez, J. *Biochem. Biophys. Acta* **1990**, *51*, 1035. Ogura, R.; Sugiyama, M.; Haramaki, N.; Hidaka, T. *Cancer Res.* **1991**, *51*, 3555.
31. Gaudiano, G.; Frigerio, M.; Sangsurasak, C.; Bravo, P.; Koch, T. H. *J. Am. Chem. Soc.* **1992**, *114*, 5546.
32. Cameron, D. W.; Feutrill, G. I.; Griffiths, P. G. *Tetrahedron Lett.* **1988**, *29*, 4629.
33. Schweitzer, B. A.; Koch, T. H. *J. Am. Chem. Soc.* **1993**, *115*, 5440.
34. Olson, J. B.; Koch, T. H. *J. Am. Chem. Soc.* **1986**, *108*, 756.

RECEIVED June 3, 1994

Chapter 8

Synthesis of Anthraquinone Analogues of Linked Anthracycline

Ping Ge[1] and Richard A. Russell[2]

[1]Department of Medicinal Chemistry, State University of New York—Buffalo, Buffalo, NY 14260
[2]School of Biological and Chemical Sciences, Deakin University, Geelong 3217, Australia

Molecules containing two anthraquinone moieties linked by an aliphatic chain and esterified by *trans*-4-aminocyclohexane carboxylic acid e.g. **34** have been prepared as simplified analogues of linked anthracyclines e.g. **1**.

Although doxorubicin was discovered several decades ago the detailed mechanism of its action has proved elusive. While most anthracyclines are potential DNA intercalators, by nature of their planar quinonoidal chromophore, this property correlates only poorly with their antitumor activity. As a consequence, multimodal mechanisms*(1)* have been advanced to account for the biological activity of this very significant group of chemotherapeutic agents.

Recently, it has been suggested that DNA intercalators subvert the action of either or both of the DNA-regulatory isoenzymes topoisomerase IIa or IIb by stabilizing the enzyme-DNA complex *(2,3)*. While little is known about the molecular details of this process, it has raised the possibility of designing drugs *(4)* that contain enhanced domains for both DNA and protein interaction.

The isolation of the bis-intercalating quinoxalines such as Triostin A and Echinomycin, which span 2 bp *(5)*, together with the related Luzopeptin, which appears to span 3 bp *(6)*, has done much to keep alive the concept of DNA bis-intercalating agents as potential therapeutic agents.

Linked anthracyclines have also been the source of interest since the pioneering work of Henry and Tong *(11)* yielded the hydrazides **1**. These compounds showed promising in vitro activity but they failed to trigger the promised rush of new clinical agents *(12)*.

Subsequent work by Reiss et al. added a range of anthracycline-based bisintercalators *(13)* which included the mixed acridine **2**. While these compounds showed various degrees of promise, at least in vitro, much of the synthetic work was accomplished on a microscale and left unanswered the problem of how to produce larger quantities of these agents for further elaboration or modification *(14)*.

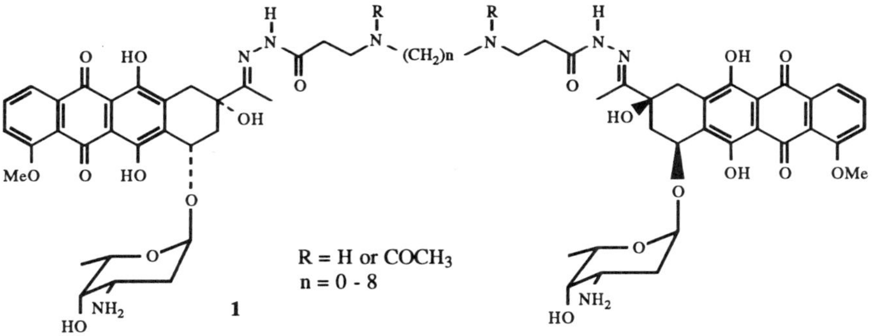

R = H or COCH3
n = 0 - 8

1

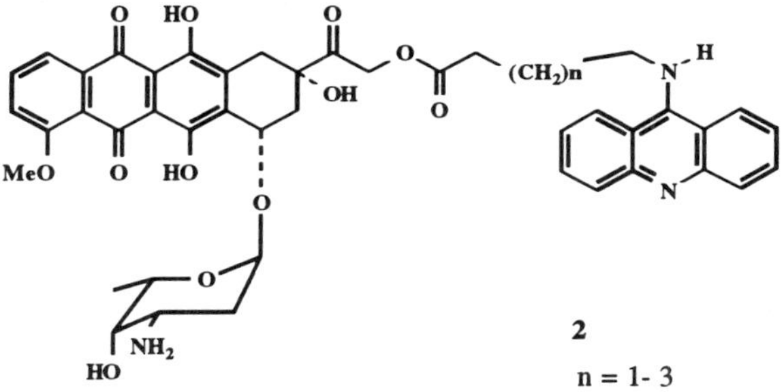

2

n = 1- 3

From a purely synthetic point of view, bis-intercalating anthracyclines have been largely ignored, no doubt in part because of the difficulties in attaching sugars efficiently to the latent aglycones. Despite many years of working with anthracyclines, we still find *(15)* that efficient glycosidation routes such as that described by Terashima *(16,17)* are sensitive to the nature of the aglycone and are frequently an experimental challenge.

The prospect of developing totally synthetic routes to linked anthracyclines appeared limited until in the work of Li, Wang and Zhang *(18,19,20)* which suggested that the daunosaminyl group of doxorubicin or idarubicin could be replaced by an ester derived from *trans* 4-aminocyclohexane carboxylic acid **3**, or by peptidyl variants **4** of this residue, and still leave high biological activity in the final molecule *(13-14)*.

At this point it seemed reasonable to pose the strategic problem as to how one might approach the synthesis of linked anthracyclines. Should two completed units be joined in a penultimate step or should the two intercalators be constructed simultaneously from a linked precursor? The latter approach was clearly distinct from previous ones and offered possibilities for applying strategies based on the phthalide anion annelation of *p*-quinone monoacetals *(21-31)* (Scheme I). This latter reaction has been an integral part of our endeavors in the anthracycline field and hence offered a good starting point.

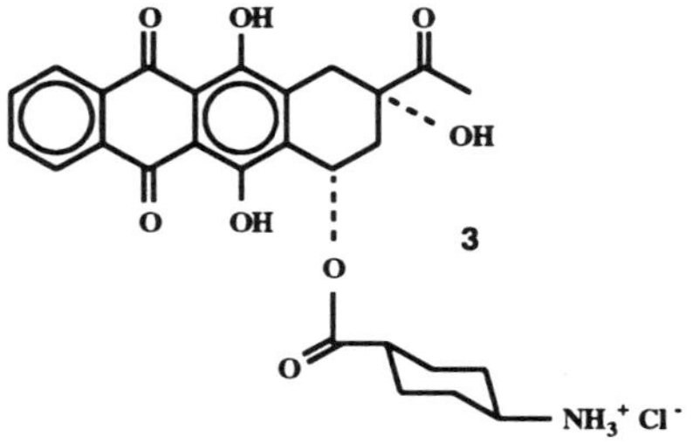

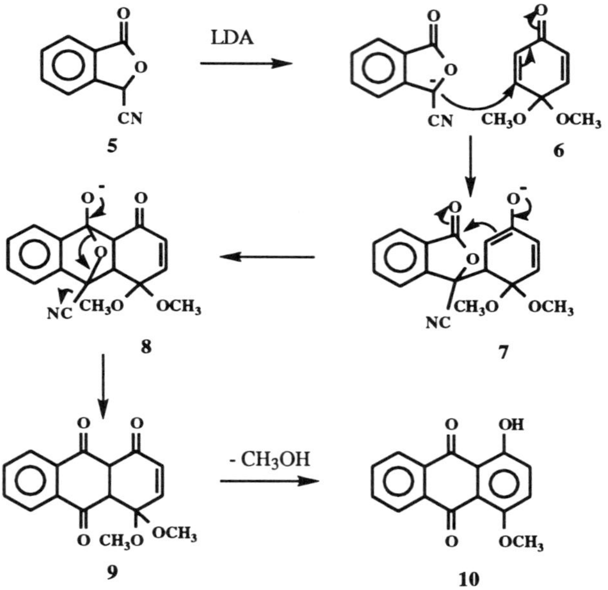

Scheme I

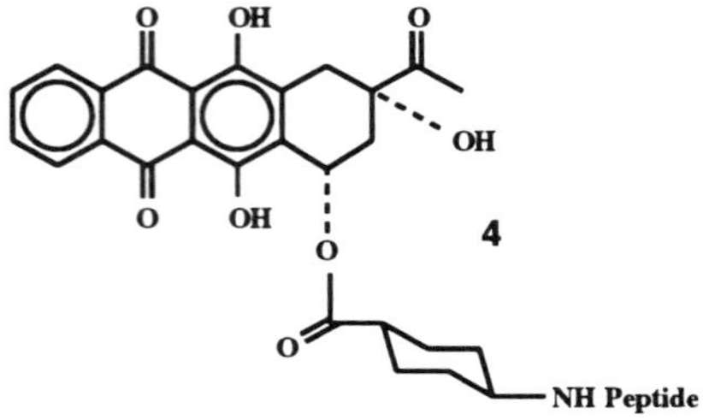

The foundation work of Swenton's group *(32,33)* on the regioselective synthesis of *p*-quinone monoacetals by anodic oxidation of 1,4-dimethoxybenzenes, followed by mild acid hydrolysis of the resulting bisacetals was readily applied to linked molecules (Scheme II).

Although the lack of total regioselectivity in the hydrolysis step lead to undesirable mixtures with hydrocarbon linking groups (**11** L = CH$_2$), the strategy provided an effective synthesis of polyether linked precursors (Scheme III) and their subsequent annelated products *(34)*.

In this sequence the dienone **17** was formed in a 9:1 ratio with its regiomer, and could be isolated in a 51% yield. Conversion to the bis-anthrquinone was

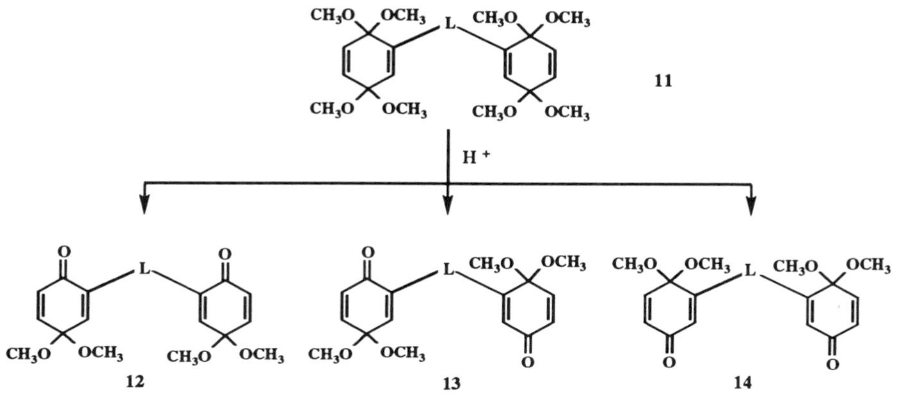

Scheme II

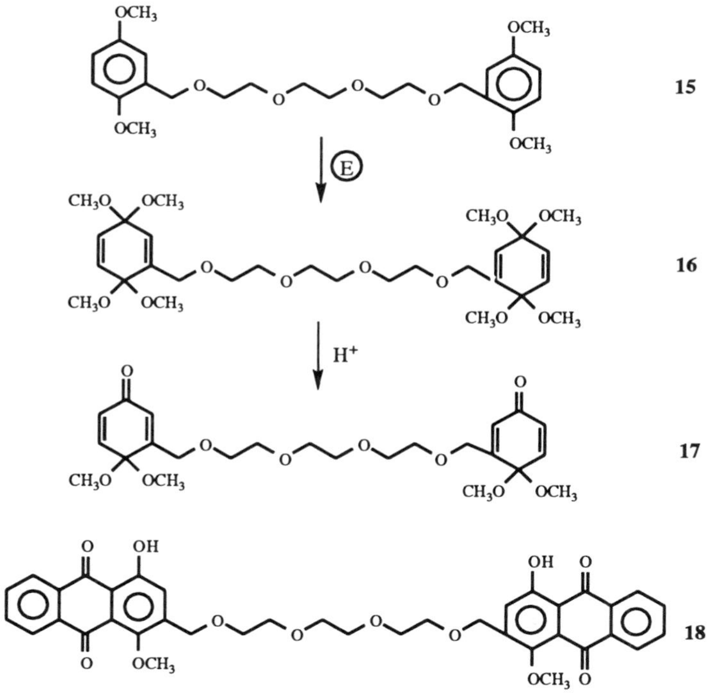

Scheme III

achieved in 50% yield, a result considerably lower than yields achieved for simple
quinone monoacetals. Similarly, some precursors containing secondary amide links
e.g **19** were prepared in good yield and with a high regiospecificity controlled by the
amide side chain *(35)*.

(a) n = 2
(b) n = 4

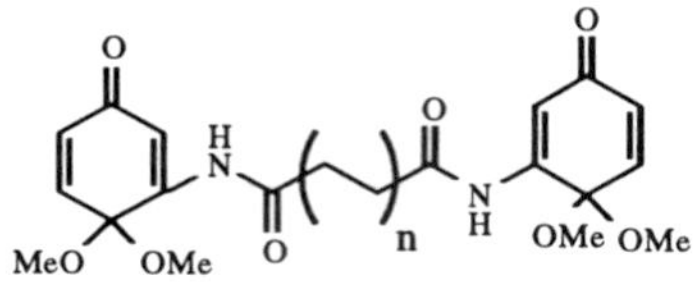

19

Notwithstanding this success, the limited solubility in methanol of amides such as **20**
severely limited the scale on which oxidations could be conducted. In addition the
anodic oxidation of amides was frequently difficult and substrate dependent. Thus
attempts to apply this approach to the N-methylated compound **22** (Scheme IV) failed
completely and yielded only polymeric materials **23** *(25)*.

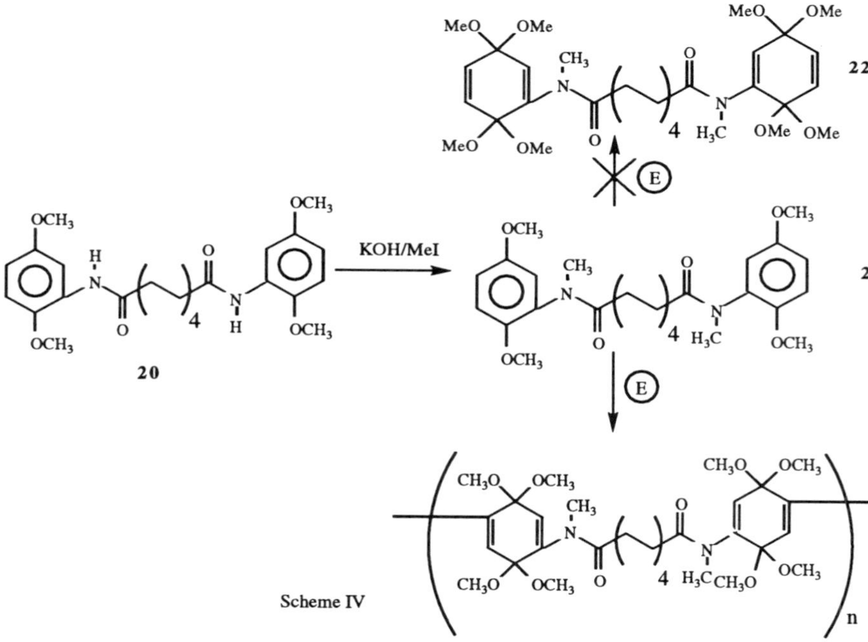

Scheme IV

A more satisfactory approach to the problem of regioselectivity lay in the
observation that a variety of aryl ethers, (e.g.. TMS,TBDS, MOM and MEM)were
labile towards anodic oxidation and could be converted to the corresponding dienone
without the need for a hydrolysis step (Scheme V) *(36)*.

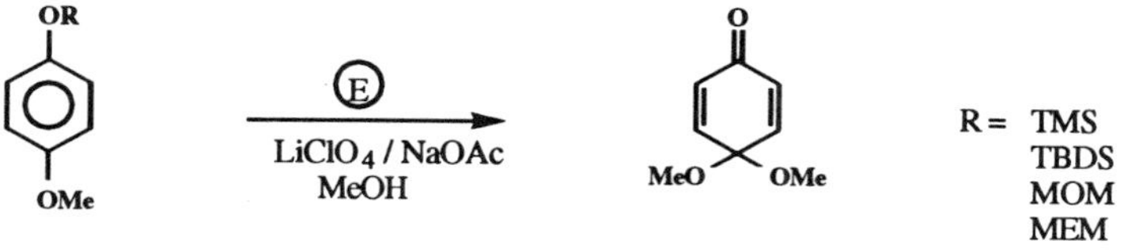

Scheme V

Bearing in mind our earlier comments concerning the complexities of glycosidation, we were in the course of developing the aminocyclohexane carboxylate ester alternative when our rather limited bioassays revealed that the simple analogues (**26** R=H or F) used as test vehicles for synthesis exhibited IC_{50} s in the range 0.02-0.04 mM in vitro against a P388 screen.

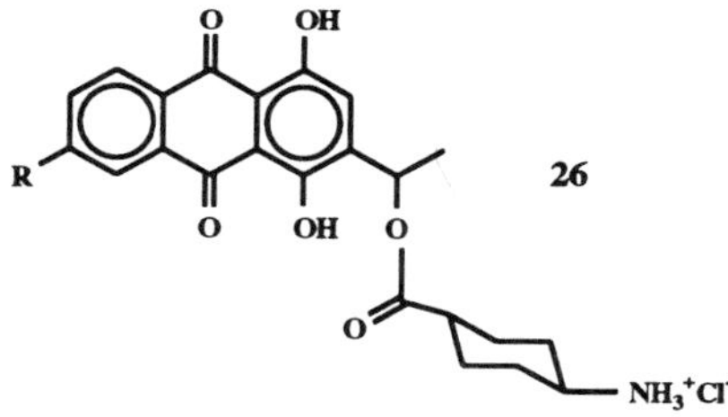

These compared with levels of 0.001-0.002 mM for comparably substituted racemic daunomycins measured under identical conditions. This surprising activity for such simple molecules raised the prospect of adapting our chemistry to construct linked molecules containing the simpler anthraquinone nucleus.

This was accomplished (Scheme VI) by reacting the aldehyde **27** with bis-Grignard reagents (n=4,6,9) to afford **28** in average yields of 50%. Subsequent protection of the resulting alcohols (MOMCl/diisopropyl ethylamine, r.t.) afforded MOM ethers **29** in quantitative yield. Anodic oxidation of these substrates in anhydrous methanol containing lithium perchlorate and anhydrous sodium acetate afforded the bis *p*-quinone monoacetals **30** which were efficiently annelated (65%) with cyanophthalide ion to afford the anthraquinones **31**. Deprotection in two steps (H_3O^+, BCl_3/-78°C) yielded **32** (86%) which was esterified by coupling with the protected (Moz) *trans*-4-aminocyclohexane carboxylic acid in the presence of DCC and DMAP, to yield ultimately the esters **33** (70%). Subsequent deprotection (HCl/dioxane) of the amides in **33** afforded the salts **34** (91%).These latter products were clearly a mixture of diastereomers as evidenced from their ^1H NMR spectra, and as such presumably contained isomers of varying activity .Accordingly the IC_{50} values in the range of 0.2-0.4 mM (P388 in vitro) provided only a crude measure of bioactivity and ignored synergistic or antagonistic effects.

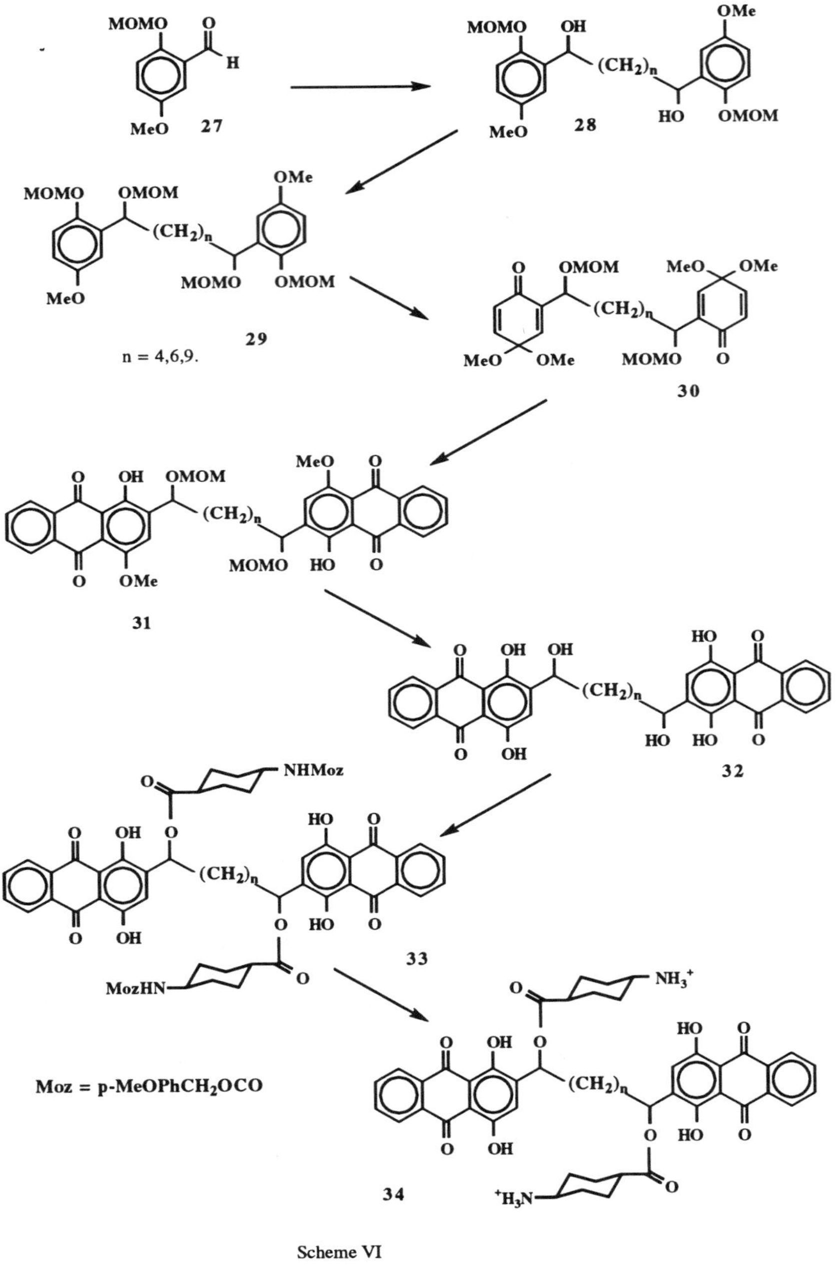

Scheme VI

Under the circumstances these initial results were not entirely discouraging. Nevertheless the low activities did not warrant the preparation of single optically pure stereoisomers. The fact that the hydrocarbon linking groups were far from optimal encouraged us to examine alternative strategies which could accommodate polyamide or polyamine linking chains. To this end we began to evaluate alternative synthetic routes which avoided anodic oxidation. Hypervalent iodine reagents e.g. $PhI(OAc)_2$ emerged as alternative oxidants to generate p-quinone monoacetals under mild conditions *(37,38)*. Our early findings suggest that there is a wealth of new chemistry which will provide easy access to precursors containing a variety of linking groups and at the same time these findings raise the prospect of incorporating minor groove binders such as distamycin *(39-41)* which may confer a better degree of site selectivity upon DNA intercalators.

In the course of our investigations we found that the use of hypervalent iodine as an oxidant for phenols provides access to many o-quinone acetals **36** (Scheme VII). The rapid nature of the oxidation coupled with the mild reaction conditions has enabled us to prepare and annelate a range of these molecules.   While some o-quinone monoacetals still dimerize too rapidly to be useful many survive long enough to act as Michael acceptors.   This finding, has bestowed on phthalide anion annelation chemistry a new significance, and a range of alternatively substituted linked anthraquinones has begun to emerge.

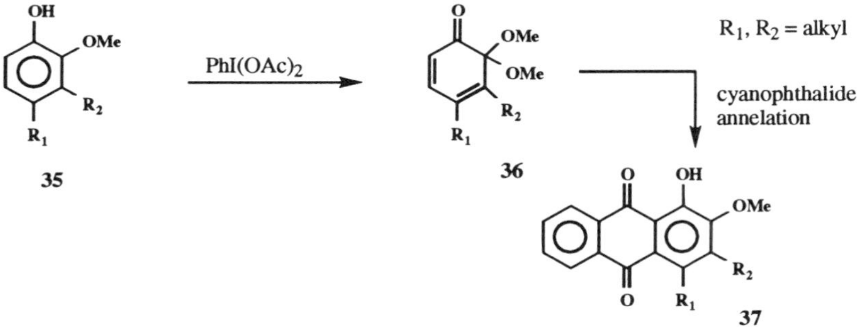

Scheme VII

Acknowledgments

We thank Mr. A. Mitchell for contributing some of his unpublished results to this paper. We also thank Professor R.N. Warrener, Central Queensland University, for his continuing interest and association with this work.

Literature Cited

1.   Tritton, T.R.; Cell surface action of adriamycin. *Pharmacol. and Ther.* **1991**, *49(3)*:293-309.
2.   Zunino, F.; Capraninco, G.; *Anti-cancer Drug Design* **1990**, *5*, 307-17.
3.   Tanako, H. *Anticancer Drugs* **1992**, *3*, 323-30.

4.   Denny, W.A. *Anticancer Drug Des.* **1989**, *4*, 241-63.
5.   Low, C.M.L.; Drew, H.R.; Waring, M.J. *Nucleic Acids Res.* **1984**, *12*, 4865-79.
6.   Crothers, D.M., *Biopolymers* **1968**, *6*, 575-84.
7.   Mikhailov, M.V.; Nikitin, S.M.; Zasedatelev, A.S.; Zhuze, A.L.; Guersky, G.V.; Gottikh, B.P. *Febs. Lett.* **1981**, *136*, 53-7.
8.   Acheson, R.M.; Constable, E.C.; Wright, R. McR.; Taylor, G.N. *J. Chem. Res., (S)* **1983**, 2-3.
9.   Corey, M.; McKee, D.D.; Kagan, J.; Henry, D.W.; Millar, J.A. *J. Am. Chem. Soc.* **1985**, *107*, 2528-36.
10. Gaugain, B.; Barbet, J.; Oberlin, R.; Roques, B.P.; Le Pecq J.B. *Biochemistry* **1978**, *17*, 5071-78.
11. Henry, D.W.; Tong G.L.; U.S. Patent 4, **1978**, *112*, 217.
12. Apple, M.A.; Yanagisawa, H.; Hut, C.A. *Amer. Assoc. Cancer Res.* Abstracts No. 675, **1978**.
13. Brownlee, R.T.C.; Cacoli, P.; Chandler, C.J.; Phillips, D.R.; Scourides, P.A.; Reiss, J.A. *J. Chem. Soc., Chem. Commun.* **1986**, 659-61.
14. Scourides, P.A.; Brownlee, R.T.C.; Phillips, D.R.; Reiss, J.A. *J. Chromatogr.* **1984**, *288*, 127-36.
15. Russell, R.A.; Irvine, R.W.; McCormick, A.S.; Warrener, R.N. *Tetrahedron* **1988**, *44*, 4591-604.
16. Kimura, Y.; Susuki, M.; Matsuomoto, T.; Abe, R.; Terashima, S. *Chem. Lett.* **1984**, 502-4.
17. Kimura, Y.; Susuki, M.; Matsumoto, T.; Abe, R.; Tershima, S. *Bull. Chem. Soc. Japan* **1986**, *59*, 423-31.
18. Wang, Z.; Feng, D.; Zhang, C.; *Acta Pharmaceutica Sinica* **1984**, *19* (5), 349-56.
19. Li, J.; Zhang, C.; *Chinese Journal of Antibiotics* **1985**, *10(4)*, 256-8.
20. Li, J.; Zhang, C.; *Pharmaceutical Industry* **1986**, *7(1)*, 15-20.
21. Hauser, F.M.; Rhee, R.P. *J. Am. Chem. Soc.* **1977**, *99* 4533-34.
22. Hauser, F.M.; Rhee, R.P. *J. Org. Chem.* **1978**, *43*, 178-80.
23. Krauss, G.A.; Sugimoto, H. *Tetrahedron Lett.* **1978**, 2263-66.
24. Krauss, G.A.; Cho, H.; Crowley, S.; Roth, B.; Sugimoto, H.; Prugh, S. *J. Org. Chem.* **1983** *48* 3439-44.
25. Russell, R.A.; Warrener, R.N. *J. Chem. Soc., Chem. Commun.* **1981**, 108-10.
26. Russell, R.A.; Pilley, B.A.; Irvine, R.W.; Warrener, R.N. *Aust. J. Chem.* **1987**,
27. Tatsuta, K.; Ozeki, H.; Yammaguchi, M.; Tanaka, M.; Okui, T. *Tetrahedron Lett.* **1990**, *31*, 5495-98.
28. Dolson, M.G.; Chennard, B.L.; Swenton, J.S. *J. Am. Chem. Soc.* **1981**, *103*, 5263-64.
29. Swenton, J.S.; Freskos, J.N.; Morrow, G.W.; Sercel, A.D. *Tetrahedron* **1984**, *40*, 4625-33.
30. Keay, B.A.; Rodrigo, R. *Can. J. Chem.* **1985**, *63*, 735-38.
31. Keay, B.A.; Rodrigo, R. *Tetrahedron* **1984**, *40*, 4597-607;
32. Swenton, J.S. *Acc. Chem. Res.* **1983**, *16*, 74-81.

33. Swenton, J.S. in *The Chemistry of the Quinonoid Compounds*; Patai, S.; Rappaport, Z. Ed.; John Wiley and Son, U.J., **1988**, p. 899.
34. Longmore, R.W.; Russell, R.A. *Aust. J. Chem.* **1991**, *44*, 1479-86.
35. Russell, R.A.; Longmore, R.W.; Warrener, R.N. *Aust. J. Chem.* **1991**, *44* 1691-704.
36. Russell, R.A.; Day, A.I.; Pilley, B.A.; Leavy, P.J.; Warrener, R.N. *J. Chem. Soc., Chem. Commun.* **1987**, 1631-3.
37. Fleck, A.E.; Hobart, S.A.; Morrow, G.W. *Synth. Commun.* **1992**, *22*, 179-87.
38. Mitchell, A.I.; Russell, R.A. *Tetrahedron Lett.* **1993** 545-48.
39. Dervan, P.B.; Sluka, J.P. New Synthetic Methodology and Functionally Interesting Compounds; Elsevier NY 1986, pp. 307-322.
40. Griffin, J.H.; Dervan, P.B. *J. Am. Chem. Soc.* **1987**, *109*, 6840-2.
41. Wade, W.S.; Mrksich, M.; Dervan, P.B.; *J. Am. Chem. Soc.* **1992**, *114*, 8783-94.

RECEIVED August 3, 1994

Chapter 9

Synthesis and Study of Structure–Activity Relationships of New Classes of Anthracyclines

Antonino Suarato, Francesco Angelucci, Alberto Bargiotti,
Michele Caruso, Daniela Faiardi, Laura Capolongo, Cristina Geroni,
Marina Ripamonti, and Maria Grandi

Pharmacia-Farmitalia Carlo Erba, R&D Oncology–Immunology,
20159 Milan, Italy

Structure-activity studies in the field of anthracyclines have been very fruitful in defining those moieties that are necessary to produce derivatives active on MDR tumor cells. We have found that substitutions on the sugar part of anthracyclines are fundamental to confer activity on MDR cells in vitro. In particular, compounds substituted at C-3' of the sugar moiety with 4-morpholino group or selected potential alkylating moieties are able to overcome resistance both in vitro and in vivo.

Over the last 20 years, the anthracyclines daunorubicin **1** and doxorubicin **2** have proved to be effective anticancer agents for the management of hematologic malignancies and a large variety of solid tumors.

1 : R=H
2 : R=OH

Doxorubicin, in particular, is currently the first drug of choice for the treatment of many tumor types. A major limitation in the clinical use of anthracyclines however is the development of resistance in chemosensitive tumors following a successful initial response (1-3). Tumor cells resistant to anthracyclines exhibit in most cases the classical multidrug resistant (MDR) phenotype. MDR cells are resistant to several classes of drugs because they have high levels of a membrane glycoprotein, p170, that is able to recognize and expel the cross-resistant compounds, therefore preventing the drug from reaching cytotoxic intracellular concentrations (4,5). As a consequence, an important pharmacological target in the synthesis of novel anthracyclines is the identification of chemical modifications capable of conferring activity against resistant cells.

In this framework, the search for molecules active on resistant tumors has been one of our major objectives in the area of new anthracycline research. In order to identify new molecules that might not be recognized by p170, we have synthesized several classes of new compounds modified on the aglycone, modified on the sugar residue, and modified on both.

The in vitro cytotoxic activity of all molecules we synthesized was evaluated using a human colon adenocarcinoma cell line, LoVo (6), and its doxorubicin-resistant subline, LoVo/DX (7). The cytotoxic activity is reported as IC_{50}, the concentration inhibiting 50% of colony formation, calculated on concentration response curves. The resistance index (R.I.) is the ratio between the IC_{50} on resistant cells and the IC_{50} on sensitive cells.

The in vivo activity was evaluated using disseminated P388 murine leukemia (10^6 cell/mouse transplanted iv in CD2F1 mice)(8) and its doxorubicin-resistant subline, P388/DX (10^5 cell/mouse transplanted iv in BD2F1 mice)(9). Treatment with drug was given iv one day after tumor transplantation. The antitumor activity of the drug was evaluated by comparing the median survival time (MST) of the treated group with that of the control group, and the results were expressed as %T/C, where:

%T/C = (MST of treated group)/(MST of control group) x 100

Both LoVo/DX and P388/DX cells exhibit the *mdr* phenotype.

Anthracyclines Modified on the Sugar Moiety

The aminosugar is a critical determinant for the pharmacological and biochemical activity of daunorubicin and doxorubicin. The basic amino group at C-3' confers water solubility to the anthracyclines and is implicated in determining the DNA affinity binding (10). It is noteworthy that chemical modifications on the sugar part of anthra-cyclines may affect their hydrophobic behavior, and in most cases, the ability to overcome MDR is related to the increased lipophilicity of the compounds and consequently to their ability to quickly enter cells and reach cytotoxic intracellular levels (*11*).

Anthracyclines modified on the sugar moiety have been prepared either by condensing the aglycone with novel synthetic sugars or by chemically modifying daunosamine aminosugar. So far, a large number of such derivatives have been described in the scientific and patent literature (*12*). Of particular interest among these are classes of compounds in which the amino group at C-3' has been placed at C-4' site, and the C-3' substituted with a hydrogen atom or hydroxy group. 4'-Amino-anthracyclines were prepared either by synthetic manipulation or by synthesis of aminosugars and subsequent condensation with aglycones. Of particular synthetic interest was the splitting of the C-3'-amino of 4'-epidaunorubicin **3** via aziridino intermediate **4,** whose regiochemical opening gave 3'-hydroxy-4'-amino anthracycline glycoside **5,** the aminosugar in *xylo* configuration. Inversion of the 3'-hydroxy group afforded isodauno **6,** which was converted to the corresponding C-14-hydroxy derivative **7** (*13*).

Similarly, *xylo* anthracycline glycoside **8** was converted, via aziridino intermediate **9**, into glycoside **10** with an *arabino* configuration (*14*).

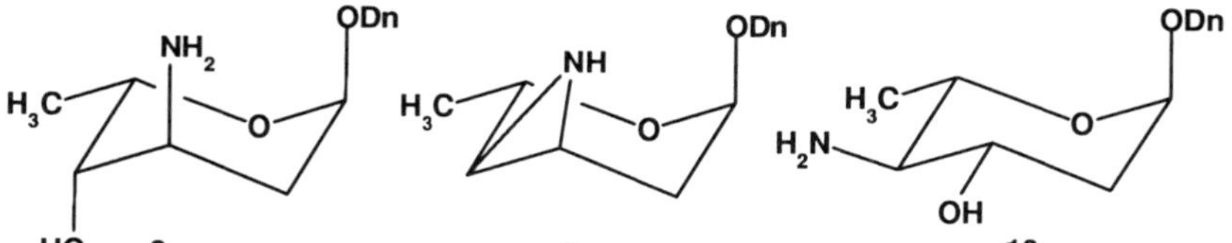

3'-Deamino-4'-amino anthracyclines **11** and **12** were prepared respectively from **5** and **10** upon substitution of 3'-hydroxy group with iodine atom and subsequent radical displacement (*15*).

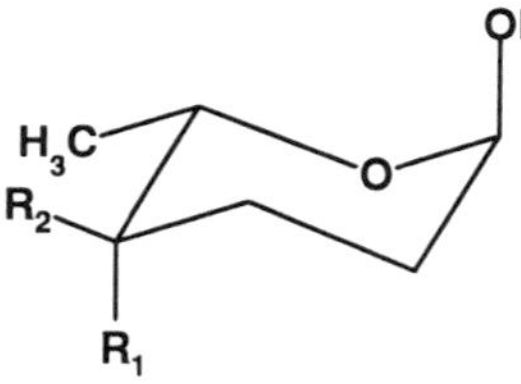

11 : $R_1 = NH_2$ $R_2 = H$

12 : $R_1 = H$ $R_2 = NH_2$

The class of 4'-aminoanthracyclines is endowed with interesting antitumor activity, particularly compound **7** (Isodoxo), which showed remarkable antitumor activity both on leukemia and solid tumors and was found to be less cardiotoxic than doxorubicin (*16,17*). These biological findings prompted us to investigate alternative ways to synthesize 4-aminosugars, in order to prepare analogues with different aglycones and to study their activity against MDR cells. Starting from L-*rhamnal* **13**, 2,4,6-trideoxy-4-amino-L-*lyxo*-exo-pyranose **14**, 2,4,6-tri-deoxy-4-amino-L-*arabino*-exopyranose **15**, 2,3,4,6-tetra-deoxy-4-amino-L-*threo*-exopyranose **16** and 2,3,4,6-tetra-deoxy-4-amino-L-*erythro*-exopyranose **17** were prepared (*12*).

These aminosugars, protected as trifluoroacetate at the

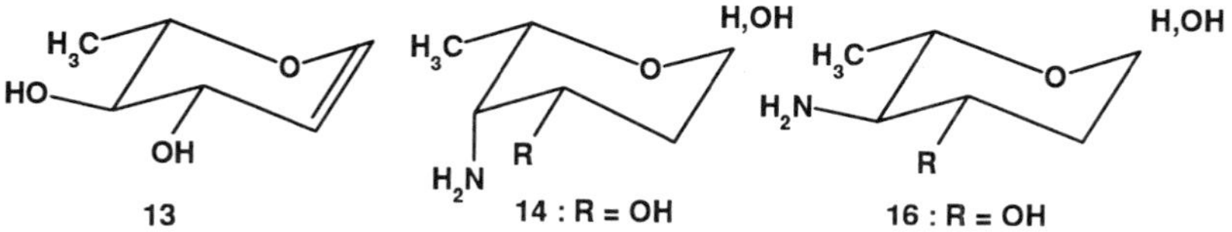

13

14 : R = OH
15 : R = H

16 : R = OH
17 : R = H

amino and hydroxy groups, were converted into 1-chloro derivatives and condensed with synthetic aglycones to produce different classes of anthracylines. Among these, the most interesting compound was 4-demethoxy-3'-deamino-4'-deoxy-4'epiaminodaunorubicin **18**, derived by condensing 4-demethoxydaunomycinone with the 1-chloro-N-trifluoroacetyl derivative of aminosugar **17** (*15*).

The cytotoxic activity of 4'-amino anthracyclines in comparison with doxorubicin is reported in Table I. Compounds **7** and **10** bearing the hydroxy group at position C-3' do not exhibit improved activity against LoVo/Dx cells,as shown by their respective R.I. values of 21.8 and 27.5. Conversely, 4'-amino derivatives **11**, **12**, **18** are active on LoVo/DX cells, with R.I. values of 2.6, 9, 0.9, respectively. Compounds **12** and **18,** in which the amino group is at position 4'-equatorial, have increased potency against both cell lines in respect to doxorubicin. These compounds are significantly more lipophilic than those bearing hydroxy group at the 3'-position (results not shown).

Table I.   Cytotoxic Activity of Anthracyclines Modified on the Sugar Moiety in Comparison with Doxorubicin

Compound	IC_{50} (ng/ml) [a]		
	LoVo	LoVo/Dx	R.I. [b]
doxorubicin	60	2180	36.3
7	40	875	21.8
10	18	495	27.5
11	100	263	2.6
12	5	45	9.0
18	11	10	0.9

[a]IC_{50} =   Concentration inhibiting colony growth by 50% :4 h treatment.
[b]R.I.=   resistence index = $(IC_{50}LoVo/Dx)/(IC_{50}LoVo)$.

Compound **18** was selected for in vivo tests (Table II). Notwithstanding its lack of cross-resistance, as evidenced in vitro on LoVo/Dx cells and in several MDR cells (data not shown), the compound was found to be inactive on disseminated P388/DX leukemia.

Table II. Activity of 4'-Epi-amino Analog of Daunorubicin 18
         Against P388 and P388/Dx Murine Leukemias

Tumor	O.D.[a]	T/C
	(mg/kg)	(%)
P388 iv	4.0	225
P388/Dx iv	5.2	121

[a]O.D.= optimal dose.

These results suggested that among "classical anthra-cyclines" a low R.I. in vitro is not always associated with maintainance of in vivo activity on MDR tumors.

Morpholino Substituted Anthracyclines

Reductive alkylation of doxorubicin with 2,2'-oxybis-acetaldehyde in the presence of sodium cyanoborohydride led to the discovery of two of the most potent anthracycline derivatives: 3'-deamino-3'-(4-morpholino)doxorubicin (**19**) and 3'-deamino-3'-(3-cyano-4-morpholino)doxorubicin (**20**), both active on sensitive and MDR cell lines (*18,19*).

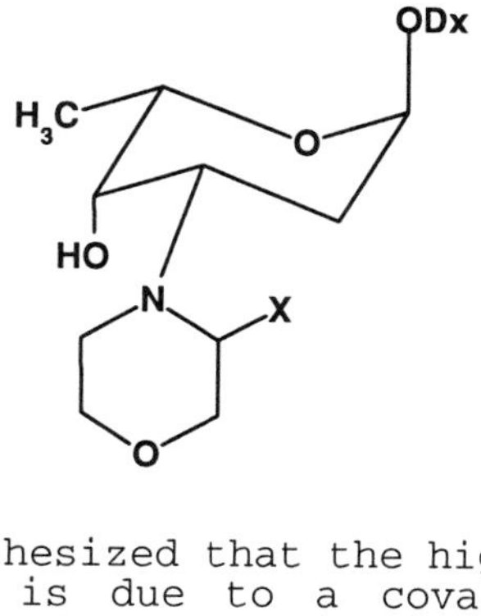

19 : X =H

20 : X = CN

It can be hypothesized that the high potency of morpholinyl anthracyclines is due to a covalent interaction of the metabolically activated morpholinyl moiety with DNA (*20,21*). However, the exact mechanism of action is still unclear. A synthetic program was launched in our laboratories to pursue optimization of this class of molecules and a better understanding of their structure-activity relationships.

When the 4-morpholino ring was shifted from the C-3' to the
C-4' position (either equatorial or axial position), the
resultant morpholino derivatives **21**, **22**, **23** were found to
be equally effective on both cell lines but to have no
activity in vivo on disseminated P388 leukemias. It is of
note that the optimal dose is significantly higher than that
of 3'-morpholino derivatives (Table III). Taken together this
findings suggest a key role for the C-3' position in the
biological activity of morpholino derivatives on MDR tumors.

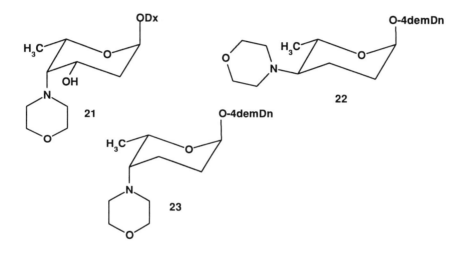

Table III. Cytotoxicity on LoVo and LoVo/Dx Cells and
           Antitumor Activity against P388 and P388/Dx
           Leukemias of 4'(morpholino)derivatives in
           comparison with 19

Compound	IC_{50} (ng/ml) [a]			P388		P388/Dx	
	LoVo	LoVo/Dx	R.I. [b]	O.D. [c] (mg/kg)	T/C (%)	O.D. [c] (mg/kg)	T/C (%)
19	171	1252	7.3	0.05	161	0.05	207
21	21	169	8.0	26[d]	235	20	100
22	852	1381	1.6	–	–	11.7	106
23	203	636	3.1	16.9[d]	180	26[d]	111

[a] IC_{50}=concentration inhibiting colony growth by 50%: 4 h
    treatment.
[b] R.I.=resistence index=(IC_{50}LoVo/Dx)/(IC_{50}LoVo).
[c] O.D.=optimal dose
[d] highest tested dose

3'-(2-Alkoxy-4-morpholino)doxorubicin Derivatives

An important class of anthracyclines is represented by those in which alkoxy residues are linked at position C-2 of the 4-morpholino ring of doxorubicin. In such compounds glycosidic linkages are introduced into the morpholino system, thus inserting additional chemical feature to the anthracycline. A number of such derivatives have been synthesized and their biological activity evaluated.

3'-(2-Alkoxy-4-morpholino)anthracyclines of general formula **24** have been prepared by condensing 1,5-diiodo-2-alkoxy-3-oxapentane **25** with doxorubicin in the presence of triethylamine. Diiodo derivatives **25** are prepared from ring opening of 1-alkyl-β-L-*arabino*-pyranosides **26** and subsequent chemical modifications (*22*).

24   25   26

This procedure allows retention of chirality at position C-1 from the starting glycoside to position C-2 of the diiodo derivatives and gives 3'-(2-alkoxy-4-morpholino) anthracyclines with defined chirality at C-2 (*23*). In addition, bis-alkylation of the amino group of anthracyclines also avoids the reduction of the 13-carbonyl group, as reported by Acton during the Borch reductive alkylation for the formation of the morpholino ring *via* bis-aldehyde (*18*). The structures of 2-alkoxy-4-morpholino anthracyclines and the cytotoxic activity on LoVo and LoVo/DX cells in comparison with 3'-(4-morpholino)doxorubicin are reported in Table IV.

Among the 3'-(2-alkoxy-4-morpholino) anthracyclines, the most active analogue was 3'-(2-methoxy-4-morpholino) doxorubicin **24a**; under the test conditions employed, this compound showed an R.I. value of 3.5 and 6 fold increase in potency versus 3'-(4-morpholino)doxorubicin. The decreased in vitro activity ranging from 10-to 20-fold, for analogues **24b-24e** in comparison with **24a** was consistent with the increased steric hindrance at C-2 of the morpholino ring. On the other hand, these compounds were active in vivo with 1.5- to 10-fold reduced potency versus **24a**. The antitumor activity measured in vivo (Table V) confirmed the lack of cross-resistance observed in vitro.

Table IV.   Cytotoxic Activity of (2-Alkoxy-4-morpholino)
             doxorubicin Derivatives against LoVo and LoVo/Dx
             cells

Compound	R	IC$_{50}$(ng/ml)[a]		
		LoVo	LoVo/Dx	R.I.[b]
24a	CH$_3$	9	32	3.5
24b	C$_2$H$_5$	104	356	3.4
24c	CH(CH$_3$)$_2$	106	517	4.9
24d	CH$_2$C$_6$H$_5$	154	530	3.4
24e	C(CH$_3$)$_3$	243	830	3.4

[a]IC$_{50}$=concentration inhibiting colony growth by 50%: 4 h
   treatment.
[b]R.I.= resistence index = (IC$_{50}$LoVo/Dx)/(IC$_{50}$LoVo).

Table V.   Antitumor Activity of (2-Alkoxy-4-morpholino)
             doxorubicin Derivatives against P388 and P388/Dx
             Leukemias

Compound	R	P388		P388/Dx	
		O.D.[a]	T/C	O.D.[a]	T/C
		(mg/kg)	(%)	(mg/kg)	(%)
24a	CH$_3$	0.09	250	0.11	208
24b	C$_2$H$_5$	0.22	219	0.16	189
24c	CH(CH$_3$)$_2$	0.50	200	0.50	167
24d	CH$_2$C$_6$H$_5$	0.50	175	0.50	156
24e	C(CH$_3$)$_3$	0.45	200	1.00	172

[a]O.D.= optimal dose.

On the basis of the above results, the effect of 3'-(2-methoxy-4-morpholino)doxorubicin was explored on different leukemias and solid tumors in mice, and the results of biological activity assays, in comparison with those of doxorubicin, are illustrated in Table VI and VII. This compound showed clear activity on MDR cells and tumors and was also found more effective than doxorubicin in experimental solid neoplasms, such as the MX1 and mammary carcinomas. The compound is now undergoing Phase I clinical trials (24).

Table VI. Comparative Activity of 3'-(2-Methoxy-4-morpholino) doxorubicin 24a and Doxorubicin

Experimental model[a]	Compound 24a[b]	Doxorubicin[b]
P388	+++	+++
P388/Dx Johnson	+++	--
P388/Dx Schabel	++	--
L1210	++	++
L1210/cDDP	+	+
L1210/L-PAM	+	+

[a]Disseminated leukemias; compounds were administered iv.
[b]T/C%: +, 130-150; ++, 150-200; +++, >200.

Table VII. Comparative Activity of 3'-(2-Methoxy-4-morpholino doxorubicin 24a and Doxorubicin

Experimental model	24a[a]	Doxorubicin[a]
MTV	++	+++
Colon 38	--	+++
Lewis Lung	+	+++
MXT	--	+/-
M5076	+++	+++
MX1	+++	++
LoVo	+	++
LoVo/Dx	--	--
A549	--	--
CX1	--	--
A592	+++	++

[a]% growth inhibition: +, 40-50; ++, 50-80; +++, >80.

3'-(2-Acyloxy-4-morpholino)doxorubicin

In order to explore the influence on MDR tumor cells of other classes of modified at 4-morpholino ring anthracyclines, we decided to substitute on that moiety at position C-2 groups with different polarity than the alkoxy. Consequently, we chose to synthesize 3'-(2-acyloxy-4-morpholino)anthracyclines. The leading compound of this class of anthracyclines is 3'-(2-O-benzoyl-4-morpholino)-doxorubicin **27**, which was prepared via Borch reductive alkylation of doxorubicin with bis-aldehyde **28** obtained from periodate oxidation of **29** (*25*).

      27   **28**   **29**

     3'-(2-O-benzoyl-4-morpholino)doxorubicin was endowed with an R.I. value of 3.9 and in vivo antitumor activity comparable to that of 3'-(2-alkoxy-4-morpholino) derivatives, with %T/C of 200 and 167 at an optimal dose of 0.5 mg/kg on both disseminated P388 and P388/DX leukemias. These results are similar to those obtained in testing 3'-(2-O-benzyl-4-morpholino)doxorubicin **24d**, indicating that steric factors, rather than electronic effects, are more important in influencing the potency and activity of anthracyclines substituted at position C-2 in the 3'-(4-morpholino) system.

Mustard Anthracyclines

Recently, a new distinct class of anthracyclines active on MDR tumor cells was developed in our laboratories. Structure-relationships data collected from different classes of anthracyclines prompted us to design new molecules capable of both intercalating and alkylating at DNA. Based on the data collected, it was worthwhile to introduce an alkylating moiety at position C-3' of the aminosugar, this idea was also supported by recent spectroscopic evidence indicating the orientation of the 3'-methoxymorpholinyl moiety within the minor groove of DNA helix, near the ATbase pairs (26). We chose the classical nitrogen mustard in which the nitrogen atom was that of the 3'-amino group of anthracycline glycoside. In addition, a hindered withdrawing group, such as a methylsulfonic group, was attached at position C-4' with the aim of reducing the basicity of the nitrogen mustard. The leading compound of this new class of molecules is 4-demethoxy-3'-bis(2-chloro-ethyl)-4'-deoxy-4'-O-sulfonyl-daunorubicin **30**, prepared by reacting 4-demethoxydaunorubicin **31** with ethylene oxide to give the 3'-N-bis(2-hydroxyethyl) derivative **32**, which is converted in one step to **30** by treatment with mesyl chloride in pyridine (27).

The compound showed an R.I. value of 3.5. Preliminary biological data, reported in Table VIII, indicate that this new type of substitution may represent a breakthrough in the field of new anthracyclines.

Table VIII.   Antitumor Activity of Compound 30 against P388 and P388/Dx Leukemias

Compound	P388			P388/Dx		
	Dose (mg/kg)	T/C (%)	LTS[a]	Dose (mg/kg)	T/C (%)	LTS[a]
30	4.8	481	3/10	4.8	233	2/19
				6.2	>410	9/19

[a]LTS: long term survivals (>60 days) after end of the experiments.

Literature cited

1.   Gottesman,M.M. *Cancer Res.* **1993**, *53*, 747-754.
2.   Kaye,S.B. *Br.J.Cancer* **1988**, *58*, 691-694.

3.   vanKalten,C.K; Pinedo,H.M; Giaccone, G. *Eur.J.Cancer*
      1991, *27*, 1481-1486.
4.   Beck,W.T. *Eur.J.Cancer* **1990**, *26*, 513-515.
5.   Dalton,W.S; Grogan,T.M; Meltzer,P.S. *J.Clin.Oncol.*
      1989, *7*, 415-424.
6.   Drewinko,B; Romsdahl,M.M; Yang,M.Y; Ahearn,M.J;
      Trujillo,J.M. *Cancer Res*. **1990**, *36*, 467-475.
7.   Grandi,M; Geroni,C; Giuliani,F.C. *Br.J.Cancer* **1986**,
      54,   515-518.
8.   Geran,R.I; Greenberg,N.H; MacDonald,M.M; Shumaker,A.M;
      Abbot,B.J. *Cancer Chem.Rep.* **1972,** *Part 3, 1-103*.
9.   Johnson,R.K; Chitnis,M.P; Embery,W.M; Gregory,E.B.
      Cancer Treat.Rep. **1978**, *62*, 1535-1547.
10. Arcamone,F. *Doxorubicin*, Medicinal Chemistry; Academic
      Press: New York, N.Y., **1981**, Vol.17.
11. Facchetti,I; Grandi,M; Cucchi,P; Geroni,C; Penco,S;
      Vigevani,A. *Anti-Cancer Drug Design* **1991**, *5*, 385-397.
12. Suarato,A; Angelucci,F; Bargiotti,A. *Chimicaoggi* **1990**,
      8, 9.
13. Bargiotti,A; Caruso,M; Suarato,A; Penco,S; Giuliani,F.
      U.S.Patent **1987**, 4,684,629.
14. Suarato,A; Caruso,M; Penco,S; Giuliani,F. *GB Patent*
      1988, 2195998A.
15. Bargiotti,A; Suarato,A; Zini,P; Grandi,M; Pezzoni,G.
      U.S.Patent **1991**, 4,987,126.
16. Suarato,A; Bordoni,T; Caruso,M; Barbieri,B; Bargiotti,
      A; Geroni,C; Arcamone,F; Giuliani,F. Proc. Amon.Assoc.
      Cancer Res. **1987**,2721.
17. Giuliani,F; Fiebig,H.H; Pezzoni,G; Bahari,B; Caruso,M;
      Bargiotti,A; Suarato,A. *NCI-EORTC Symposium on New
      Drugs in Cancer Therapy*, Amsterdam, **1989.**
18. Acton,E.M; Tong,G.L; Mosher,C.W; Wolgemuth,R.L.
      J.Med.Chem. **1984**, *27*, 638-645.
19. Johnston,J.B; Habernicht,B; Acton,E.M. *Biochem.
      Pharmacol.* **1983**, *32, 3255-3258.
20. Westendorf,J; Groth,G; Steinheider,G; Marquardt,H.
      J.Cell Biol.Toxicol. **1985**, *1*, 87-101.
21. Westendorf,J; Aydin,M; Groth,G; Weller,O; Marquardt,H.
      Cancer Res. **1989**, *49, 5262-5266.
22. Suarato,A; Angelucci,F; Bargiotti,A; Grandi,M;Pezzoni,
      G. *G.B. Patent* **1989**, 8905668
23. Faiardi,D;   Bargiotti,A;   Grandi,M;   Suarato,A.
      G.B.Patent   **1989**, 8928654.
24. Bargiotti,A; Faiardi,D; Grandi,D; Ripamonti,M;Suarato,
      A. *XVIth International Carbohydrate Symposium* **1992**,
      Paris, *A164*.
25. Ripamonti,M; Pezzoni,G; Pesenti,E; Pastori,A; Farao,M;
      Bargiotti,A; Suarato,A; Spreafico,F; Grandi,M. *Br.J.
      Cancer* **1992**, *65*, 703-707.
26. Faiardi,D; Bargiotti,A; Suarato,A. *G.B. Patent* 1991,
      9019934.

27. Odefey,C; Westendorf,J; Johannes,D; Dieckmann,T; Oshkinat,T. *Chem.-Biol.Interact.* **1992**, *85(2-3)*, 117-122.
28. Bargiotti,A; Caruso,M; Faiardi,D; Suarato,A; Mongelli, N. *G.B. Patent* **1991,** 9114549

RECEIVED June 3, 1994

Chapter 10

Molecular Recognition of DNA by Daunorubicin

Jonathan B. Chaires

Department of Biochemistry, University of Mississippi Medical Center, 2500 North State Street, Jackson, MS 39216–4505

Solution studies of the interaction of the potent anticancer drug daunorubicin (daunomycin) with DNA have converged with structural and theoretical studies to provide a coherent picture of the preferential interaction of the antibiotic with the DNA sequences 5'(A/T)GC and 5'(A/T)CG, where the notation (A/T) indicates that either A or T may occupy the position. Daunorubicin is unique among monointercalating compounds both in its recognition of a 3 bp site and in its binding to DNA by a mixed mode. The anthraquinone ring system of daunorubicin intercalates, with concomitant unwinding and lengthening of the DNA helix, while the daunosamine moiety is bound within the minor groove. The sequence preference for daunorubicin binding arises from a combination of specific hydrogen bond formation between the drug and DNA base pairs and from a favorable stereochemical fit of the daunosamine in the minor groove in the vicinity of an AT base pair. This chapter summarizes the results of macroscopic studies used to characterize the physical chemistry of the binding of daunorubicin to DNA and discusses newer results obtained by DNAse I footprinting methods that have characterized the preferred DNA binding sites of daunorubicin, which in turn have begun to characterize the microscopic binding properties of the drug. The observed binding behavior is discussed in light of the extensive structural and theoretical results now available.

The anthracycline antibiotics have been, and continue to be, important weapons in the chemical arsenal used against cancer. As recently as 1987, doxorubicin (adriamycin) was the leading anticancer drug in terms of sales in the United States (1). Because of their proven utility in cancer chemotherapy, the anthracycline antibiotics have been intensively studied in attempts to understand the underlying chemical, biochemical, and pharmacological mechanisms that make them effective anticancer agents. Several excellent monographs have summarized much of the wealth of information obtained

0097–6156/95/0574–0156$08.00/0

about these important drugs over the last two decades of research (2,3). In spite of this intensive study, the fundamental mechanism(s) by which the anthracycline antibiotics act remain unclear. Further study is necessary to understand how these important drugs work at the molecular level.

It is now clear that the mechanisms of action of the parental anthracycline antibiotics daunorubicin (daunomycin) and doxorubicin are pleiotropic (4). Convincing arguments may be made to implicate either DNA or membranes as targets for these drugs or to point to the ability of these compounds to generate free radicals as an integral part of their mode of action. Topoisomerase II has been identified as perhaps the key target for the anthracycline antibiotics (5–11), although the precise molecular mechanism by which inhibition of this key enzyme occurs remains to be defined.

Historically, DNA has received considerable attention as an important target for doxorubicin and daunorubicin. Both compounds bind avidly to DNA by the process of intercalation (12–14). DNA binding results in inhibition of both DNA replication and RNA transcription, effects that may be observed both in vivo and in vitro (15–18). It has been argued that the magnitude of observed macroscopic DNA binding constants ($\approx 10^6$ M^{-1}) is too low to account for the physiological concentrations at which the anthracycline antibiotics exert their effects (19), and that something other than DNA alone must be the key target of the drugs. More will be said about this later on. Strong evidence implicating DNA as an important target for the anthracyclines has appeared. Valentini, et al, have found a positive correlation between DNA binding affinity and biological activity in studies utilizing 26 different anthracycline antibiotics (20). Microspectrofluorometry has shown that doxorubicin is quantitatively localized into the nuclei of living cells (21). Interestingly, it has been shown that doxorubicin selectively displaces a unique set of nuclear proteins from nuclei, a set distinct from those displaced by actinomycin, mitoxantrone, and amenatrone (22). Finally, recent studies have shown that DNA binding is necessary (but not sufficient) for the inhibition of topoisomerase II by anthracycline antibiotics (9,10).

Whether or not DNA is the ultimate cellular target of the anthracyclines, their DNA binding properties are still of intense, fundamental interest to the physical biochemist and the medicinal chemist. One reason for this is that the anthracycline antibiotics have been (until just last year) the only monointercalators for which atomic level crystallographic structural information was available for their complexes with DNA oligonucleotides (23–27). Such structural information makes these compounds, at the least, important models for how intercalators bind to DNA. Knowledge of the structural details of their binding has provided information that may guide the rational design of new compounds with improved DNA binding properties. Kinetic and thermodynamic studies in solution are essential complements to these structural studies, along with synthetic efforts to produce new compounds with systematically altered binding properties. Such solution studies of anthracycline antibiotic binding to DNA have been the primary focus of this laboratory and will be summarized here.

MACROSCOPIC BINDING STUDIES OF THE DAUNORUBICIN – DNA INTERACTION

Figure 1 shows a binding isotherm for the interaction of daunorubicin with calf thymus DNA. This figure is a composite, and was constructed by combining data from four different laboratories that had studied the binding reaction under identical solution conditions (28–32; Satyanarayana, S., Chaires, J. B., unpublished data). A variety of experimental approaches have been used to obtain these data, including fluorescence and absorption spectroscopy, equilibrium dialysis, and phase partition methods. The combined data, it is clear in Figure 1, are in excellent agreement, at least at binding ratios (r) above 0.05. Analysis of the data of Figure 1 (neglecting the data below r = 0.05) by the simple neighbor exclusion model of McGhee and von Hippel (33) yields a

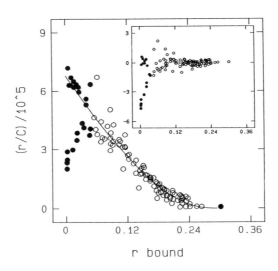

FIGURE 1.   Composite binding isotherm for the interaction of daunorubicin with calf thymus DNA at 20°C, pH 7.0, 0.2 M Na⁺. Data were taken from reports from four different laboratories (28–32). Data above r=0.05 were fit to the neighbor exclusion model (33), yielding the best fit shown as the solid line. The filled symbols indicate the points omitted when fitting the data. The inset shows the residuals (experimental data − fit) plotted as a function of the binding ratio r.

binding constant (K) for the interaction of daunorubicin with an isolated DNA site of 6.6 ($\pm$0.2) x 10^5 M^{-1} and an exclusion parameter (n) of 3.3 $\pm$ 0.1 bp. The latter value is consistent with structural studies (23–27) that show that daunorubicin physically covers 3 bp when bound to DNA. It must be emphasized that the binding isotherm shown in Figure 1 is a *macroscopic* characterization of binding and that the K value obtained by the analysis using the neighbor exclusion model is obtained under the assumption of identical and noninteracting binding sites along the DNA lattice. Footprinting studies, to be summarized in a later section, reveal that this assumption is probably invalid. The macroscopic K value is, nonetheless, still a useful quantitative measure of binding, if it is kept in mind that it represents a complicated average binding constant that masks a distribution of binding constants for the interaction of daunorubicin at different sequences along the DNA lattice.

By studying daunorubicin binding to DNA as a function of temperature, ionic strength, and base composition, it is possible to derive a complete *macroscopic* thermodynamic profile for the binding interaction. Table I shows such a profile. Several points emerge from Table I. Daunorubicin binding to calf thymus DNA is energetically favorable, with $\Delta G^0 = -8.0$ kcal mol^{-1}. The favorable binding free energy results largely from a favorable binding enthalpy ($\Delta H^0 = -12.0$ kcal mol^{-1}). The direct measurement, by calorimetry, of the binding enthalpy of daunorubicin for its interaction with a number of natural DNA samples and polynucleotides is an area of recent progress (34–35). The negative entropy that accompanies daunorubicin binding must arise from the complicated interplay of many contributions, including changes in both DNA and antibiotic hydration, ion release, the loss of translational and rotational freedom by the antibiotic upon binding, and DNA conformational changes upon intercalation.

TABLE I. Thermodynamic Profile for the Interaction of Daunorubicin with Calf Thymus DNA[a]

$K = 6.6$ ($\pm$0.2) x 10^5 M^{-1} (20°C)

n = 3.3 ($\pm$0.1) bp

$\Delta G^0 = -RT\ln K = -7.9$ kcal mol^{-1} (20°C)

$\Delta H^0 = -12.8$ ($\pm$2.0) kcal mol^{-1} (van't Hoff)

-11.4 ($\pm$0.7) kcal mol^{-1} (calorimetry)

$\Delta S^0 = -14$ cal mol^{-1} deg^{-1} (20°C)

$(\delta\ln K/\delta\ln[Na^+]) = -1.25 \pm 0.2$

$\Delta G_{obs} = \Delta G_0 + \Delta G_{el} = -6.9 + (0.738)\ln[Na^+]$

$K_{obs} = (37.0$ x $10^5)(2f^2 - f^3)$ f = fraction GC bp

[a] The values reported refer to standard solution conditions of 0.2 M NaCl, pH 7.0.

Table I also shows that daunorubicin binding to DNA is strongly dependent upon NaCl concentration, as reflected by the quantity $(\delta\ln K/\delta\ln[Na^+])$. This salt dependence has been analyzed in detail (36). Briefly, the observed salt dependence arises from the coupling of the binding positively charged drug molecule with Na^+ binding. The overall contribution to the binding free energy arising from polyelectrolyte effects is comparatively minor, approximately -1.0 kcal in 0.2 M NaCl. The implication of this is that the main noncovalent interactions stabilizing the daunorubicin–DNA complex are hydrogen bonds and van der Waals interactions.

Finally, Table I shows that the magnitude of the daunorubicin binding constant for DNA is strongly dependent upon the base composition of the DNA to which it is binding. The functional dependency is complex and goes as at least the *square* of the fractional GC content. Such a dependency reflects site specific drug binding and may be explained by a model in which daunorubicin preferentially binds to sites containing adjacent GC base pairs (37).

The kinetics of daunorubicin binding to DNA have been thoroughly studied (38–43). These studies reveal that daunorubicin binding is both slow and complex. These features distinguish daunorubicin from simple intercalators like ethidium and proflavin, whose binding is complete on the millisecond time scale and which seem to bind by a simple bimolecular reaction mechanism. The lifetime of the daunorubicin–DNA complex is approximately 1 second, and at least three steps are required to describe the time course of its DNA binding reaction. These complicated kinetics most probably arise from its preferential binding to certain DNA sequences.

MICROSCOPIC BINDING STUDIES

The advent of "footprinting" methodology (44–48) has made it possible to study the *microscopic* binding of antibiotics to specific DNA sites. Both DNase I footprinting (49–53) and a high resolution transcription assay (54,55) have been used to identify preferred anthracycline antibiotic binding sites in DNA. The results from these methods reveal unambiguously that the anthracyclines do not bind randomly along the DNA lattice, but rather bind preferentially to particular DNA sequences. Aspects of quantitative footprinting titration studies of daunorubicin will be highlighted here.

Footprinting titration studies of the interaction of daunorubicin with a 165 bp restriction fragment containing the tyrosine tRNA operator region (the "*tyr T* fragment") revealed four responses upon addition of drug (52). One class showed no effect of the antibiotic on the DNase I cleavage rate. Such sites are interpreted as sites that do not bind antibiotic. A second class showed protection from DNase I cleavage upon addition of drug and represents sites near where antibiotic binds. Another class of sites shows protection, but only after a threshold drug concentration has been reached. These sites may arise from cooperative binding interactions in the vicinity of high affinity binding sites. A final class of sites shows enhanced cleavage by DNase I in the presence of drug. These may arise from structural distortion of the DNA helix near drug binding sites. From the point of view of the DNA sequence preference of daunorubicin binding, the unprotected and simple protected sites are of the most interest and will be discussed here.

Table II lists the sites whose rate of cleavage by DNase I is unaffected by the addition of daunorubicin. The information content of these aligned sequences may be evaluated by using an algorithm presented by Stormo (56). Such an analysis reveals that the aligned sequences in Table II are not entirely random, and that certain positions contain a modest information content. In particular, these data reveal that cleavage of sequences to the 5' side of contiguous AT base pairs tends to be unaffected by the addition of daunorubicin. Thus, the sequence 5'(A/T)(A/T)N, where (A/T) means either A or T and N means any nucleotide, does not appear to bind daunorubicin. Macroscopic binding studies have shown that daunorubicin binds poorly to the synthetic polynucleotide poly dA–poly dT (57), a finding that is consistent with this conclusion from footprinting results.

TABLE II. Analysis of Sequences Surrounding Sites Whose Rate of DNase I Digestion Was Unaffected by the Addition of Daunorubicin[a]

5'			—POSITION—			3'
	-3	-2	-1	$+1$	$+2$	$+3$
A	0.318	0.409	0.227	0.409	0.455	0.316
T	0.136	0.136	0.318	0.409	0.273	0.227
G	0.227	0.273	0.227	0.091	0.227	0.273
C	0.318	0.182	0.227	0.091	0.045	0.182
I_{sq}:	0.07	0.12	0.01	**0.30**	**0.27**	0.02

[a] The table shows the frequency of bases surrounding the unprotected sites, along with the calculated information content (I_{sq}) obtained by using the algorithm of Stormo (56). The number of sites analyzed was 22. I_{sq} ranges from 0 (no information) to 2.0 (maximum information).

In contrast to these sites unaffected by daunorubicin, several sequences show striking protection from DNase I digestion in the presence of daunorubicin. The best protected sites, revealed by quantitative microdensitometry of autoradiograms used to record the footprinting experiment, are shown in Table III. These sites appear to be completely saturated at a daunorubicin concentration of $\leq$ 0.25 μM. Inspection of these sites reveals that the majority of these sequences contain a common triplet sequence, 5'(A/T)CG or 5'(A/T)GC, i.e., an AT base pair flanked by contiguous GC base pairs in an alternating purine–pyrimidine (or pyrimidine–purine) configuration. Since macroscopic binding studies, and x–ray crystallographic structures, have shown that daunorubicin binds to a 3 bp site, it is reasonable that its preferred binding sequence is a triplet. The results of these footprinting studies are fully consistent with the prediction of a preferred triplet daunorubicin binding site, 5'TCG, made by Pullman and coworkers (58–60), based on their computational studies. This is one of the few cases in which computational chemistry has successfully predicted the behavior of a DNA–ligand interaction that has been subsequently proven correct by experiment (60).

While daunorubicin most certainly binds preferentially to these triplet sequences, it must be emphasized that its binding specificity is not absolute and that the drug can and does bind to other triplet sequences, albeit with lower affinity. Footprinting titration studies allow for lower limit estimates to be made of individual site binding constants at each position where protection is registered. Figure 2 show the extremes of behavior observed for daunorubicin binding to the *tyr T* fragment. Binding constants at the two sites shown vary by a factor of 60, corresponding to a difference in binding free energy of about 2 kcal mol[-1]. Figure 3 shows the distribution of experimentally determined binding constants for all of the protected sites observed in the footprinting titration experiments. The preferred sites listed in Table III are the high affinity sites, and their binding constants differ from the mean value by about a factor of 5. Thus, daunorubicin discriminates strongly *against* certain sequences as binding sites, the 5'(A/T)A motif, but will bind to others with a range of binding affinities.

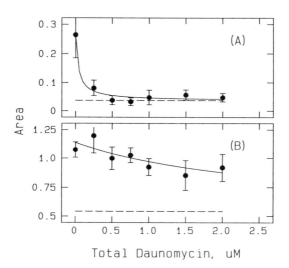

FIGURE 2. Individual site binding isotherms for the interaction of daunorubicin at two specific sequences within the *tyr T* fragment. Panel A shows the highest affinity site observed in footprinting titration experiments; panel B shows the lowest affinity site that is protected. Nonlinear least squares analysis yields a binding constant of 24.4×10^6 M^{-1} for the data in panel A, and 0.4×10^6 M^{-1} for the data in panel B (values that correspond to a difference in binding free energy between the two sites of about 2 kcal mol^{-1}).

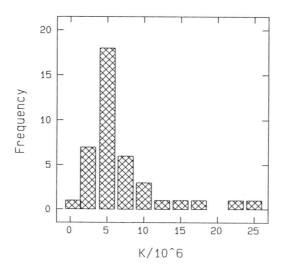

FIGURE 3. Distribution of the binding constants observed for the 40 protected sites monitored in footprinting studies. The mean for this distribution is $5.8 \ (\pm 5.2) \times 10^6$ M^{-1}.

To test the results emerging from footprinting experiments, DNA oligonucleotides were designed and synthesized for use in fluorescence titration binding studies. Hexanucleotide sequences were used, one containing the putative preferred triplet sequence (5'ACG) and others of simpler repeating CG or TA sequences to serve as controls. The first study of this type was reported by Rizzo and coworkers (61) and utilized hexameric duplex DNA samples. We have reproduced their findings (Satyanarayana, S., Chaires, J. B., unpublished data), but have in addition synthesized hairpin molecules for binding studies to avoid the necessity of conducting binding studies in 1 M NaCl, a salt concentration required to stabilize the DNA hexamers. The results of studies conducted with the hairpin molecules under more standard conditions (0.2 M NaCl) are shown in Table IV. Binding is tighter to the hairpin sequence containing the triplet 5'ACG than to either of the two control hairpins. This finding is consistent with the published work utilizing hexameric duplexes (61) and strongly supports the conclusions from our footprinting titration studies.

Table III. Sites and Surrounding Sequences Showing the Greatest Protection Upon Addition of Daunorubicin[a]

POSITION	SEQUENCE
67	5'CACTT^TACAG
70	5'TTTAC^*AGC*GG
59	5'AA*ACG*^TAACA
119	5'GA*CGA*^GGCCA
95	5'GA*TGC*^GCCCC
54	5'TTTCT^CAACG
36	5'TT*ACG*^CAACC
38	5'AC*GCA*^ACCAG
100	5'GCCCC^*GCT*TC
64	5'TAACA^CTTTA

[a] The symbol (^) shows the protected bond in each sequence. The common triplet sequence is italicized.

MOLECULAR DETERMINANTS OF THE DAUNORUBICIN SEQUENCE PREFERENCE

The preference of daunomycin binding to the triplets 5'(A/T)CG and 5'(A/T)GC may be rationalized using the extensive structural (23–27) and computational (58–60) data available. First, daunorubicin is unique in that it binds to DNA by a *mixed* mode. The anthraquinone ring *intercalates* into DNA, with its long axis nearly perpendicular to the long axes of the base pairs comprising the intercalation site. The daunosamine and C13 and C14 portions of daunorubicin lie in the minor groove, providing *groove binding* interactions.

The key determinants of the sequence preference of daunorubicin binding appear to be the O9 hydroxyl and, surprisingly, the daunosamine. The O9 hydroxyl forms two hydrogen bonds with the central guanine in the preferred triplet sequence. Only a single hydrogen bond could be formed if an AT base pair were to occupy that position. The AT base pair at the 5' end of the triplet is preferred because it allows for a better stereochemical fit of the daunosamine in the minor groove. If a GC base pair were at this position, the N2 of guanine would protrude into the minor groove and hinder the fit of the daunosamine. In addition, the N3 amino group in daunosamine can form a direct hydrogen bond with thymine at the 5' end of the triplet. These interactions can explain the sequence preference emerging from the footprinting studies and the somewhat unusual mixed triplet motif of the preferred site. The computational, structural, and solution studies thus provide a mutually consistent picture of daunorubicin's sequence preference and its underlying molecular basis.

TABLE IV. Binding Constants for the Interaction of Daunorubicin with Hexameric Hairpin Molecules or Duplex Hexamers

SEQUENCE	$K_1/10^5$ M^{-1}
5'CGTACG^{T}T GCATGC$_T$T	40.0 ($\pm$3.5)[a]
5'CGCGCG^{T}T GCGCGC$_T$T	20.0 ($\pm$1.8)[a]
5'TATATA^{T}T ATATAT$_T$T	11.0 ($\pm$1.0)[a]
5'CGTACG GCATGC	4.6 ($\pm$0.4)[b]
5'CGCGCG GCGCGC	2.3 ($\pm$0.6)[b]

[a] Satyanarayana, S.; Chaires, J. B., unpublished data. Solution conditions: 20°C, pH 7.0, 0.2 M NaCl.
[b] Data taken from Ref. 61. Solution conditions: 20°C, pH 7.0, 1 M NaCl.

BIOLOGICAL IMPLICATIONS OF DAUNORUBICIN BINDING TO PREFERRED SITES

A recent study (62) showed that doxorubicin, in the absence of excision repair, induced highly specific deletion and base substitution mutations in *Escherichia coli.* A striking finding was that 80% of the deletion mutation were found at the end of the triplet sequences 5'(A/T)CG or 5'(A/T)GC, the very triplets identified as preferred binding sites in our footprinting studies. Thus, the sequence preference inferred in vitro persists in vivo.

Our documentation of a *distribution* of binding sites along the DNA lattice has several implications. First, it has been argued (19) that the magnitude of observed *macroscopic* DNA binding constants ($\approx 10^6$ M^{-1}) is too low to account for the physiological concentrations ($\approx 10^{-8}$ M) at which the anthracycline antibiotics exert their effects and that something other than DNA alone must therefore be the key target of the drugs. This view must be reconsidered in light of our studies. Our footprinting results show that selected DNA sites can bind daunorubicin with binding constants at least an order of magnitude greater that the overall macroscopic binding constant. These sites will generally not be observed in macroscopic binding studies due to their low frequency, and due to the inherent difficulty in measuring tight binding by the optical methods typically employed in such studies. High affinity sites with binding constants of greater than 10^7 M^{-1} exist on DNA and would be selectively occupied under "physiological" drug concentrations.

Second, it is common to attempt to correlate the inhibition of some biological activity (DNA replication, topoisomerase or helicase activity, etc.) with some macroscopic measure of binding, i.e. K or ΔT_m values. Since binding to DNA is best described by a distribution of binding constants, such attempts are likely to be unsatisfactory and will neglect the possible preferential binding of antibiotic to certain sequences. Since many biological activities utilizing the DNA template rely on specific regulatory sequences, a better correlation between anthracycline binding to DNA and these activities might be made by recognizing the broad distribution of affinities along the DNA lattice and by using the appropriate microscopic drug binding constant.

ACKNOWLEDGEMENTS

Work in the author's laboratory has been generously supported by grant CA35635 from the National Cancer Institute. Dr. Julio Herrera participated in quantitative footprinting studies, and his substantial contributions are gratefully acknowledged.

LITERATURE CITED

1. Fukushima, M. *Nature* **1989**, *342*, 850 –851.
2. Arcamone, F. *Doxorubicin*; Academic Press: New York, 1981.
3. Lown, J. W. (ed.) *Anthracycline and Anthracenedione Based Anticancer Agents*; Elsevier: Amersterdam, The Netherlands, 1988.
4. Gianni, L.; Corden, B. J.; Myers, C. F. *Rev. Biochem. Toxicol.* **1983**, *5*, 1 – 82.
5. Drlica, K.; Franco, R. J. *Biochemistry* **1988**, *27*, 2253 – 2259.
6. Liu, L. F. *Ann. Rev. Biochem.* **1989**, *58*, 351 –375.
7. D'Arpa, P.; Liu, L. F. *Biochem. Biophys. Acta* **1989**, *989*, 163 – 177.
8. Zunino, F; Capranico, G. *Anti–Cancer Drug Design* **1990**, *5*, 307 – 317.
9, Bodley, L.; Liu, L. F.; Israel, M.; Seshadri, R.; Koseki, Y.; Giuliani, F. C.; Kirschenbaum, S.; Silber, R.; Potmesil, M. *Cancer Res.* **1989**, *49*, 5969 – 5978.
10. Capranico, G.; Zunino, F.; Kohn, K; Pommier, Y. *Biochemistry* **1990**, *29*, 562 – 569.
11. Capranico, G.; Kohn, K; Pommier, Y. *Nucleic Acids Res.* **1991**, *18*, 6611 – 6619.
12. Fritzche, H.; Berg, H. *Gazz. Chim. Ital.* **1987**, *117*, 331 – 352.

13. Fritzche, H., Walter, A. **In:** *Chemistry & Physics of DNA-Ligand Interactions*; Kallenbach, N. R. (*ed.*); Adenine Press: Schenectady, N. Y., **1989**, pp. 1 – 35.
14. Chaires, J. B. *Biophys. Chem.*, **1990**, *35*, 191 – 202.
15. Myers, C. E.; Mimnaugh, E. G.; Yeh, G. C.; Sinha, B. K. **In:** *Anthracycline and Anthracenedione Based Anticancer Agents*; Lown, J. W. (*ed.*); Elsevier: Amersterdam, The Netherlands, 1988, pp. 527 – 569.
16. Gelvan, D.; Samuni, A. *Biochem. Pharm.* **1986**, *35*, 3267 – 3275.
17. Kriebardis, T.; Meng, D.; Aktipis, S. *J. Biol. Chem.* **1987**, *262*, 12632 – 12640.
18. Goodman, M. F.; Lee, G. M.; Bachur, N. R. *J. Biol. Chem.* **1977**, *252*, 2670 –2674.
19. Myers, C. E.; Chabner, B. A. **In:** *Cancer Chemotherapy: Principles and Practice*; Chabner, B. A.; Collins, J. M.; J. B. Lippincott Co.: Philadelphia, **1990**, chapter 14.
20. Valentini, L.; Nicolella, V.; Vannini, E.; Menozzi, M.; Penco, S.; Arcamone, F. *Il Farmaco* 1985, *40*, 377 – 390.
21. Gigli, M.; Doglia, S. M.; Millot, J. M.; Valentinin, L; Manfait, M *Biochim. Biophys. Acta* **1988**, *950*, 13 – 20.
22. Bartkowiak, J.; Kapuscinski, J.; Melamed, M. R.; Darzynkiewicz, Z. *Proc. Natl. Acad. Sci. USA* **1989**, *86*, 5151 – 5154.
23. Quigley, G. J.; Wang, A. H.–J.; Ughetto, G.; van der Marel, G.; van Boom, J. H.; Rich, A. *Proc. Natl. Acad. Sci. USA* **1980**, *77*, 7204 – 7208.
24. Wang, A. H.–J.; Ughetto, G.; Quigley, G. J.; Rich, A. *Biochemistry* **1987**. *26*, 1152 – 1163.
25. Moore, M. H., Hunter, W. N.; Langlois d'Estaintot, B.; Kennard, O *J. Mol. Biol.* **1989**, *206*, 693 – 705.
26. Frederick, C. A., Williams, L. D. Ughetto, G.; van Boom, J. H.; Rich, A.; Wang, A. H.–J. *Biochemistry* **1990**, *29*, 2538 – 2549.
27. Nunn, C. M.; van Meervelt, L.; Zhang, S.; Moore, M. H. ; Kennard, O. *J. Mol. Biol.* **1991**, *222*, 167 – 177.
28. Chaires, J. B.; Dattagupta, N. D.; Crothers, D. M. *Biochemistry* **1982**, *21*, 3933 –3940.
29. Chaires, J. B. *Biopolymers* **1985**, *24*, 403 – 419.
30. Graves, D. E.; Krugh, T. R. *Biochemistry* **1983**, *22*, 3941 – 3947.
31. Barcelo, F.; Martorell, J.; Gavilanes, F.; Gonzalez–Ros, J. M. *Biochem. Pharmacol.* **1988**, *37*, 2133 – 2138.
32. Schutz, H.; Gollmick, F. A.; Stutter, E. *Studia Biophysica* **1979**, *75*, 147 – 159.
33. McGhee, J. D.; von Hippel, P. H. *J. Mol. Biol.* **1974**, *86*, 469 – 489.
34. Remeta, D. P.; Mudd, C. P.; Berger, R. L.; Breslauer, K. J. *Biochemistry* **1991**, *30*, 9799 – 9809.
35. Remeta, D. P.; Mudd, C. P.; Berger, R. L.; Breslauer, K. J. *Biochemistry* **1993**, *32*, 5064 – 5073.
36. Chaires, J. B.; Priebe, W.; Graves, D. E.; Burke, T. G. *J. Am. Chem. Soc* **1993**, *115*, 5360 – 5364.
37. Chaires, J. B. **In:** *Advances in DNA Sequence Specific Agents, Vol. 1*; Hurley, L. H., *Ed.*; Jai Press, Inc.: Greenwich, CT, **1992**, pp. 3 – 23.
38. Stutter, E.; Forster, W. *Studia Biophysica* **1979**, *75*, 199 – 208.
39. Forster, W.; Stutter, E. *Int. J. Biol. Macromol.* **1984**, *6*, 114 – 124.
40. Chaires, J. B.; Dattagupta. N.; Crothers, D. M. *Biochemistry* **1985**, *24*, 260 – 267.
41. Krishnamoorthy, C. R.; Yen, S.–F.; Smith, J. C.; Lown, J. W.; Wilson, W. D. *Biochemistry* **1986,** *25*, 5933 – 5940
42. Phillips, D. R.; Greif, P. C.; Boston, R. C. *Mol. Pharmacol.* **1988,** *33*, 225 – 230.
43. Rizzo, V.; Sacchi, N.; Menozzi, M. *Biochemistry* **1989**, *28*, 274 – 282.

44. Van Dyke, M. w.; Hertzberg, R. P.; Dervan, P. B. *Proc. Natl. Acad. Sci. USA* **1982**, *79*, 5470 –5474.
45. Lane, M. J.; Dabrowiak, J. C.; Vournakis, J. N. *Proc. Natl. Acad. Sci. USA* **1983**, *80*, 3260 – 3264.
46. Low, C. M. L.; Drew, H. R.; Waring, M. J. *Nucleic Acids Res.* **1984**, *12*, 4865 – 4879.
47. Portugal, J. *Chem. – Biol. Interactions* **1989**, *71*, 311 – 324.
48. Goodisman, J.; Dabrowiak, J. C. In: *Advances in DNA Sequence Specific Agents, Vol. 1*; Hurley, L. H. *Ed.*; Jai Press, Inc.: Greenwich, CT, **1992**, pp. 25 – 49.
49. Chaires, J. B.; Fox, K. R.; Herrera, J. E.; Britt, M.; Waring, M. J. *Biochemistry* **1987**, *26*, 8227 – 8236.
50. Skorobogaty, A.; White, R. J.; Phillips, D. R.; Reiss, J. A. *FEBS Lett.* **1988**, *227*, 103 – 106.
51. Skorobogaty, A.; White, R. J.; Phillips, D. R.; Reiss, J. A. *Drug Design and Delivery* **1988**, *3*, 125 – 152.
52. Chaires, J. B.; Herrera, J. E.; Waring, M. J. *Biochemistry* **1990**, *29*, 6145 – 6153.
53. Chaires, J. B. In: *Molecular Basis of Specificity in Nucleic Acid – Drug Interactions*; Pullman, B.; Jortner, J., *Eds.*; Kluwer Academic Press: Dordrecht, The Netherlands, **1990**, pp. 123 – 136.
54. Phillips, D. R.; Crothers, D. M. *Biochemistry* **1986** *25*, 7355 – 7362.
55. Phillips, D. R.; Cullinane, C.; Trist, H.; White, R. J. In: *Molecular Basis of Specificity in Nucleic Acid – Drug Interactions*; Pullman, B.; Jortner, J., *Eds.;* Kluwer Academic Press: Dordrecht, The Netherlands, **1990**, pp. 137 – 155.
56. Stormo, G. D. *Meth. Enzymol.* **1991**, *208* 458 – 468.
57. Herrera, J. E.; Chaires, J. B. *Biochemistry* **1989**, *28*, 1993 – 2000.
58. Chen, K.–X.; Gresh, N.; Pullman, B. *J. Biomol. Struc. Dyn.* **1985**, *3*, 445 – 466.
59. Pullman, B. *Adv. Drug Res.* **1989**, *18*, 1 – 113.
60. Pullman, B. *Anti–Cancer Drug Design* **1990**, *7*, 95 – 105.
61. Rizzo, V.; Battistini, C.; Vigevani, A.; Sacchi, N.; Razzano, G.; Arcamone, F.; Garbesi, A.; Colonna, F. P.; Capobianco, M.; Tondelli, L. *J. Mol. Recognit.* **1989**, *2*, 132 – 141.
62. Anderson, R. D.; Veigl, M. L.; Baxter, J.; Sedwick, W. D. *Cancer Res.* **1991**, *51*, 3930 – 3937.

RECEIVED June 3, 1994

Chapter 11

Adducts of DNA and Anthracycline Antibiotics
Structures, Interactions, and Activities

Jasmine Y.-T. Wang[1], Mark Chao[2], and Andrew H.-J. Wang[1,3]

[1]Department of Cell and Structural Biology and [2]Department of Microbiology, University of Illinois at Urbana–Champaign, Urbana, IL 61801

The molecular interactions between the anthracycline anticancer drugs daunorubicin and doxorubicin and their cellular receptor molecule, DNA, have been elucidated by high resolution x-ray diffraction analyses. These structural studies allow us to understand the possible biological functions associated with different parts of the anthracycline molecules. Some parts are essential for their binding to DNA, while other parts provide the necessary binding specificity or interference with relevant cellular enzymes such as polymerases or topoisomerases. Interestingly, drug molecules are capable of adjusting their conformation so as to strategically place those various functional parts in DNA double helix to achieve optimal binding affinity and specificity. A stable covalent cross-link between daunorubicin/doxorubicin and DNA can be formed as mediated by formaldehyde (HCHO). The cross-linking reaction suggests a different route in making new antitumor compounds. Therefore, we have tested the biological activities (against L1210 cells) of daunorubicin cross-linked by HCHO to a $(dG-dC)_n$ molecule and the adduct indeed exhibits cytotoxicity. Together these studies of structure-function relationship may be useful in designing new anthracycline drugs.

The anthracycline antibiotics daunorubicin (DNR) and doxorubicin (DOX) (Figure 1) are presently in widespread clinical use for cancer chemotherapy (1-3). While the precise cellular targets for these important drugs remain to

[3]Corresponding author

0097–6156/95/0574–0168$08.00/0

be elucidated, the biological activities of DNR and DOX are likely associated with their DNA-binding properties. Unfortunately, their effectiveness is somewhat hampered by their significant cardiotoxicity and the drug resistance that develops in some cancer cells. These problems have prompted extensive searches for new drugs that may overcome those shortcomings. An emerging useful approach is to design new compounds rationally on the basis of the structure-function relationships of the existing drugs. A full understanding of how anthracycline drug molecules interact with their DNA receptor (e.g., the binding affinity and specificity toward DNA) is an important first step. Toward this goal, the three dimensional structural analyses by x-ray diffraction of the molecular complexes between several anthracycline drugs and DNA oligomers have provided valuable information regarding the role of various functional groups of the drug molecules (4).

Although DNA intercalation using the aglycone chromophore is a requirement for the DNA binding of anthracycline antibiotics, and most probably is needed for their biological activities, it is not sufficient since the aglycone alone is not an active anticancer agent. Other components besides the intercalator chromophore play essential roles in deciding whether the compounds possess antitumor activity or not. Presumably, these components contribute in conforming on the compounds different DNA binding affinity, DNA sequence specificity, or other required properties such as membrane transport characteristics. In addition, they may affect the ways in which proteins (e.g., polymerases, helicase or topoisomerases) interact with the drug-DNA complexes (1-3). Indeed, many new anthracycline derivatives, involving modifications in either the aglycone or the carbohydrate moieties, have been synthesized and tested for biological activities. The structural analyses of the drug-DNA complexes allow us to gain insights into the roles of various functional components in those compounds.

DNR/DOX and Their Interactions with DNA

The archetypal three dimensional structure of the complex of DNR/DOX and DNA was revealed in detail by the 1.2 Å resolution structure of the 2:1 complex between DNR and d(CGTACG) (5). That structure showed that the anthracycline drug molecules bind to DNA by intercalating the chromophore between the CpG steps at both ends of a distorted B-DNA double helix. The elongated aglycon chromophore (rings A-D) penetrates the DNA double helix with the ring D protruding into the major groove and the amino sugar lying in the minor groove.

That and other subsequent structural analyses allowed us to identify three major functional components of anthracyclines: the aglycone intercalator (ring B-D), the anchoring function associated with ring A, and the sugars. Each component plays a different, important role in the biological activity of the drug. Aglycone, by intercalating into DNA, causes a distortion in the double helix that may be recognized by enzymes (e.g., polymerases, topoisomerase II, topoisomerase I, helicase). The sugars are essential as they lie in the grooves of DNA for additional interactions with relevant enzymes. The O^9 hydroxyl in the DNR-DOX series provides key hydrogen bonds to DNA, anchoring the drug firmly in the double helix and favoring a guanine base sequence. Finally, the configuration at the C^7 position in ring A is important, as it joins the amino sugar to the aglycone with a right-handed chirality such that the drug can position the amino sugar in the minor groove of a right-handed B-DNA double helix.

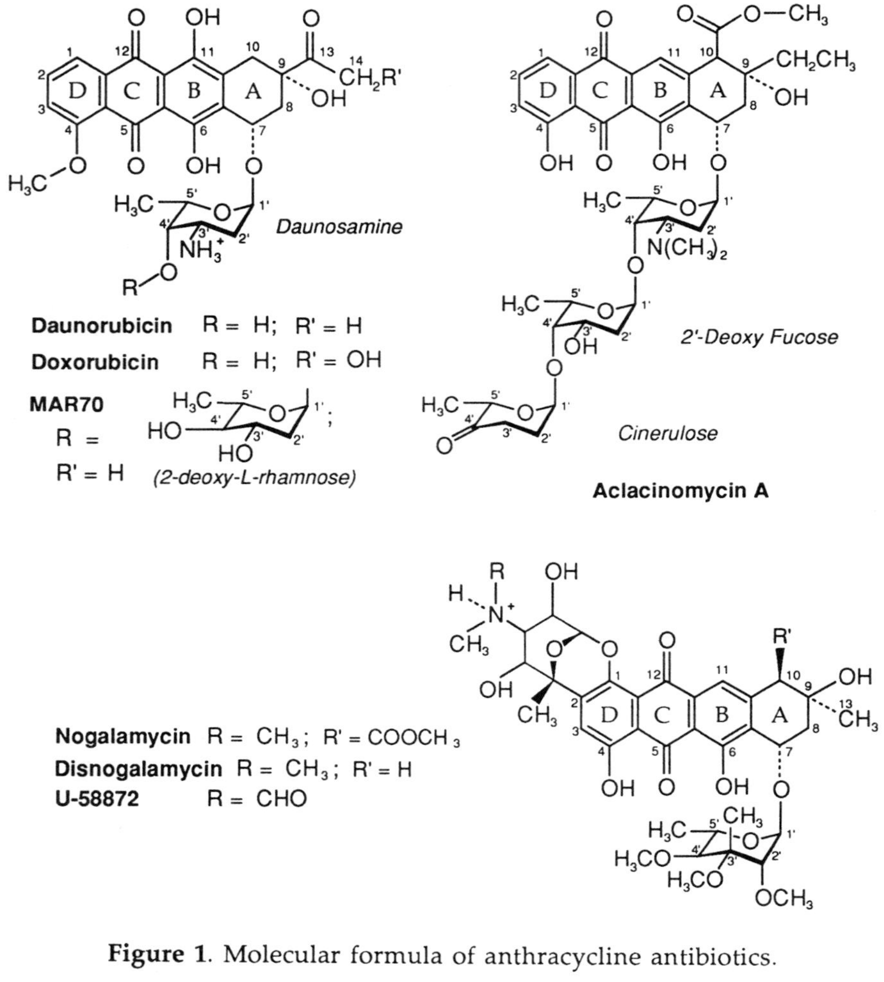

Figure 1. Molecular formula of anthracycline antibiotics.

An interesting observation about the structure is that the $N3'$ amino group of the sugar in the drug molecule approaches the edge of the base pairs in the minor groove. Furthermore, if there is a guanine at the fourth (and the symmetry-related tenth) sequence position in a hexamer such as $d(C_1G_2C_3G_4C_5G_6)$, the two amino groups ($N3'$ from DNR and $N2$ of G4) in the non-adduct complex would be in close contact, which may slightly destabilize the binding of DNR to a sequence of 5'-GCG in DNA. This is consistent with the footprinting data (6) and the prediction from theoretical studies (7) of the binding of DNR to DNA, which show a sequence preference of 5'-(A/T)CG over 5'-GCG.

It is interesting to note that the intercalative binding of DNR to DNA brings those two amino groups ($N2$ from guanine and $N3'$ from DNR) into close proximity, rigidifying them in a somewhat hydrophobic environment of the minor groove. This creates an ideal situation for a nucleophilic attack on amino groups (which become highly reactive in this shielded environment) by an agent such as HCHO. In fact, the cross-linking reaction transforms the unfavorable contact between the two amino groups in the noncovalently bonded complex to a favorable covalent insertion of a methylene group as discussed below.

At this point, we would like to point out that the information on the interactions of anthracycline drug-DNA complexes have been based on high resolution crystal structures. One possible concern has been whether the lattice packing forces might influence the conformation of these complexes. This concern has been removed by at least two lines of studies. First, some complexes have been crystallized in different crystal lattices, and their structures are substantially similar to one another. Second, the solution structure revealed by NMR corroborates the crystal structure. For example, the solution structure of nogalamycin complexed to d(CGTACG) has been analyzed by a quantitative 2D-NOESY refinement procedure (8,9). The root mean square deviation of this structure from that of the 1.3 Å resolution crystal structure (10) is only 0.3 Å. Another concern is that the DNA oligomer is too short, having an end-effect, to represent a polymeric DNA system. The only structural information that was not available from the x-ray diffraction studies of oligonucleotides is the helix unwinding angle associated with the base pairs beyond the immediate base pair adjacent to the aglycone ring. However, this affects relatively little our understanding of the interactions between anthracycline drugs and polymeric DNA. We therefore concluded that the structural features seen in crystals can be extrapolated into the solution state, and used safely as a logical framework for rational drug design work.

Formaldehyde Cross-linking of DNR/DOX to DNA

We recently discovered that formaldehyde (HCHO) can very efficiently cross-link DNR to DNA of certain sequences (11,12). We have shown by HPLC and x-ray diffraction analyses that, when DNR is mixed with d(CGCGCG) in the presence of HCHO, stable covalent adducts of DNA are formed. These adducts contain a covalent methylene bridge between the

$N^{3'}$ of daunosamine and the N^2 of a G. The reason for this efficient cross-linking reaction is that the two amino groups in the minor groove of the complex are perfectly juxtaposed, providing an ideal template for a nucleophilic addition of HCHO. The methylene bridge does not perturb the conformation of the drug-DNA complex, when compared with the structure of DNR-d(CGTACG).

The common structural features of the anthracycline-DNA complex are represented by the DNR-d(CGCGCG) cross-linked-complex in Figure 2. Two DNRs are intercalated in the d(CpG) sequences of the B-DNA hexamer duplex. Most of the structural features seen in the noncovalent DNR-d(CGTACG) complex (5) are preserved. The result of the x-ray structure analysis of the cross-linked adduct suggests that the reaction is sequence specific. Only a DNA sequence like 5'-GCG has the proper drug-binding conformation to place the N^2 amino group of the G in the triplet sequence near the $N^{3'}$ of DNR.

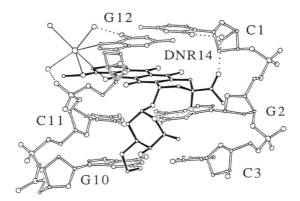

Figure 2. The three dimensional structure of the d(CGCGCG)-DNR formaldehyde cross-linked complex. Two DNRs are bound to a hexamer duplex. The aglycone chromophores are intercalated between the two CpG steps at the ends of the helix. The amino sugar lies in the minor groove. A methylene bridge is formed between the $N^{3'}$ of daunosamine and N^2 of guanine G10.

More recently, we have shown that DNR and DOX can also be cross-linked to araC-containing hexamers d(CG[araC]GCG) and d(CA[araC]GTG), respectively by formaldehyde respectively (13). These two adducts provide useful information regarding the influence of another anticancer drug, araC, on the conformation of anthracycline-DNA complex. The latter complex, DOX-d(CA[araC]GTG), crystallized in a different space group (monoclinic C2) and its three dimensional structure is substantially similar to that of the DAU-d(CG[araC]GCG) in the tetragonal $P4_12_12$ space group. This observation reinforces the argument pointed out earlier that the crystal lattice forces have only a small influence on the structure of the anthracycline drug-DNA complexes.

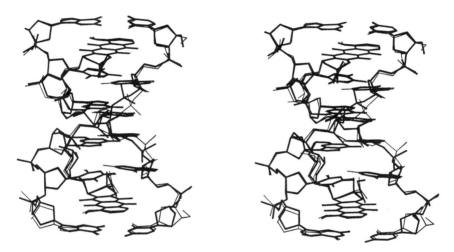

Figure 3. Superimposition of the two complexes, DNR-d(CG[araC]GCG) (thin bond) and DOX-d(CA[araC]GTG) (thick bond). They are superimposed by a least-square fitting of all common atoms.

Figure 3 compares the two structures. It is interesting to note that in the DOX-d(CA[araC]GTG) structure one of the DOX O^{14} hydroxyl groups forms a hydrogen bond to the phosphate oxygen of DNA. It is tempting to speculate that the different biological activities associated with DNR and DOX (DNR for leukemia and DOX for wide-spectrum use) may be related, in part, to the additional interactions involving O^{14} hydroxyl group.

The observation that DNR/DOX can be cross-linked to DNA may be of significant relevance, since the cross-linking ability of a number of natural antibiotics is well established. For example, several potent antitumor antibiotics act by forming covalent adducts between the drug and DNA. Anthramycin, mitomycin C, saframycin (14), and ecteinacidins (15) most likely form covalent adducts with guanine at the N^2 position. Interestingly, a highly potent anthracycline antibiotic, barminomycin/SN07 (Figure 4), contains an active aldehyde group attached to $O^{4'}$ of the daunosamine sugar (16). This aldehyde serves as a crosslinking functional group in ways very similar to the exogenic HCHO discussed above. SN07 has been cross-linked to different DNA polymers (e.g., poly[dG-dC].poly[dG-dC]), and the resulting drug-DNA adducts appeared to have higher anticancer activities (17).

It is also informative to point out that a highly potent $N^{3'}$-modified derivative of DOX, 3'-(3-cyano-4-morpholinyl)-3'-deamino-DOX (MR-CN-DOX), forms a covalent adduct to DNA in vivo with the loss of the cyano group (18-20). While the exact nature of this adduct is yet to be determined, the mechanism associated with the aldehyde-mediated adduct found in the structure of DNR-d(CGCGCG) is noteworthy. We have recently solved the crystal structure of both 3'-(4-morpholinyl)-3'-deamino-DOX and MR-CN-DOX complexed to d(CGATCG), which revealed that the C^5 and C^6 of the morpholinyl group are in the proximity of bases in the minor groove (unpublished data). It is conceivable that for certain DNA sequences the C5/C6 position of the morpholino ring may alkylate the N^2 of guanine via the imine intermediate, as shown in Figure 4.

Cross-linked DNR-DNA as a Potential Antitumor Drug

An interesting example of a cross-linked drug-DNA complex as cytotoxic agent is the SN07-DNA complex mentioned above (16). When SN07 is complexed with poly(dG-dC), the complex has significantly higher antitumor activity than SN07 alone, almost 10-fold higher. But no such effect was oberved when poly(dI-dC) was used, suggesting that the N^2 of guanine is involved. This led us to believe that a DNA-DNR formaldehyde crosslinked adduct (DNA-DNR-HCHO) may have a similar effect and may be used as a new anticancer agent.

The use of DNA as an enhancing element of cytotoxicity was explored earlier by Trouet & Jolles (21) who showed that the DNA non-covalent complexes of DNR enhanced chemotherapeutic activities on tested mice. One premise for such an approach is that the DNA-DNR complex may have a lower cardiotoxicity. However, non-covalent association of the drug-DNA complex tends to dissociate in solution. Therefore, some way to keep the drug in the DNA lattice may be a useful property. Covalent cross-link, such as the DNA-DNR-HCHO adduct may be one effective means.

DNA, in this case, can act as a carrier for DNR molecules with latent alkylating (aldehyde) function. Since DNR is intercalated and is cross-linked by formaldehyde of DNA, the DNR is not free to dissociate. It is likely that cells may respond to the crosslinked drug differently. However, in the adduct, DNR molecules can be released from the DNA lattice, because the link between Dau and DNA is a aminal group. Once the cross-linking reaction is reversed, the released drugs should have a cytotoxic effect. Thus, the adduct could have a prolonged cytotoxic effect due to the relatively stable covalent linkage. To test this, we have carried out a number of biological studies. The results are summarized below and have been described in more detail elsewhere (22).

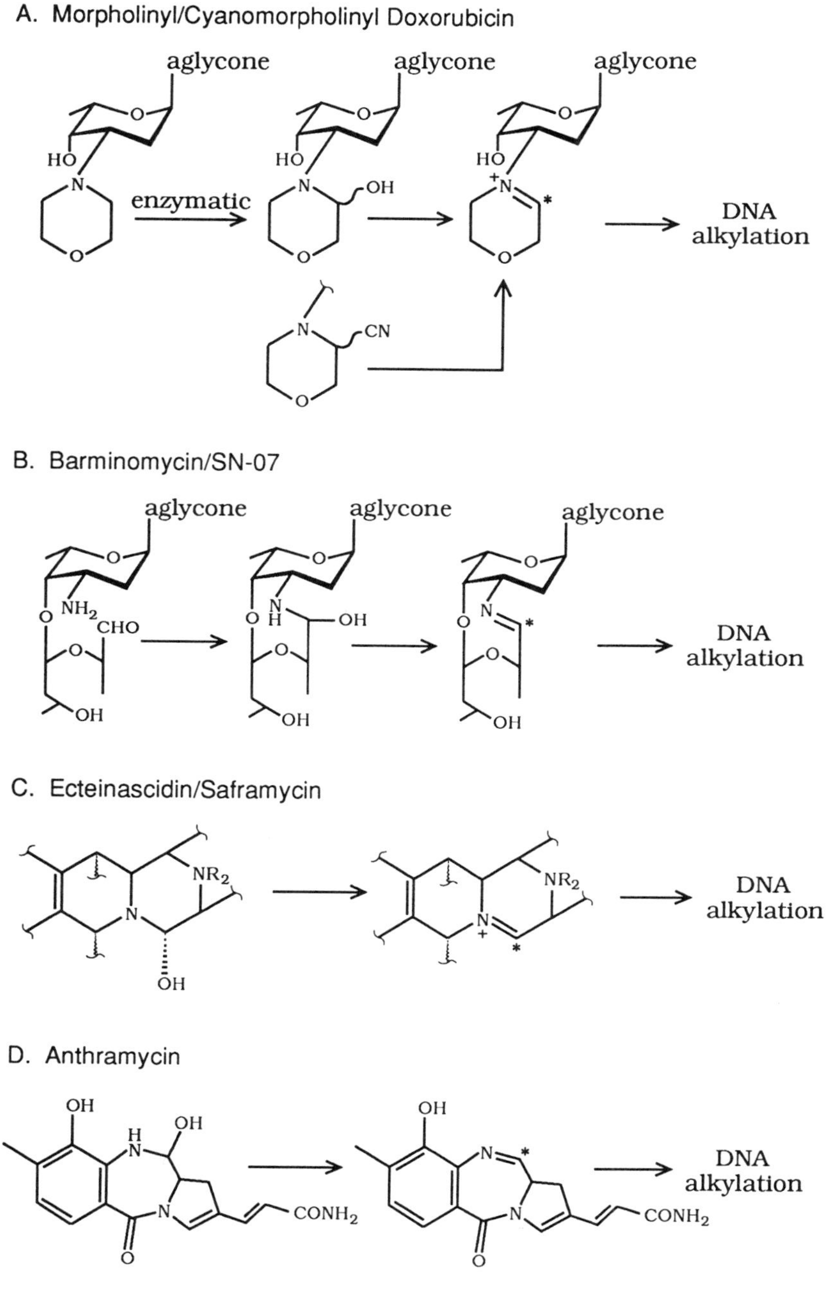
A. Morpholinyl/Cyanomorpholinyl Doxorubicin
aglycone
HO
enzymatic
OH
CN
DNA alkylation
B. Barminomycin/SN-07
NH2
CHO
OH
DNA alkylation
C. Ecteinascidin/Saframycin
NR2
DNA alkylation
D. Anthramycin
CONH2
DNA alkylation

Figure 4. Antitumor compounds that use iminium ion as activated intermediate for the alkylation of DNA.

Preparation of DNA-DNR Adduct. A DNA molecule, either poly(dG-dC) or oligonucleotide $(GC)_{20}$, was used for the adduct preparation. A 100 mM Na_2HPO_4/100 mM NaCl solution, pH 7.0, was used to resuspend the DNA. The cross-linking reaction was carried out by mixing 45 µL of 20 mM DNR with 1 mL of 4.5 mM (by base pair) $(GC)_{20}$ or poly(dG-dC) by vortexing, then adding 9 µL of 1 M formaldehyde. The reaction was incubated overnight at room temperature and subsequently dialyzed against buffer to remove any non-crosslinked (free) DNR. Samples were centrifuged briefly before taking the UV spectra. The spectral absorption A_{500} was measured before and after dialysis. No loss of A_{500} was observed, suggesting that the crosslinked DNR-DNA adduct is quite stable under these conditions. Both cross-linked and non-cross-linked adducts were lyophylized and stored at -20 °C. Fluorescence spectra of the adducts showed that the DNR is completely quenched in the cross-linked adduct, indicating that the chromophore is firmly anchored in the DNA double helix as expected. The cross-linking density is one DNR per five base pairs, slightly lower than the maximum loading density of one DNR per three base pairs. Figure 5 is a schematic diagram of the cross-linked DNR-DNA adduct. It should be pointed out that any anthracycline antibiotic may be used for this purpose, as long as there is an $N^{3'}$ amino group in the molecule. Therefore DOX, idarubicin, or other newer geneartion drugs may be incorporated in a similar manner.

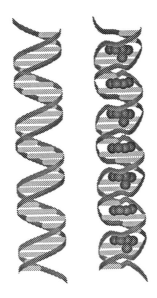

Figure 5. Schematic drawings showing free DNA (left) and DNR-loaded DNA (right).

Morphological Observation in Adduct-treated L1210 Cells. Mouse leukemia L1210 cells were cultured (37 °C, 10% CO_2) in MEM10C medium (GIBCO). Uniform amounts of DNR, either contained in cross-linked adduct, non-cross-linked complex, or as free DNR, were placed into a 96-well disposable tissue culture plate (Corning) and air dried in the culture hood. Cells then were taken from a culture in exponential growth phase and resuspended to 5×10^3 cells/mL prior to being aliquoted into the wells. The drug-containing plates were seeded with 1,000 cells for each test concentration. By limiting the cell numbers, we avoided the density-induced inhibition of cell growth.

Cells were examined by inverted microscope over several days. The cell morphology in the drug treated samples was intriguing (Figure 6). Also, strikingly different uptake of DNA-DNR-HCHO by the L1210 cells at various dosages and different retention durations of the adduct by these cells were observed. At high drug doses (e.g., > 1 μg/mL), L1210 cells shrank to about 5 times smaller than the control cells and died by the end of 48 hours. At low drug doses, the the DNA-DNR-HCHO treated cells grew at least 3 times larger than control cells. Cells appeared to be viable until day 5. At a concentration of 64 ng/mL of free DNR, most of the treated cells were destroyed at day 3. In contrast, the DNA-DNR-HCHO treated cells displayed a less rigid appearance. Starting at day 4, some cells showed fragmented morphology and were not viable. The cytosol of the remaining cells were filled with many different sized vacuoles that made the cells lumpy and more transparent. These observations suggested that the cross-linked adduct indeed has a significant cytotoxic effect on the L1210 cancer cells. This has prompted us to perform a more quantitative analysis of the cytotoxicity of the adduct.

Cytotoxicity Measurement. The cytotoxicity of the DNA-DNR-HCHO adduct against L1210 cells was tested using [methyl-^{3}H]-thymidine incorporation. Plates containing L1210 cells were incubated at 37 °C in CO_2 for 48 hours. Then, the cultures were spiked with 1 μCi/μL of tritiated [methyl-^{3}H]-thymidine (ICN, Inc.) into each well and incubated with cells at 37 °C for 24 hours. At the end of 72 hours, we observed the morphologies in four different test wells, including control wells, under an inverted microscope. Cells were then harvested with a 96-well cell harvester (Cambridge Technology, Inc.) onto glass filters, dried at room temperature in a chemical hood for 48 hours, and suspended in 5 mL scintillation liquid (ScintVerse E, Fisher Scientific). The tritium radioactivity was measured using a Beckman liquid scintillation counter.

The IC_{50} cytotoxicity assay of DNA-DNR-HCHO measures cell death, in terms of decreased incorporation of radiolabeled markers ([methyl-^{3}H]-thymidine) by the cell culture. The results are shown in Figure 7. We observed that both cross-linked and non-cross-linked adducts retained a cytotoxic effect on cancer cells, and differed only in the dosage. It is interesting to note that at low dosage (<60 ng/ml) the cross-linked adduct prolonged the cytotoxic effect and continued to kill cells after 5 days to a greater extent than the non-crosslinked or the free DNR (data not shown). This supports our hypothesis that the drug-DNA conjugate may have a longer-lasting effect and also reflects the potent anticancer activity of DNA-DNR-HCHO.

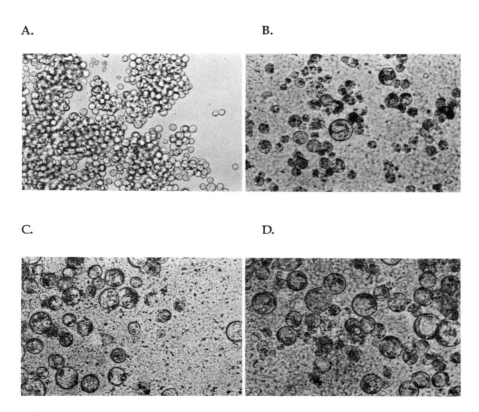

Figure 6. Cell morphologies of L1210 cells 3 days after treatment with drugs. (A) Cells without any added drug. (B) Cells treated with DNR (64 ng/mL). (C) Cells treated with DNR complexed to $(GC)_{20}$ without cross-link (64 ng/mL of DNR-equivalent). (D) Cells treated with DNR cross-linked to $(GC)_{20}$ with (64 ng/mL of DNR-equivalent).

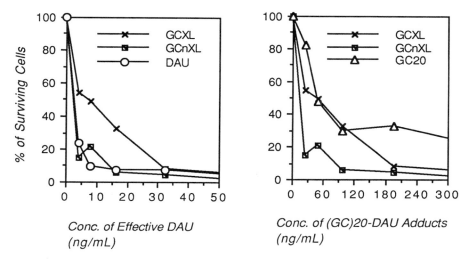

Figure 7. Cytotoxicity of different compounds measured as % of cell survival. Abbreviations used are DAU for free daunorubicin, GCXL for DNR cross-linked to $(GC)_{20}$, and GCnXL for DNR complexed to $(GC)_{20}$ without cross-link.

The results (Figure 7) show that no cells survived at concentrations of 1 μg/mL or higher in crosslinked, non-cross-linked and free DNR. The estimated IC_{50} was ~10 ng/mL (effective DNR concentration) for cross-linked adducts (GCXL), ~5 ng/mL (effective DNR concentration) for non-cross-linked adducts (GCnXL), and 5 ng/mL for DNR alone. The relative values among the three compounds suggest that the crosslinked adduct is still active. In comparison, in Oki's report (23), the IC_{50} of DOX is 20 ng/mL as measured by the inhibition of growth and macromolecular synthesis. In the presence of $(GC)_{20}$ only, cells are also inhibited. Though we still do not know exactly what mechanism is causing this effect, it has been shown previously by others in the studies of antisense DNA that DNA (polymer and oligonucleotides) have certain toxic effects on cells. Based on these preliminary data, we are optimistic that using this unique approach the adduct may hold promise as a new chemotherapeutic probe. However, more experiments need to be carried out to determine the molecular and cellular mechanisms of cytotoxicity in order to better envision the development of this new potential antitumor drug.

Immunogenicity of a Poly(dG-dC)-DNR Cross-linked Adduct. One concern in using drug-DNA conjugate as a drug is that the conjugate may induce immune response, a possible serious side effect. We therefore carried out immunological studies to see whether the cross-linked adduct can elicit strong immune response in mice. DNA is known to be a weak antigen in inducing antibody. Until now, only anti-ssDNA antibodies could be readily induced by immunization with ssDNA polymer complexed with methylated bovine serum albumin (BSA) (24,25). Various lengths of natural and synthetic single-stranded polynucleotides are potent immunogens when mixed with methylated BSA or covalently conjugated to albumin. On the other hand, the induction of antibodies against native double-stranded B-DNA form has yet to succeed. Recently a novel intraspleen immunization method has been developed (26) and found to be effective in inducing immune response to horse serum albumin in mice.

We attempted this intraspleen immunization method (26) to directly stimulate the spleen with the cross-linked adduct. Poly(dG-dC)-DNR cross-linked complex was used as antigen. It was coated on nylon membrane and surgically inserted directly into the spleen. Five BALB/c mice were treated with antigen. Two sequential boosters were given at intervals of about one month. Sera (100-150 microliters) were collected and used for ELISA measurement of anti-single-stranded and anti-double-stranded antibody activities (27). The results of the ELISA screening of the four surviving mice revealed no specific immune response against the DNR-DNA adduct antigens. In fact, it appeared that, *despite the distorted DNA conformation, the adduct was still non-immunogenic.* This is not highly surprising as it is well known that double helical DNA is extremely nonimmunogenic. In contrast, the antibodies against the Z-DNA conformation of poly(dG-dC) have been readily produced (28). Our results suggest that a possible undesirable immunological reaction is unlikely.

Interestingly, Ballard and Voss (29) found that a monoclonal anti-dsDNA antibody (BV17-45) generated from a (NZBxNZW) F1 lupus-prone mouse bound specifically to B-form DNA, but not to A-form or to Z-form DNA. Anti-dsDNA antibodies are important in that the level of activity tends to correspond to the intensity of the disease (30), and such autoantiboies have been found in lupus patients. It seemed reasonable to ask whether one can treat (NZBxNZW) F1 autoimmune mouse with DNA-DNR-HCHO complex. We propose that the anti-dsDNA antibody on the T or B cell surface should bind specifically to the DNA of the complex. The binding could trigger the endocytosis of DNA-DNR-HCHO complex into this specific T or B cell. Eventually, the DNR released inside could induce the cytotoxic effect in this specific T or B cell, thereby decreasing the body's level of anti-dsDNA antibody. Therefore we tested two groups of (NZBXNZW) F1 autoimmune mice. The DNA-DNR-HCHO adduct was used both alone and in conjunction with a number of carriers, including poly-L-lysine.

Thus far, no apparent favorable survival pattern among the treated mice could be observed. Nonetheless the experiments showed that the DNA-DNR-HCHO adduct did not have a toxic effect toward those mice when compared with the mice in the control group. More work is needed to test this hypothesis further.

Conclusions

High resolution x-ray diffraction analyses of several anthracycline-DNA complexes provide the fine details of the molecular interactions between those anticancer drugs and their cellular receptor molecule, DNA. These structural studies allow us to understand the possible biological functions associated with different parts of the anthracycline molecules. Some parts are essential for their binding to DNA, while other parts provide the necessary binding specificity, or interference with cellular enzymes such as polymerases or topoisomerases. The cross-link between DNR/DOX with DNA mediated by formaldehyde is suggestive for a different direction in making new antitumor compounds. We have tested the biological activities of daunorubicin formaldehyde-crosslinked to $(dG-dC)_n$ molecule. In conclusion, our results suggested that there are several potential beneficial properties of such kind of drug-DNA cross-link as anticancer drug.

   1. The conjugate can be prepared readily and efficiently with very high loading of the drug (to a maximum of one DOX per three base pairs for poly(dG-dC)).

   2. The conjugate may be long-lived in the blood stream, serving as a slow-releasing drug reservoir.

   3. The cytotoxic agent is hidden in the DNA lattice, avoiding the attack by enzyme which causes the formation of the free radical species of DOX. This may reduce the cardiotoxicity side effect.

   4. If the conjugate can be picked up by the cancer cells directly, it may be effective against resistant cells.

   5. The conjugate should be resistant to nuclease, since the DNA conformation is severely distorted by the intercalated drug.

   6. Sequence-specific DNA-DOX adduct may be prepared, as long as some guanine nucleotides are present.

Acknowledgments

This work was supported by National Institutes of Health grants (GM-41612 and CA-52506) to AHJW. We thank Dr. G. Wilson for assistance in the initial cytotoxicity measurement, Professor W.-C. Chang of National Taiwan University for conducting the intraspleen immunization experiment, Professor E. Voss, Jr. for providing the facility for the immunological studies. The contributions of Mr. Y.-G. Gao, Mr. M. Sriram, Mr. Y. Guan, Ms. H. Zhang, and Dr. H. Robinson are grateful acknowledged.

Literature Cited

1.   *Anthracycline and Anthacenedione-based Anticancer Agents;* Lown, J. W. Ed.; Elsevier, New York, NY, 1988.

2.   Denny, W. A. *Anti-cancer Drug Design* **1989**, *4*, 241-263.
3.   *Molecular Basis of Specificity in Nuceic Acid-Drug Interactions;* Pullman, B.; Jortner, J. Eds.; Kluwer Academic Publishers, Dordrecht, 1990.
4.   Wang, A. H.-J. *Current Opinin. Struct. Biol.* **1992**, *2*, 361-368.
5.   Wang, A. H.-J.; Ughetto, G.; Quigley, G. J.; Rich, A. *Biochemistry* **1987**, *26*, 1152-1163.
6.   Chaires, J. B.; Fox, K. R.; Herrera, J. E.; Britt, M.; Waring, M. J. *Biochemistry* **1987**, *26*, 8227-8236.
7.   Chen, K. X.; Gresh, N.; Pullman, B. *J. Biomol. Struct. Dyn.* **1985**, *3*, 445-466.
8.   Robinson, H.; Wang, A. H.-J. *Biochemistry* **1992**, *31*, 3524-3533.
9.   Robinson, H.; Yang, D.; Wang, A. H.-J. *Gene* **(1994, in press)**.
10. Liaw, Y.-C., Gao, Y.-G., Robinson, H., van der Marel, G. A., van Boom, J. H., & Wang, A. H.-J. (1989) *Biochemistry* **1989**, *28*, 9913-9918.
11. Wang, A. H.-J.; Gao, Y.-G.; Liaw, Y.-C.; Li, Y. K. *Biochemistry* **1991**, *30*, 3812-3815.
12. Gao, Y.-G.; Liaw, Y.-C.; Robinson, H.; van der Marel, G. A.; van Boom, J. H.; Wang, A. H.-J. *Proc. Natl. Acad. Sci. USA.* **1991**, *88*, 4845-4849.
13. Zhang, H.; Gao, Y.-G.; van der Marel, G. A.; van Boom, J. H.; Wang, A. H.-J. *J. Biol. Chem.* **1993**, *268*, 10095-10101.
14. Warpehoski, M. A.; Hurley, L. H. *Chemical Res. Toxicol.* **1988** , *1*, 315-333.
15. Sakai, R.; Rinehart, K. L.; Guan, Y.; Wang, A. H.-J. *Proc. Natl. Acad. Sci. USA* **1992**, *89*, 11456-11460.
16. Kimura, K.; Takahashi, H.; Nakayama, S.; Miyata, N.; Kawanishi, H. *Agri. Biol. Chem.* **1989**, *53*, 1797-1803.
17. Kimura, K.; Takahashi, H.; Takaoka, H.; Miyata, N.; Nakayama, S.; Miyata, N.; Kawanishi, H. *Agri. Biol. Chem.* **1990**, *54*, 1645-1650.
18. Westendorf, J.; Aydin, M.; Groth, G.; Weller, O.; Marquardt, H. *Cancer Res.*, **1989**, *49*, 5262-5266.
19. Cullinane, C., & Phillips, D. R. *Biochemistry*, **1992**, *31*, 9513-9519.
20. Lau, D. H. M.; Duran, G. E.; Sikic, B. I. *J. Natl. Cancer Inst.* **1992**, *84*, 1587-1592.
21. Trouet, A.; Jolles, G. *Seminars in Oncology.* **1984**, *11*, 64-72.
22. Wang, J. Y.-T. *M.S. Thesis,* University of Illinois, Urbana, IL., 1993.
23. Oki. T. In *Anthracycline and Anthacenedione-based Anticancer Agents;* Lown, J. W. Ed.; Elsevier, New York, NY, 1988; pp103-127.
24. Plescia, O. J., Braun, W. & Palczuk, N. C. *Proc. Nat. Acad. Sci. USA* **1964**, *52*, 279-283.
25. Tan, E. M.; Natali, P.G. *J. Immunol.* **1970**, *104*, 902-906
26. Hong, T.-H.; Chen, S.-H.; Tang, T.-K.; Wang, S.-C.; Chang, T. H. *J. Immunol. Methods* **1989**, *120*, 151-157.
27. Lacy, M. J.; Voss, E. W. Jr. *J. Immunol. Methods* **1989**, *116*, 87-98.
28. Stollar, B. D. *CRC Rev. Biochem.* **1986**, *20*, 1-36.
29. Ballard, D. W.; Voss, E. W. Jr. *J. Immunol.* **1985**, *135*, 3372-3380.
30. Hughes, G. R. V. *The Connective Tissue Diseases;* Blackwell, Oxford, 1980.

RECEIVED June 3, 1994

Chapter 12

DNA Topoisomerases and Their Inhibition by Anthracyclines

Yves Pommier

Laboratory of Molecular Pharmacology, Developmental Therapeutics Program, Division of Cancer Treatment, National Cancer Institute, National Institutes of Health, Bethesda, MD 20892

Several of the most active anticancer drugs, including anthracyclines, poison cellular DNA topoisomerase II (top2). The genomic sites of top2 inhibition differ for each class of drug, thereby providing a rationale that may help to explain the mechanisms of drug inhibition ("stacking model") and the differential activity of top2 inhibitors in the clinic. Resistance to anthracyclines is often associated with top2 alterations and increased drug efflux by the P-glycoprotein[MDR]. Different parts of the anthracycline molecules are probably involved in top2 inhibition, recognition by the P-glycoprotein[MDR], and generation of free radicals. Therefore, it should be possible to design more selective top2 inhibitors. Since one anthracycline, morpholinyldoxorubicin, inhibits topoisomerase I (top1) rather than top2, it may be also be possible to find top1 inhibitors among anthracyclines.

DNA topoisomerases represent a major focus of research not only for cancer chemotherapy but also for gene regulation, cell cycle, mitosis, and chromosome structure. A number of reviews have been written on the subject (1-9).

DNA Topoisomerases: Molecular Biology and Functions

The length of eukaryotic DNA and its anchorage to nuclear matrix attachment regions limits the free rotation of one strand around the other as the two strands of the DNA double-helix are separated for DNA metabolism (transcription, replication, recombination, repair). DNA topoisomerases catalyze the unlinking of the DNA strands by making transient DNA strand breaks and allowing the DNA to rotate around or traverse through these breaks. There are two types of topoisomerases known in eukaryotes, type 1 and type 2 topoisomerases (top1 & top2) and three types in yeast where topoisomerase III has recently been identified (3-6,10-12). DNA gyrase is the bacterial equivalent of topoisomerase II. Antibacterial quinolones (nalidixic acid, ciprofloxacin, norfloxacin, and derivatives) are DNA gyrase inhibitors with no or very limited effect on the host human top2. Topoisomerase-mediated DNA breaks correspond to transesterification reactions where a DNA phosphoester bond is transferred to a specific enzyme tyrosine residue.

In the case of top1, the enzyme becomes linked to the 3'-terminus of a DNA single-strand break. In the case of top2, each molecule of an enzyme homodimer becomes linked to the 5'-terminus of a DNA double-strand break (Figure 1).

Both top1 and top2 can remove DNA supercoiling by catalyzing DNA swiveling and relaxation. They can complement each other in this function at least in yeast, where the absence of one can be compensated by the presence of the other topoisomerase. However, yeast top2 mutants are not viable and die at mitosis because top2 is essential for chromosome condensation and structure (3-6,10-16) and for the proper segregation of mitotic (13-19) and meiotic (20,21) chromosomes. This is due to the fact that, in addition to its DNA relaxing activity, top2 is essential for the separation of chromatin loops (decatenation of replicated DNA), and condensation of chromosomes as well as proper segregation of sister chromatids (17-19,22-26). The accumulation of top2 at the end of S-phase and during G2 and its concentration in the chromosome scaffold is consistent with the enzyme's roles during mitosis. The role of top2 in maintaining the structural integrity of mitotic chromosomes has recently been disputed (16). However, a fraction of top2 remains selectively associated with the telophase chromosomes, indicative of an important function during mitosis (27).

A relationship is possible between top2 and cell cycle-associated kinases/cyclins since top2 phosphorylation increases during G2/M (28), resulting in enhanced catalytic activity (see below). Top2 is also the major chromosome protein recognized by the mitotic phosphoprotein antibody MPM2 (29). The present evidence indicates, however, that top2 phosphorylation during G2/M in yeast is carried out by casein kinase II rather than cdk1-cyclin B (30).

Two top2 isoenzymes have been isolated, top2-alpha and top2-beta. They differ in their molecular mass (31), enzymatic properties (31), genes (31-35), cell cycle regulation (36-38), and cellular distribution (39,40). In addition, the top2 inhibitor teniposide (VM-26) has been reported to be 3- to 8-fold more active on the top2-alpha isoform (31).

Molecular Interactions between Drugs and Topoisomerases

Most topoisomerase inhibitors induce topoisomerase-mediated DNA strand breaks. These lesions are commonly referred to as cleavable complexes because the breaks are sealed by the topoisomerase enzymes and are detectable after protein denaturation (commonly in the presence of sodium dodecyl sulfate [SDS]). The cleavable complexes can be detected in cells as protein-linked DNA breaks and DNA-protein cross-links by alkaline elution and also as protein-DNA complexes by SDS-KCl precipitation assays (8,12,41). Inhibitors are generally specific either for top1 or for top2. However, there are a small number of compounds which can induce cleavable complexes with both enzymes, such as saintopin, intoplicine, and indolocarbazole derivatives (see Tables II & III).

Topoisomerase I inhibitors. The top1-linked DNA strand breaks (cleavable complexes) are single-stranded and correspond to a transesterification intermediate between an enzyme tyrosine and the 3'-DNA terminus of the cleaved DNA (Figure 1).

Camptothecin (CPT) and derivatives (topotecan, CPT-11, 9-aminocamptothecin) are the only top1 poisons in clinical trial (42). These drugs inhibit top1 by inhibiting the resealing of top1-mediated DNA single-strand breaks (43,44). Structure-activity studies indicate that the stereochemistry of the lactone ring is essential for enzyme inhibition and that ring opening through nucleophilic attack from the enzyme or the DNA may be required for activity (6,45-49).

The discovery of top1 inhibition by CPT has stimulated the rapid discovery of a number of other top1 inhibitors including one anthracycline [morpholinyldoxorubicin (47)], indolocarbazoles (50-52), bulgarein (53), DNA minor groove ligands (54), and fagaronine (55) (Table I).

Some topoisomerase inhibitors can induce cleavable complexes with both topoisomerases [actinomycin D (56,57), azaIQD (58), saintopin (59), intoplicine (60), and 6-amino-substituted benzo[c]phenanthridines (61)]. Interestingly, the DNA sequence selectivity of top1 cleavage is different for each chemical class of drug.

Topoisomerase II inhibitors. There are many top2 inhibitors and a large number of them, including the anthracyclines had been used as anticancer agents before the identification of top2 as their target   (Table II).   The antitumor top2 inhibitors presently used in the clinic poison the enzyme by stabilizing cleavable complexes, presumably by inhibiting DNA religation and preventing enzyme catalytic activity (12,41,62,63).

A large number of top2 inhibitors from diverse chemical families have been identified in the last 4 years (Table II) [amonafide (48,64), flavones, genistein and derivatives (65-67), the nitroimidazole Ro 15-0216 (68), withangulatin (69), streptonigrin (70), terpenoids (71,72), azatoxins (73,74), quinolones (75-79), anthraquinones (80), menogaril (81), naphtoquinones (82), and the polyaromatic quinone antibiotic UCE6 (52)], without mentioning the dual top1 and top2 inhibitors listed in the previous section. New anthracyclines have also been introduced in the clinics [4-demethoxydaunorubicin (Idarubicin), 4'-epidoxorubicin (Epirubicin)].   4-demethoxydaunorubicin [Idarubicin] exhibits greater potency against purified top2 than daunorubicin, indicating that removal of the 4-methoxy group yields better top2 inhibitors (83).   All the anthracyclines studied so far retain the same DNA sequence selective inhibition of top2 (83).

Drugs exhibit specific DNA cleavage patterns in the presence of top2. Also some, such as terpenoids (71,72) and anthraquinones with alkylating groups (80) produce irreversible cleavable complexes. This last class of compounds might be useful in determining the drug binding sites on top2 (and/or DNA).

Inhibition of top2 catalytic activity without trapping of cleavable complexes can also be observed (Table II). This is the case for strong DNA

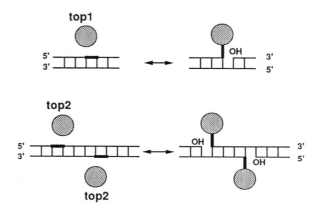

Figure 1. Cleavable complexes.

Table I.	Topoisomerase I Inhibitors			
Drug Type	*Trapping of cleavable complexes*	*refs.*	*Suppression of cleavable complexes*	*refs.*
Camptothecins	Topotecan CPT-11 (SN-38) 9-Aminocamptothecin 10,11-MDO-CPT	(6,179) (6,179) (6,179) (6,179)		
Other non-DNA Binders	AzaIQD [a] Indolocarbazoles	(58) (50)	Heparin Corilagin, Chebulagic acid Beta-Lapachone Diethylstilbestrols	(180) (181, 182) (183) (184)
DNA binders	Actinomycin D [a] Saintopin [a] Intoplicine [a] Morpholino- doxorubicin Bulgarein Fagaronine	(56,57) (59) (60) (57) (53) (55)	 Saintopin [a] Intoplicine [a] Morpholino- doxorubicin Bulgarein Fagaronine	 (59) (60) (57) (53) (55)
DNA groove binders	Hoechst 33342	(185)	Distamycin Netropsin	(186)

[a] Dual top1 and top2 inhibitor

Table II.   Topoisomerase II Inhibitors

Drug Type	Induction of cleavable complexes		refs.	Suppression of cleavable complexes		refs.
Intercalators, DNA binders	Doxorubicin		(12,62)	Doxorubicin		(12,62)
	Daunorubicin		(12,62)	Daunorubicin		(12,62)
	Epirubicin		(12,62)	Epirubicin		(12,62)
	Idarubicin		(12,62)	Idarubicin		(12,62)
	Amsacrine		(12,62)			
	Mitoxantrone		(12,62)	Mitoxantrone		(12,62)
	Anthrapyrazoles		(12,62)	Anthrapyrazoles		(12,62)
	Elliptinium		(12,62)	Elliptinium		(12,62)
	Actinomycin D	a	(12,62)			
	Alkylating anthraquinones	b	(80)			
	Menogaril		(81)	Menogaril		(81)
	Intoplicine	a	(60)	Intoplicine	a	(60)
	Saintopin	a	(59)	Saintopin	a	(59)
	Amonafide		(187)	Amonafide		(187)
	Streptonigrin		(70)	Bulgarein		(53)
				Ethidium bromide		(87)
				Ditercalinium		(88)
				Distamycin Netropsin		(89)
Non-intercalators	VP-16, VM-26		(12,62)	Merbarone		(90)
	AzaIQD	a	(58)	Bis(2,6-dioxopiperazine)		(91,92)
	Flavones - Flavonones Isoflavones [genistein]		(65,66,70)	Fostreicin (?)		(94)
				Suramin		(93)
	Nitroimidazole (Ro 15-0216)		(68)			
	Terpenoids (terpentecin, clerocidin, UCT4B)	b	(71,188)			
	Naphthoquinones		(82)			
	Whithangulatin		(69)			
	Polyaromatic quinone (UCE6)		(52)			
	Quinolones		(76,77,189)			
	Azatoxin		(73,74)			

a Dual top1 & top2 inhibitor
b not reversible

intercalators, such as anthracyclines and ellipticines at concentrations that saturate the DNA and alter its structure (*84-89*). Some non-DNA binders can also suppress cleavable complexes (Table II), e.g. merbarone (*90*), dioxopiperazine derivatives (*91,92*), suramin (*93*), and possibly fostreicin, for which contradictory results have been published (*94-96*) (Table II).

Hence, three types of dose-response curves can be found for top2 (and top1) inhibitors: 1) a monotonal increase of cleavable complexes with drug concentration in the case of weak (or non-) DNA binders (amsacrine, VP-16, VM-26); 2) a bell-shaped curve in the case of anthracyclines and other DNA intercalators; and 3) a monotonal decrease of cleavable complexes in the case of bulky intercalators (ethidium bromide, ditercalinium, aclarubicin) (*87,88,97*) or non-DNA binders that inhibit the catalytic activity without trapping cleavable complexes (see above).

It is well known that top2 inhibitors have different clinical potencies and activity spectra. Also, drug cytotoxic potencies are not well correlated with the frequency of cleavable complexes. Even at the top2 level, drugs exhibit very different effects. The kinetics of cleavable complex formation and reversal vary in drug-treated cells. They are slow in the case of doxorubicin but very rapid in the case of VP-16 or amsacrine (*98,99*). Since the duration of exposure to cleavable complexes probably determines cytotoxicity, this may explain the greater cytotoxicity of doxorubicin over VP-16.

Drugs induce not only the top2-mediated DNA double-strand breaks but also top2-mediated single-strand breaks, the ratio of which varies widely among drugs. Anthracyclines and ellipticine produce almost exclusively DNA double-strand breaks, while VP-16 and amsacrine produce 10-20 single-strand breaks per double-strand break (*12,98,99*). Hence, the greater cytotoxicity of anthracyclines compared with that of amsacrine or VP-16 may be due to the greater frequency of double-strand breaks, which may be more lethal than single-strand breaks (*12*).

Recent work by Osheroff and coworkers indicates that drugs may act differently in the top2 catalytic cycle (*63,75,100,101*). While etoposide (VP-16) severely inhibits cleavable complex religation and has little effect on strand passage and ATP hydrolysis, genistein and quinolones have little effect on top2-mediated religation but impair the ability of top2 to carry out its strand passage event and ATP hydrolysis. Amsacrine is unique since it inhibits similarly religation, strand passage, and ATP hydrolysis. These observations strongly suggest that the drugs interact with different top2 protein domains, which is consistent with the finding that some drug-resistant top2 mutant enzymes are not cross-resistant to all inhibitors (*102-105*). Enzyme deletion mutants may prove useful in delineating the top2 domain(s) that interact with the drugs (*106*).

The DNA sequence and genomic localization of top2 cleavable complexes varies among drugs (*83,107-110*). Usually, drugs from the same chemical family produce closely related patterns of top2 cleavage, whereas compounds structurally and electronically unrelated produce different patterns both in purified DNA and in drug-treated cells. These differences may play a key role in the differential cytotoxicity and spectrum of antitumor activity of top2 inhibitors.

Base Sequence Preference of Topoisomerase Inhibition; Stacking Model.
The development of a method to align and analyze DNA sequences around
topoisomerase cleavage sites enabled us to demonstrate that drug differences
were correlated with base sequence preferences and to propose the stacking
model (*107,110-113*).

Top1 cleavage sites do not occur randomly and exhibit a strong
preference for T at the 3'-terminus of the DNA breaks (position -1). In
addition, in the presence of CPT, a strong preference for G at the 5'-DNA
terminus (position +1 relative to the break) (Figure 2) is found especially at
the most intense sites. This is consistent with the fact that CPT enhances only
a subset of top1 cleavage sites (those having G+1) (*43,110,111,113*). We have
demonstrated further the strong preference for guanine at the 5'-DNA
terminus of the top1 breaks induced by CPT using oligonucleotides (*111*).
More recently, using UV photoactivated camptothecin, we have obtained
evidence of a selective interaction of CPT with guanine even in the absence
of top1 (*114*). Taken together, these data are consistent with our stacking
model that proposes that CPT binds inside the top1 cleavage sites bearing a
guanine at their 5'-terminus and stacks along this base, thereby preventing
resealing of the cleavable complexes (*115*).

Similarly, in the case of top2, each class of inhibitor tends to enhance
cleavage at sites with different base sequence preference, either at the 3'-
(position -1) or 5'-terminus (position +1) relative to the observed cleavage site
(see Figure 2). This base preference was first demonstrated for doxorubicin
with A-1 (*107*) (Figure 2), and then later for amsacrine with A+1, etoposides
(VP-16, VM-26) with C-1, and for ellipticines and mitoxantrone with
pyrimidine-1 (*110,112,116-119*)]. Taking advantage of the strong drug-
induced cleavage sites that we had mapped in SV40 DNA (*108*), we designed
experiments with oligonucleotides in which the bases flanking the cleavage
sites were mutated. The base sequence preferences were similar to those
found in longer DNA fragments (*116*). These strong drug-selective
preferences for certain bases immediately flanking the cleavage sites suggest
that the drugs interact directly with these bases. Since all top2 inhibitors --
whether intercalator or not -- have a planar aromatic portion, the simplest
explanation is that the drugs stack inside the cleavage sites. Depending on
the drug structure, the preferential stacking would take place either at the 3'-
or the 5'-terminus with a specific base; in the case of anthracyclines with an
adenine at the 3'-terminus (position -1) (Figure 2). This hypothesis implies
that topoisomerases first cleave the DNA at many sites and that the drugs
then bind specifically to some sites and prevent DNA religation (*41*).
However, Osheroff and coworkers have obtained evidence that some drugs
induce the formation rather than inhibit the resealing of top2 cleavable
complexes (*63,75,100,101*). The base sequence analysis data for
anthracyclines (*107*) strongly suggests that stacking with an adenine at the 3'-
end of one break site is sufficient for the trapping of top2 double-strand
breaks, which is consistent with the concerted action of both enzyme
subunits during catalysis.

Determinants of Cellular Sensitivity to Topoisomerase Inhibitors

Figure 3 outlines the factors that determine antitumor cytotoxic activity.

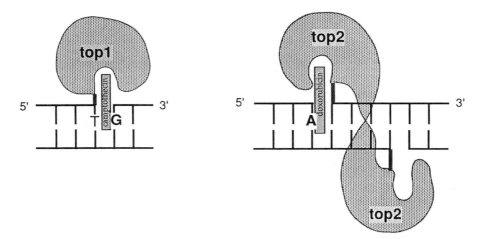

Figure 2. Stacking model for topoisomerase inhibitors.

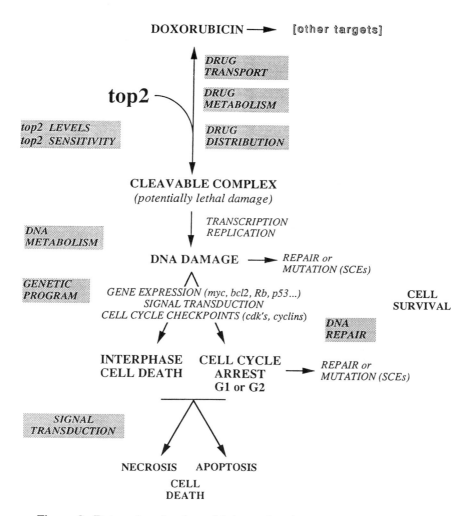

Figure 3. Determinants of sensitivity and resistance to topoisomerase inhibitors.

At the plasma membrane level, it is important to note that most anticancer top2 inhibitors are substrates for the P-glycoprotein^{MDR} that actively extrudes drugs from the cells (120-122). Therefore, cells overexpressing the P-glycoprotein^{MDR} are resistant to most top2 inhibitors. P-glycoprotein^{MDR} is a critical determinant of cellular sensitivity in the case of anthracyclines and other top2 inhibitors, such as mitoxantrone and etoposides (VP-16, VM-26). In addition, top2 and P-glycoprotein^{MDR} alterations are commonly associated in cells resistant to adriamycin and mitoxantrone (123,124). However, top2 is not altered in vinblastin-resistant cell lines with P-glycoprotein^{MDR} alterations (125). Trifluoroperazine can revert resistance by acting both on the P-glycoprotein^{MDR} and top2-induced DNA lesions (125).

A better knowledge of the P-glycoprotein^{MDR} pharmacophore in association with rational drug synthesis and testing against P-glycoprotein^{MDR} and top2 may be a powerful way to design compounds that overcome P-glycoprotein^{MDR}-mediated resistance without compromising the anti-top2 and antitumor activities. In the case of anthracyclines, the amino group on the daunosamine sugar may be involved in the recognition of the drug by the P-glycoprotein^{MDR} but not in the top2 interaction since the deamino derivative, hydroxyrubicin, is less subject to drug resistance while retaining top2 inhibitory activity (126,127). Among top1 inhibitors, topotecan is also a P-glycoprotein^{MDR} substrate (128,129). By contrast, CPT does not appear to be a P-glycoprotein^{MDR} substrate and should be more active than topotecan in multidrug-resistant cells.

Another multidrug-resistance-associated protein (MRP) has also been identified in cells that do not overexpress the P-glycoprotein^{MDR} (130,131). A recent study shows that MRP confers resistance to doxorubicin and VP-16 in transfected cells (132).

Intracellular distribution is different among drugs and possibly between cell lines. It is well known that DNA intercalators such as doxorubicin tend to concentrate in nuclei and are active at very low concentrations.

Assuming that the same amount of drug reaches the topoisomerase targets, then the higher the levels of top1 or top2, the more sensitive the cells. More cleavable complexes are formed in the presence of abundant topoisomerases, and resistant cells usually exhibit low enzyme levels. In addition, the difference in genomic distribution of cleavable complexes between drugs (see the Stacking Model section above) (12,109) may be accentuated in chromatin due to the effect of nucleosomes which restrict topoisomerase access to the linker regions (133). As a result, amsacrine forms very prominent cleavable complexes in the human c-myc P2 promoter in drug-treated cells, while VP-16 has a more diffuse effect (104). Hence, not only is the total number of cleavable complexes critical for drug effects but also the genomic distribution of these cleavable complexes. Differential effects on selective oncogenes may play an essential role in determining the sensitivity of a cell line to a certain class of agent.

There is clear evidence that cleavable complexes are only potentially lethal and not sufficient for cell killing. Inhibition of DNA synthesis at the time of CPT treatment abolishes the CPT-induced cytotoxicity without changing the frequency of top1 cleavable complexes (134,135). This is because CPT generates DNA double-strand breaks upon collision of a replication fork with a top 1 cleavable complex. Such lesions are very persistent after CPT removal even though the cleavable complexes are reversible within minutes after drug removal (136,137). The clinical relevance of this finding is that, because CPT cytotoxicity is highly dependent on DNA replication and on the number of cells in S phase (at least in the cell lines that have been studied), the longer the drug exposure, then the greater the cytotoxicity.

Interaction of RNA transcription with cleavable complexes may play a role in the activity of top2 inhibitors. The dependence of top1 and top2 inhibitor cytotoxicity on replication and transcription probably explains why simultaneous treatments with CPT and VP-16 are antagonistic (138); VP-16 probably suppresses CPT effects by inhibiting DNA replication and CPT suppresses the effects of VP-16 by inhibiting RNA transcription. Other conditions have been described where the cytotoxicity of topoisomerase inhibitors can be abrogated without effect on cleavable complex formation. These include intracellular calcium depletion by EDTA (139) and protein synthesis inhibition by cycloheximide (12,62). Poly(adenosine diphosphoribose) synthesis may also be important for cell killing since poly(adenosine diphosphoribose)-deficient Chinese hamster cells are resistant to VP-16 and hypersensitive to CPT (140).

These observations validate the term "cleavable complexes" since there is usually no (or very rare) functional DNA interruption by topoisomerases. Indeed, the enzyme-induced DNA breaks are sealed tightly within the enzyme complexes and reverse readily upon drug removal without subsequent DNA recombination. DNA damage would be produced when the cleavable complexes become "cleaved" or disrupted upon drug stabilization and interference with dynamic cellular processes such as a moving replication or transcription forks (see middle part of Figure 3).

DNA damage induces recombinations that have been detected as sister chromatid exchanges after treatments of cells with inhibitors of top1 (141-144) and inhibitors of top2 (22,145-149). Frameshift mutations, duplications, deletions, and chromosomal damage are also induced by top2 inhibitors (150,151). Top2-mediated DNA damage may be responsible for the chromosomal translocations (152) associated with etoposide-induced secondary malignancies (153-156).

DNA repair can correct drug-induced and topoisomerase-mediated DNA damage. Yeast cells are usually resistant to topoisomerase inhibitors unless they are RAD52 mutants, e.g., deficient in DNA double-strand break repair (157). Also, DNA repair-deficient cell lines (ataxia telangiectasia and Cockayne's syndrome) are hypersensitive to CPT (158,159). Unrepaired DNA damage may lead to the accumulation of genetic alterations, such as sister chromatid exchanges and recombinations (7,147-149,151).

Cell cycle control may also play a pivotal role in the cytotoxicity of topoisomerase inhibitors. Lack of arrest in G1 or G2 may not provide the cell with the time required for DNA repair and may lead to an accumulation of further damage. Hence, deregulation of cyclins, cell cycle regulated kinases and phosphatases, and p53 mutations may be involved in the cytotoxicity of topoisomerase inhibitors toward cancer cells (160-164).

Altered regulation of DNA repair and of cell cycle events that appear quite common in neoplastic cells may explain the selectivity of chemotherapy (*164*). Such defects may also provide a basis for carcinogenesis, and it is well established that cancer-prone diseases are characterized by DNA-repair deficiencies [xeroderma pigmentosum, ataxia telangiectasia, Bloom syndrome (*165*)] and tumor suppressor gene mutations [retinoblastoma, Li-Fraumeni syndrome (p53)].

Another determinant of sensitivity to topoisomerase inhibitors is the preexisting genetic program of the cell. Some cells such as HL-60 cells are known to be hypersensitive to a variety of aggressions, including DNA damage by topoisomerase inhibitors. The underlying mechanism for this hypersensitivity may be the facile induction of apoptosis (*166-171*). Overexpression of the c-*myc* proto-oncogene and down regulation of the *bcl*-2 gene have been involved in committing the cells to an apoptosis-prone phenotype. A recent report (*163,172*) suggests that p53 could play a key role in the case of VP-16 by mediating apoptosis in response to DNA damage without regulating apoptosis induced by glucocorticoids. Interestingly, p53 is not expressed in HL-60 cells, indicating that other DNA damage-dependent pathways may direct apoptosis in these cells. Further analysis is required to unravel apoptosis and will provide molecular and clinical pharmacologists with probes that will enable the detection of drug-sensitive and resistant status. Apoptosis may also play a role in drug-induced side effects such as hematopoietic toxicity. Indeed, hematopoietic progenitors may be prone to apoptosis. Hence, studies on the pharmacological regulation of apoptosis may prove useful. We have recently identified several classes of pharmacological agents that can suppress topoisomerase inhibitor-induced apoptosis (*169*) and shown that *bcl*-2 overexpression renders cells resistant to VP-16 (*173*).

Mechanisms of Resistance to Topoisomerase Inhibitors and Anthracyclines

There are at least two well-characterized mechanisms of resistance to topoisomerase inhibitors:

1) limited drug accumulation and access to the topoisomerase target, and
2) reduced formation of cleavable complexes.

Limited drug accumulation is usually due to P-glycoprotein[MDR] overexpression and associated with top2 alterations in doxorubicin-resistant cells (*123,124*); for review see (*120-122*). Non-P-glycoprotein[MDR] resistance is also important (*130,131*). A recent study shows that the MRP (Multidrug Resistance-associated Protein) gene confers resistance to doxorubicin and VP-16 in transfected cells (*132*).

Reduction of topoisomerase cleavable complexes can either be due to decreased enzyme levels, or enzyme mutations. Reduction of enzyme levels are more common than enzyme mutations; for review see (*8,122*). The responsibility of enzyme mutations for drug resistance has been established unambiguously for top1 mutants since the recombinant proteins from two of these mutants are CPT resistant (*174,175*). Analyses of drug-resistant enzyme mutants (Table III) provide important information on the functional domains of topoisomerases and the sites of drug interaction with their protein targets.

Table III.   **Drug-Resistant Topoisomerase II Mutants**

Cell line	Mutation(s)	Refs.
HL-60/AMSA	Arg-486-Lys	(190)
KBM-3/AMSA	Arg-486-Lys Lys-479-Glu Lys-519-stop codon	(191)
CEM/VM-1	Arg-449-Glu Prol-802-Ser	(192) (193)
CEM/VM-1-5	Arg-449-Gln Prol-802-Ser	(192) (193)
CEM/VP-1	Lys-797-Asn	(194)
VpmR-5	Arg-493-Gln	(195)

Several CPT-resistant cell lines show collateral hypersensitivity to top2 inhibitors (176,177) and some cell lines that are resistant to top2 inhibitors are hypersensitive to CPT (140). Therefore, it is possible that functional deficiency in one topoisomerase can be compensated for by an increase in the other topoisomerase. Although this observation may suggest that association of top2 and top1 inhibitors could be useful in cancer chemotherapy, cell culture experiments indicate that treatments with top1 and top2 inhibitors have to be separated by several hours in order to avoid drug antagonism (178). Recent data suggest, however, that dual inhibitors of top1 and top2 may exhibit interesting antitumor activity (60).

Acknowledgments

The author wishes to thank Dr. Kurt W. Kohn, Chief of the Laboratory of Molecular Pharmacology, NCI, for his support and exciting discussions.

Literature Cited

1. Wang, J. C. *Annu. Rev. Biochem.* **1985**, *54*, 665-697.
2. Wang, J. C. *Biochim. Biophys. Acta* **1987**, *909*, 1-9.
3. Wang, J. C. *J. Biol. Chem.* **1991**, *266*, 6659-6662.
4. Caron, P. R. Wang, J. C. In *DNA Topoisomerases as Target of Therapeutics: A Structural Overview* ; Andoh, T., Ikeda, H. and Oguro, M., Eds; Molecular Biology of DNA topoisomerases and its application to chemotherapy; CRC Press: Boca Raton, FL, 1993; 1-18.
5. Champoux, J. In *Mechanistic Aspects of Type-I Topoisomerases* ; Wang, J. C. and Cozarelli, N. R., Eds; DNA topology and its biological effects; Cold Spring Harbor Laboratory: Cold Spring Harbor, 1990; 217-242.

6. Pommier, Y. Tanizawa, A. In *Mammalian DNA Topoisomerase I and its Inhibitors* ; Hickman, J. and Tritton, T., Eds; Cancer Chemotherapy; Blackwell Scientific Publications LTD: Oxford, 1993; 214-250.

7. Pommier, Y. Bertrand, R. In *The mechanisms of formation of chromosomal aberrations: role of eukaryotic DNA topoisomerases* ; Kirsch, I. R., Eds; The causes and consequences of chromosomal aberrations; CRC Press: Boca Raton, 1993; 277-309.

8. Pommier, Y. *Cancer Chemother. Pharmacol.* **1993**, *32*, 103-108.

9. Pommier, Y.; Leteurtre, F.; Fesen, M. R.; Fujimori, A.; Bertrand, R.; Solary, E.; Kohlhagen, G. Kohn, K. W. *Cancer Investigations* **1994**, *in press,*

10. Osheroff, N.; Zechierich, E. L. Gale, K. C. *BioEssays* **1991**, *13*, 269-275.

11. Kohn, K. W.; Pommier, Y.; Kerrigan, D.; Markovits, J. Covey, J. M. *NCI Monographs* **1987**, *4*, 61-71.

12. Pommier, Y. Kohn, K. W. In *Topoisomerase II Inhibition by Antitumor Intercalators and Demethylepipodophyllotoxins* ; Gazer, R. I., Eds; Developments in Cancer Chemotherapy; CRC Press, Inc.: Boca Raton, FA, 1989; 175-196.

13. Adachi, Y.; Luke, M. Laemmli, U. K. *Cell.* **1991**, *64*, 137-148.

14. Earnshaw, W. C. Heck, M. M. *J. Cell. Biol.* **1985**, *100*, 1716-1725.

15. Gasser, S. M.; Laroche, T.; Falquet, J.; Boy de la Tour, E. Laemmli, U. K. *J. Mol. Biol.* **1986**, *188*, 613-629.

16. Hirano, T. Mitchison, T. J. *J. Cell. Biol.* **1993**, *120*, 601-612.

17. DiNardo, S.; Voelkel, K. Sternglanz, R. *Proc. Natl. Acad. Sci. U.S.A.* **1984**, *81*, 2616-2620.

18. Holm, C.; Stearns, T. Botstein, D. *Mol.Cell.Biol.* **1989**, *9*, 159-168.

19. Uemura, T.; Ohkura, H.; Adachi, Y.; Morino, K.; Shiozaki, K. Yanagida, M. *Cell* **1987**, *50*, 917-925.

20. Rose, D. Holm, C. *Mol. Cell. Biol.* **1993**, *13*, 3445-3455.

21. Moens, P. B. Earnshaw, W. C. *Chromosoma.* **1990**, *98*, 317-322.

22. Shamu, C. E. Murray, A. M. *J. Cell Biol.* **1992**, *117*, 921-934.

23. Wood, E. R. Earnshaw, W. C. *J. Cell. Biol.* **1990**, *111*, 2839-2850.

24. Newport, J. *Cell* **1987**, *48*, 205-217.

25. Newport, J. Spann, T. *Cell* **1987**, *48*, 219-230.

26. Downes, C. S.; Mullinger, A. M. Johnson, R. T. *Proc. Natl. Acad. Sci. U.S.A.* **1991**, *88*, 8895-8899.

27. Swedlow, J. R.; Sedat, J. W. Agard, D. A. *Cell* **1993**, *73*, 97-108.

28. Heck, M. M. S.; Hittelman, W. N. Earnshaw, W. C. *J. Biol. Chem.* **1989**, *264*, 15161-15164.

29. Taagepera, S.; Rao, P. N.; Drake, F. H. Gorbsky, G. J. *Proc. Natl. Acad. Sci. U.S.A.* **1993**, *90*, 8407-8411.

30. Cardenas, M. E.; Dang, Q.; Glover, C. V. C. Gasser, S. M. *EMBO. J.* **1992**, *11*, 1785-1796.

31. Drake, F. H.; Hofmann, G. A.; Bartus, H. F.; Mattern, M. R.; Crooke, S. T. Mirabelli, C. K. *Biochemistry* **1989**, *28*, 8154-8160.

32. Tsai-Pflugfelder, M.; Liu, L. F.; Liu, A. A.; Tewey, K. M.; Whang-Peng, J.; Knutsen, T.; Huebner, K.; Croce, C. M. Wang, J. C. *Proc. Natl. Acad. Sci. U.S.A* **1988**, *85*, 7177-7181.

33. Jenkins, J. R.; Ayton, P.; Jones, T.; Davies, S. L.; Simmons, D. L.; Harris, A. L.; Sheer, D. Hickson, I. D. *Nucleic Acids Res.* **1992**, *20*, 5587-5592.

34. Patel, S. Fisher, L. M. *Br. J. Cancer* **1993**, *67*, 456-463.

35. Tan, K. B.; Dorman, T. E.; Falls, K. M.; Chung, T. D. Y.; Mirabelli, C. K.; Crooke, S. T. Mao, J.-I. *Cancer Res.* **1992**, *52*, 231-234.
36. Tsutsui, K.; Tsutsui, K.; Okada, S.; Watanabe, M.; Shohmori, T.; Seki, S. Inoue, Y. *J. Biol. Chem.* **1993**, *268*, 19076-19083.
37. Woessner, R. D.; Mattern, M. R.; Mirabelli, C. K.; Johnson, R. K. Drake, F. H. *Cell Growth Differentiation* **1991**, *2*, 209-214.
38. Capranico, G.; Tinelli, S.; Austin, C. A.; Fisher, M. L. Zunino, F. *Biochimical et Biophysica Acta* **1992**, *1132*, 43-48.
39. Negri, C.; Scovassi, A. I.; Braghetti, A.; Guano, F. Astaldi Ricotti, G. C. *Exp. Cell Res.* **1993**, *206*, 128-33.
40. Petrov, P.; Drake, F.; Loranger, A.; Huang, W. Hancock, R. *Exp. Cell Res.* **1993**, *204*, 73-81.
41. Liu, L. F. D'Arpa, P. *Important Adv. Oncol.* **1992**, 79-89.
42. Slichenmyer, W. J.; Rowinsky, E. K.; Donehower, R. C. Kaufmann, S. H. *J. Natl. Cancer Inst.* **1993**, *85*, 271-287.
43. Porter, S. E. Champoux, J. J. *Nucleic Acids Res.* **1989**, *17*, 8521-8532.
44. Kjeldsen, E.; Svejstrup, J. Q.; Gromova, I.; Alsner, J. Westergaard, O. *J. Mol. Biol.* **1992**, *228*, 1025-1030.
45. Pommier, Y.; Jaxel, C.; Heise, C.; Kerrigan, D. Kohn, K. W. In *Structure-Activity Relationship of Topoisomerase I Inhibition by Camptothecin Derivatives: Evidence for the Existence of a Ternary Complex* ; Potmesil, M. and Kohn, K. W., Eds; DNA topoisomerases in cancer; Oxford University Press: New York, 1991; 121-132.
46. Jaxel, C.; Kohn, K. W.; Wani, M. C.; Wall, M. E. Pommier, Y. *Cancer Res.* **1989**, *49*, 1465-1469.
47. Hertzberg, R. P.; Caranfa, M. J.; Holden, K. G.; Jakas, D. R.; Gallagher, G.; Mattern, M. R.; Mong, S. M.; Bartus, J. O.; Johnson, R. K. Kingsbury, W. D. *J. Med. Chem.* **1989**, *32*, 715-720.
48. Hsiang, Y.-H.; Liu, L. F.; Wall, M. E.; Wani, M. C.; Nicholas, A. W.; Manikumar, G.; Kirschenbaum, S.; Silber, R. Potmesil, M. *Cancer Res.* **1989**, *49*, 4385-4389.
49. Kingsbury, W. D.; Boehm, J. C.; Jakas, D. R.; Holden, K. G.; Hecht, S. M.; Gallagher, G.; Caranfa, M. J.; McCabe, F.; Faucette, L. F.; Johnson, R. K. Hertzberg, R. P. *J. Med. Chem.* **1991**, *34*, 98-107.
50. Yamashita, Y.; Fujii, N.; Murakata, C.; Ashizawa, T.; Okabe, M. Nakano, H. *Biochemistry* **1992**, *31*, 12069-12075.
51. Yoshinari, T.; Yamada, A.; Uemura, D.; Nomura, K.; Arakawa, H.; Kojiri, K.; Yoshida, E.; Suda, H. Okura, A. *Cancer Res.* **1993**, *53*, 490-494.
52. Fujii, N.; Yamashita, Y.; Chiba, S.; Uosaki, Y.; Saitoh, Y.; Tuji, Y. Nakano, H. *J. Antibiot.* **1993**, *46*, 1173-1174.
53. Fujii, N.; Yamashita, Y.; Saitoh, Y. Nakano, H. *J. Biol. Chem.* **1993**, *268*, 13160-13165.
54. Chen, A. Y.; Yu, C.; Gatto, B. Liu, L. F. *Proc. Natl. Acad. Sci. U.S.A.* **1993**, *90*, 8131-8135.
55. Larsen, A. K.; Grondard, L.; Couprie, J.; Desoizé, B.; Comof, L.; Jardillier, J.-C. Riou, J.-F. *Biochem. Pharmacol.* **1993**, *19*, 1403-1412.
56. Trask, D. K. Muller, M. T. *Proc. Natl. Acad. Sci. U.S.A.* **1988**, *85*, 1417-1421.
57. Wassermann, K.; Markovits, J.; Jaxel, C.; Capranico, G.; Kohn, K. W. Pommier, Y. *Mol. Pharmacol.* **1990**, *38*, 38-45.
58. Riou, J.-F.; Helissey, P.; Grondard, L. Giorgi-Renault, S. *Mol. Pharmacol.* **1991**, *40*, 699-706.

59. Yamashita, Y.; Kawada, S.-Z.; Fujii, N. Nakano, H. *Biochemistry* **1991**, *30*, 5838-5845.
60. Poddevin, B.; Riou, J.-F.; Lavelle, F. Pommier, Y. *Mol. Pharmacol.* **1993**, *44*, 767-774.
61. Janin, Y. L.; Croisy, A.; Riou, J.-F. Bisagni, E. *J. Med. Chem.* **1993**, *36*, 3686-3692.
62. Liu, L. F. *Annu. Rev. Biochem.* **1989**, *58*, 351-375.
63. Corbett, A. H. Osheroff, N. *Chem. Res. Toxicol.* **1993**, *6*, 585-597.
64. Andersson, B. S.; Beran, M.; Bakic, M.; Silberman, L. E.; Newman, R. A. Zwelling, L. A. *Cancer Res.* **1987**, *47*, 1040-1044.
65. Markovits, J.; Linassier, C.; Fosse, P.; Couperie, J.; Pierre, J.; Jacquemin-Sablon, A.; Saucier, J.-M.; Le Pecq, J. B. Larsen, A. K. *Cancer Res.* **1989**, *49*, 5111-5117.
66. Austin, C. A.; Patel, S.; Ono, K.; Nakane, H. Fisher, L. M. *Biochem. J.* **1992**, *282*, 883-889.
67. Yamashita, Y.; Kawada, S.-Z. Nakano, H. *Biochem. Pharmacol.* **1990**, *39*, 737-744.
68. Sorensen, B. S.; Jensen, P. S.; Andersen, A. H.; Christiansen, K.; Alsner, J.; Thomsen, B. Westergaard, O. *Biochemistry* **1990**, *29*, 9507-9515.
69. Juang, J.-K.; Huang, H. W.; Chen, C.-M. Liu, H. J. *Biochem. Biophys. Res. Commun.* **1989**, *159*, 1128-1134.
70. Yamashita, Y.; Kawada, S.-Z.; Fujii, N. Nakano, H. *Cancer Res.* **1990**, *50*, 5841-5844.
71. Kawada, S.-Z.; Yamashita, Y.; Fujii, N. Nakano, H. *Cancer Res.* **1991**, *51*, 2922-2925.
72. Kawada, S.-Z.; Yamashita, Y.; Uosaki, Y.; Gomi, K.; Iwasaki, T.; Takiguchi, T. Nakano, H. *J. Antibiot.* **1992**, *45*, 1182-1184.
73. Leteurtre, F.; Madalengoitia, J.; Orr, A.; Guzi, T. J.; Lehnert, E.; Macdonald, T. Pommier, Y. *Cancer Res.* **1992**, *52*, 4478-4483.
74. Solary, E.; Leteurtre, F.; Paull, K. D.; Scudiero, D.; Hamel, E. Pommier, Y. *Biochem. Pharmacol.* **1993**, *45*, 2449-2456.
75. Robinson, M. J.; Martin, B. A.; Gootz, T. D.; McGuirk, P. R.; Moynihan, M.; Sutcliffe, J. A. Osheroff, N. *J. Biol. Chem.* **1991**, *266*, 14585-14592.
76. Yamashita, Y.; Ashizawa, T.; Marimoto, M.; Hosomi, J. Nakano, H. *Cancer Res.* **1992**, *52*, 2818-2822.
77. Elsea, S. H.; Osheroff, N. Nitiss, J. L. *J. Biol. Chem.* **1992**, *267*, 13150-13153.
78. Froelich-Ammon, S. J.; McGuirk, P. R.; Gootz, T. D.; Jefson, M. R. Osheroff, N. *Antimicrob. Agents Chemother.* **1993**, *37*, 646-651.
79. Yoshinari, T.; Mano, E.; Arakawa, H.; Kurama, M.; Iguchi, T.; Nakagawa, S.; Tanaka, N. Okura, A. *Jpn. J. Cancer Res.* **1993**, *84*, 800-806.
80. Kong, X. B.; Rubin, L.; Chen, L. I.; Ciszewska, G.; Watanabe, K. A.; Tong, W. P.; Sirotnak, F. M. Chou, T. C. *Mol. Pharmacol.* **1992**, *41*, 237-244.
81. Ono, K.; Ikegami, Y.; Nishizawa, M. Andoh, T. *Jpn. J. Cancer Res.* **1992**, *83*, 1018-1023.
82. Fujii, N.; Yamashita, Y.; Arima, Y.; Nagashima, M. Nakano, H. *Antimicrob. Agents Chemother.* **1992**, *36*, 2589-2594.
83. Capranico, G.; Zunino, F.; Kohn, K. W. Pommier, Y. *Biochemistry* **1990**, *29*, 562-569.
84. Tewey, K. M.; Rowe, T. C.; Yang, L.; Halligan, B. D. Liu, L. F. *Science* **1984**, *226*, 466-468.

85. Pommier, Y.; Minford, J. K.; Schwartz, R. E.; Zwelling, L. A. Kohn, K. W. *Biochemistry* **1985**, *24*, 6410-6416.
86. Pommier, Y.; Schwartz, R. E.; Zwelling, L. A. Kohn, K. W. *Biochemistry* **1985**, *24*, 6406-6410.
87. Rowe, T.; Kupfer, G. Ross, W. *Biochem. Pharmacol.* **1985**, *34*, 2483-2487.
88. Markovits, J.; Pommier, Y.; Mattern, M. R.; Esnault, C.; Roques, B. P.; Le Pecq, J. B. Kohn, K. W. *Cancer Res.* **1986**, *46*, 5821-5826.
89. Fesen, M. Pommier, Y. *J. Biol. Chem.* **1989**, *19*, 11354-11359.
90. Drake, F. H.; Hofmann, G. A.; Mong, S. M.; Bartus, J. O.; Hertzberg, R. P.; Johnson, R. K.; Mattern, M. R. Mirabelli, C. K. *Cancer Res.* **1989**, *49*, 2578-2583.
91. Ishida, R.; Miki, T.; Narita, T.; Yui, R.; Sato, M.; Utsumi, K. R.; Tanabe, K. Andoh, T. *Cancer Res.* **1991**, *51*, 4909-4916.
92. Tanabe, K.; Ikegami, R. Andoh, T. *Cancer Res.* **1991**, *51*, 4903-4908.
93. Bojanowski, K.; Lelievre, S.; Markovits, J.; Couprie, J.; Jacquemin-Sablon, A. Larsen, A. K. *Proc. Natl. Acad. Sci. U.S.A.* **1992**, *89*, 3025-3029.
94. Gedik, C. M. Collins, A. R. *Nucleic Acids Res.* **1990**, *18*, 1007-1013.
95. Frosina, G. Rossi, O. *Carcinogenesis* **1992**, *13*, 1371-1377.
96. Boritzki, T. J.; Wolfard, T. S.; Besserer, J. A.; Jackson, R. C. Fry, D. W. *Biochem. Pharmacol.* **1988**, *37*, 4063-4068.
97. Jensen, P. B.; Jensen, P. S.; Demant, E. J. F.; Friche, E.; Sorensen, B. S.; Sehested, M.; Wassermann, K.; Vindelov, L.; Westergaard, O. Hansen, H. H. *Cancer Res.* **1991**, *51*, 5093-5099.
98. Long, B. H.; Musial, S. T. Brattain, M. G. *Cancer Res.* **1985**, *45*, 3106-3112.
99. Zwelling, L. A.; Michaels, S.; Erickson, L. C.; Ungerleider, R. S.; Nichols, M. Kohn, K. W. *Biochemistry* **1981**, *20*, 6553-6563.
100. Robinson, M. J.; Corbett, A. H. Osheroff, N. *Biochemistry* **1993**, *32*, 3638-3643.
101. Robinson, M. J. Osheroff, N. *Biochemistry* **1991**, *30*, 1807-1813.
102. Zwelling, L. A.; Hinds, M.; Chan, D.; Mayes, J.; Sie, K. L.; Parder, E.; Silberman, L.; Radcliffe, A.; Beran, M. Blick, M. *J. Biol. Chem.* **1989**, *264*, 16411-16420.
103. Mayes, J.; Hinds, M.; Soares, L.; Altschuler, E.; Kim, P. Zwelling, L. A. *Biochem. Pharmacol.* **1993**, *46*, 599-507.
104. Huff, A. C.; Ward, R. E. Kreuzer, K. N. *Mol. Gen. Genet.* **1990**, *221*, 27-32.
105. Huff, A. C. Kreuzer, K. N. *J. Biol. Chem.* **1990**, *265*, 20496-20505.
106. Crenshaw, D. G. Hsieh, T.-S. *J. Biol. Chem.* **1993**, *28*, 21328-21334.
107. Capranico, G.; Kohn, K. W. Pommier, Y. *Nucleic. Acids. Res.* **1990**, *18*, 6611-6619.
108. Pommier, Y.; Capranico, G.; Orr, A. Kohn, K. W. *J. Mol. Biol.* **1991**, *222*, 909-924.
109. Pommier, Y.; Orr, A.; Kohn, K. W. Riou, J. F. *Cancer Res.* **1992**, *52*, 3125-3130.
110. Pommier, Y.; Kohn, K. W.; Capranico, G. Jaxel, C. In *Base sequence selectivity of topoisomerase inhibitors suggests a common model for drug action* ; Andoh, T., Ikeda, H. and Oguro, M., Eds; Molecular Biology of DNA Topoisomerases and its Application to Chemotherapy; CRC Press: Boca Raton, FL, 1993; 215-227.
111. Jaxel, C.; Capranico, G.; Kerrigan, D.; Kohn, K. W. Pommier, Y. *J. Biol. Chem.* **1991**, *266*, 20418-20423.

112. Pommier, Y.; Capranico, G. Kohn, K. W. *Proc. Am. Assoc. Cancer Res.* **1991**, *32*, 335.

113. Tanizawa, A.; Kohn, K. W. Pommier, Y. *Nucleic Acids Res.* **1993**, *21*, 5157-5156.

114. Leteurtre, F.; Fesen, M.; Kohlhagen, G.; Kohn, K. W. Pommier, Y. *Biochemistry* **1993**, *32*, 8955-8962.

115. Svejstrup, J. Q.; Christiansen, K.; Gromova, I. I.; Andersen, A. H. Westergaard, O. *J. Mol. Biol.* **1991**, *222*, 669-678.

116. Capranico, G.; Tinelli, S.; Zunino, F.; Kohn, K. W. Pommier, Y. *Biochemistry* **1993**, *32*, 145-152.

117. Fosse, P.; Rene, B.; Le Bret, M.; Paoletti, C. Saucier, J.-M. *Nucleic Acids Res.* **1991**, *19*, 2861-2868.

118. Freudenreich, C. H. Kreuzer, K. N. *EMBO J.* **1993**, *12*, 2085-2097.

119. Capranico, G.; De Isabella, P.; Tinelli, S.; Bigioni, S. Zunino, F. *Biochemistry* **1993**, *32*, 3032-3048.

120. Ling, V. *Cancer* **1992**, *69*, 2603-2609.

121. Chabner, B. C. Horwitz, S. B. *Cancer Chemother. Biol. Response Modif.* **1990**, *11*, 74-81.

122. Beck, W. T. Danks, M. K. *Semin. Cancer Biol.* **1991**, *2*, 235-244.

123. Ganapathi, R.; Grabowski, D.; Ford, J.; Heiss, C.; Kerrigan, D. Pommier, Y. *Cancer Commun.* **1989**, *1*, 217-224.

124. Sinha, B. K.; Haim, N.; Dusre, L.; Kerrigan, D. Pommier, Y. *Cancer Res.* **1988**, *48*, 5096-5100.

125. Kamath, N.; Grabowski, D.; Ford, J.; Drake, F.; Kerrigan, D.; Pommier, Y. Ganapathi, R. *Cancer Commun.* **1991**, *3*, 37-44.

126. Priebe, W.; Van, N. T.; Burke, T. G. Perez-Soler, R. *Anti-Cancer Drugs* **1993**, *4*, 37-48.

127. Solary, E.; Ling, Y.-H.; Perez-Soler, R.; Priebe, W. Pommier, Y. *Int. J. Cancer* **1994**, *In Press*,

128. Hendricks, C. B.; Rowinsky, E. K.; Grochow, L. B.; Donchower, R. C. Kaufmann, S. H. *Cancer Res.* **1992**, *52*, 2268-2278.

129. Chen, A. Y.; Yu, C.; Potmesil, M.; Wall, M. E.; Wani, M. C. Liu, L. F. *Cancer Res.* **1991**, *51*, 6039-6044.

130. Cole, S. P.; Chanda, E. R.; Dicke, F. P.; Gerlach, J. H. Mirski, S. E. *Cancer Res.* **1991**, *51*, 3345-3352.

131. Cole, S. P.; Bhardwaj, G.; Gerlach, J. H.; Mackie, J. E.; Grant, C. E.; Almquist, K. C.; Stewart, A. J.; Kurz, E. U.; Duncan, A. M. Deeley, R. G. *Science* **1992**, *258*, 1650-1654.

132. Grant, C. E.; Valdimarsson, G.; Hipfner, D. R.; Almquist, K. C.; Cole, S. P. Deeley, R. G. *Cancer Res.* **1994**, *54*, 357-361.

133. Capranico, G.; Jaxel, C.; Roberge, M.; Kohn, K. W. Pommier, Y. *Nucleic Acids Res.* **1990**, *18*, 4553-4559.

134. Hsiang, Y.-H.; Lihou, M. G. Liu, L. F. *Cancer Res.* **1989**, *49*, 5077-5082.

135. Holm, C.; Covey, J. M.; Kerrigan, D. Pommier, Y. *Cancer Res.* **1989**, *49*, 6365-6368.

136. Ryan, A. J.; Squires, S.; Strutt, H. L. Johnson, R. T. *Nucleic Acids Res.* **1991**, *19*, 3295-3300.

137. Covey, J. M.; Jaxel, C.; Kohn, K. W. Pommier, Y. *Cancer Res.* **1989**, *49*, 5016-5022.

138. Bertrand, R.; O'Connor, P. M.; Kerrigan, D. Pommier, Y. *Proc. Am. Assoc. Cancer Res.* **1991**, *32*, 335.
139. Bertrand, R.; Kerrigan, D.; Sarang, M. Pommier, Y. *Biochem. Pharmacol.* **1991**, *42*, 77-85.
140. Chatterjee, S.; Cheng, M. F. Berger, N. A. *Cancer Commun.* **1990**, *2*, 401-407.
141. Huang, C. C.; Han, C. S.; Yue, X. F.; Shen, C. M.; Wang, S. W.; Wu, F. G. Xu, B. *J. Natl. Cancer Inst.* **1983**, *71*, 841-847.
142. Degrassi, F.; De Salvia, R.; Tanzarella, C. Palitti, F. *Mutat. Res.* **1989**, *211*, 125-130.
143. Tobey, R. A. *Cancer Res.* **1972**, *32*, 2720-2725.
144. Zhao, J. H.; Tohda, H. Oikawa, A. *Mutat. Res.* **1992**, *282*, 49-54.
145. Lim, M.; Liu, L. F.; Jacobson-kram, D. Williams, J. R. *Cell Biology and Toxicology* **1986**, *2*, 485-494.
146. Dillehay, L. E.; Jacobson-Kram, D. Williams, J. R. *Mutat. Res.* **1989**, *215*, 15-23.
147. Pommier, Y.; Zwelling, L. A.; Kao-Shan, C. S.; Whang-Peng, J. Bradley, M. O. *Cancer Res.* **1985**, *45*, 3143-3149.
148. Pommier, Y.; Kerrigan, D.; Covey, J. M.; Kao-Shan, C. S. Whang-Peng, J. *Cancer Res.* **1988**, *48*, 512-516.
149. Berger, N. A.; Chatterjee, S.; Schmotzer, J. A. Helms, S. R. *Proc. Natl. Acad. Sci. U.S.A.* **1991**, *88*, 8740-8743.
150. Masurekar, M.; Kreuzer, K. N. Ripley, L. S. *Genetics* **1991**, *127*, 453-462.
151. Han, Y.-H.; Austin, M. J. F.; Pommier, Y. Povirk, L. F. *J. Mol. Biol.* **1993**, *229*, 52-66.
152. Negrini, M.; Felix, C. A.; Martin, C.; Lange, B. J.; Nakamura, T.; Canaani, E. Croce, C. M. *Cancer Res.* **1993**, *53*, 4489-4492.
153. Albain, K. S.; Le Beau, M. M.; Ullirsch, R. Schumacher, H. *Genes Chromosom. Cancer* **1990**, *2*, 53-58.
154. Smith, M. A.; Rubinstein, L.; Cazenave, L.; Ungerleider, R. S.; Maurer, H. M.; Heyn, R.; Khan, F. M. Gehan, E. *J. Natl. Cancer Inst.* **1993**, *85*, 554-558.
155. Ratain, M. J. Rowley, J. D. *Ann. Oncol.* **1992**, *3*, 107-111.
156. Nichols, C. R.; Breeden, E. S.; Loehrer, P. J.; Williams, S. D. Einhorn, L. H. *J. Natl. Cancer Inst.* **1993**, *85*, 36-40.
157. Nitiss, J. Wang, J. C. *Proc. Natl. Acad. Sci. U.S.A* **1988**, *85*, 7501-7505.
158. Squires, S.; Ryan, A. J.; Strutt, H. L. Johnson, R. T. *Cancer Res.* **1993**, *53*, 2012-2019.
159. Smith, P. J.; Makinson, T. A. Watson, J. V. *Int. J. Radiat. Biol.* **1989**, *55*, 217-231.
160. Tsao, Y.-P.; D'Arpa, P. Liu, L. F. *Cancer Res.* **1992**, *52*, 1823-1829.
161. Lock, R. B. Ross, W. E. *Cancer Res.* **1990**, *50*, 3761-3766.
162. Lock, R. B. Ross, W. E. *Cancer Res.* **1990**, *50*, 3767-3771.
163. Lane, D. P. *Nature* **1993**, *362*, 786-787.
164. O'Connor, P. M. Kohn, K. W. *Sem. Cancer Biol.* **1992**, *3*, 409-416.
165. Pommier, Y.; Runger, T. M.; Kerrigan, D. Kraemer, K. H. *Mutat. Res.* **1991**, *254*, 185-190.
166. Solary, E.; Bertrand, R.; Jenkins, J. Pommier, Y. *Exp. Cell Res.* **1993**, *203*, 495-498.
167. Solary, E.; Bertrand, R.; Kohn, K. W. Pommier, Y. *Blood* **1993**, *81*, 1359-1368.

168. Bertrand, R.; Sarang, M.; Jenkin, J.; Kerrigan, D. Pommier, Y. *Cancer Res.* **1991**, *51*, 6280-6285.

169. Bertrand, R.; Solary, E.; Jenkins, J. Pommier, Y. *Exp. Cell Res.* **1993**, *207*, 388-397.

170. Bertrand, R.; Solary, E.; Kohn, K. W. Pommier, Y. *Proc. Am. Assoc. Cancer Res.* **1993**, *34*, 292.

171. Pommier, Y.; Bertrand, R. Solary, E. *Leukemia & Lymphoma* **1994**, *in press,*

172. Clarke, A. R.; Purdie, C. A.; Harrison, D. J.; Morris, R. G.; Bird, C. C.; Hooper, M. L. Wyllie, A. H. *Nature* **1993**, *362*, 849-852.

173. Kamesaki, S.; Kamesaki, H.; Jorgensen, T. J.; Tanizawa, A.; Pommier, Y. Cossman, J. *Cancer Res.* **1993**, *53*, 4251-4256.

174. Tamura, H.-O.; Kohchi, C.; Yamada, R.; Ikeda, T.; Koiwa, O.; Patterson, E.; Keene, J. D.; Okada, K.; Kjeldsen, E.; Nishikawa, K. Andoh, T. *Nucleic Acids Res.* **1991**, *19*, 69-75.

175. Tanizawa, A.; Tabuchi, A.; Bertrand, R. Pommier, Y. *J. Biol. Chem.* **1993**, *268*, 25463-25468.

176. Sugimoto, Y.; Tsukahara, S.; Oh-hara, T.; Liu, L. F. Tsuruo, T. *Cancer Res.* **1990**, *50*, 7962-7965.

177. Gupta, R. S.; Gupta, R.; Eng, B.; Lock, R. B.; Ross, W. E.; Hertzberg, R. P.; Caranfa, M. J. Johnson, R. K. *Cancer Res.* **1988**, *48*, 6404-6410.

178. Bertrand, R.; O'Connor, P.; Kerrigan, D. Pommier, Y. *Eur. J. Cancer.* **1992**, *28A*, 743-748.

179. Liu, S. Y.; Hwang, B. D.; Liu, Z. C. Cheng, Y. C. *Cancer Res.* **1989**, *49*, 1366-1370.

180. Ishii, K.; Katase, A.; Andoh, T. Seno, N. *Biochem. Biophys. Res. Commun.* **1987**, *104*, 541-547.

181. Berry, D. E.; MacKenzie, L.; Shultis, E. A. Chan, J. A. *J. Org. Chem.* **1992**, *57*, 420-422.

182. Hecht, S. M.; Berry, D. E.; MacKenzie, L. J.; Busby, R. W. Nasuti, C. A. *J. Nat. Prod.* **1992**, *55*, 401-413.

183. Li, C. J.; Averboukh, L. Pardee, A. B. *J. Biol. Chem.* **1993**, *268*, 22463-22468.

184. Oda, T.; Sato, Y.; Kodama, M. Kaneko, M. *Biol. Pharm. Bull.* **1993**, *16*, 708-710.

185. Chen, A. Y.; Yu, C.; Bodley, A.; Peng, L. F. Liu, L. F. *Cancer Res.* **1993**, *53*, 1332-1337.

186. Mortensen, U. H.; Stevnsner, T.; Krogh, S.; Olesen, K.; Westergaard, O. Bonven, B. J. *Nucleic Acids Res.* **1990**, *18*, 1983-1989.

187. Hsiang, Y.-H.; Jiang, J. B. Liu, L. F. *Mol. Pharmacol.* **1989**, *36*, 371-376.

188. McCullough, J. E.; Muller, M. T.; Howells, A. J.; Maxwell, A.; O'Sullivan, J.; Summerhill, R. S.; Parker, W. L.; Wells, S.; Bonner, D. P. Fernandes, P. B. *J. Antibiot.* **1993**, *46*, 526-530.

189. Wentland, M. P.; Lesher, G. Y.; Reuman, M.; Gruett, M. D.; Singh, M. D.; Aldous, S. C.; Dorff, P. H.; Rake, J. B. Coughlin, S. A. *J. Med. Chem.* **1993**, *36*, 2801-2809.

190. Hinds, M.; Deisseroth, K.; Mayes, J.; Altschuler, E.; Jansen, R.; Ledley, F. D. Zwelling, L. A. *Cancer Res.* **1991**, *51*, 4729-4731.

191. Lee, M. S.; Wang, J. C. Beran, M. *J. Mol. Biol.* **1992**, *223*, 837-843.

192. Bugg, B. Y.; Danks, M. K.; Beck, W. T. Suttle, D. P. *Proc. Natl. Acad. Sci. U.S.A.* **1991**, *88*, 7654-7658.

193. Danks, M. K.; Warmoth, M. R.; Friche, E.; Granzen, B.; Bugg, B. Y.; Harker, W. G.; Zwelling, L. A.; Futscher, B. W.; Suttle, D. P. Beck, W. T. *Cancer Res.* **1993**, *6*, 1373-1379.

194. Patel, S.; Austin, C. A. Fisher, L. M. *Anticancer Drug. Des.* **1990**, *5*, 149-157.

195. Chan, V. T. W.; Ng, S.-W.; Eder, P. Schnipper, L. E. *J. Biol. Chem.* **1993**, *268*, 2160-2165.

RECEIVED June 3, 1994

Chapter 13

Anthracycline Antihelicase Action
New Mechanism with Implications for Guanosine–Cytidine Intercalation Specificity

Nicholas R. Bachur[1–3], Robin Johnson[2], Fang Yu[2,3], Robert Hickey[2,4], and Linda Malkas[2,3]

[1]Department of Medicine, [2]Program of Oncology, [3]Department of Pharmacology and Experimental Therapeutics, University of Maryland School of Medicine, and [4]Department of Pharmaceutical Sciences, University of Maryland School of Pharmacy, Baltimore, MD 21201

Natural anthracycline antibiotics bind to double stranded DNA (*1*). Several biochemical consequences of the interaction of anthracycline antibiotics and DNA have been described and range from the inhibition of DNA synthesis, RNA synthesis, and DNA repair to the more recently described inhibition of topoisomerase II (topo II). Although mechanisms for the binding of some anthracyclines and DNA are well described and established, the exact biochemical events and specificity for cancer cells that result from the physical binding of the substances to DNA are controversial and not fully understood.

In living cells, double stranded or duplex DNA is the fundamental information-storing structure. In order to replicate the DNA or to transcribe the base sequence information to RNA, the duplex DNA strands must be separated to expose the DNA single strands for biochemical and molecular processing. An important aspect of these mechanisms, central to replication or transcription, lies in the nature of double-stranded DNA. Base paired, single DNA strands bind into double-stranded duplex formation through a thermodynamically favored reaction, so that the double stranded (duplex) form of DNA is the thermodynamically and kinetically favored structure (*2*). Since the duplex form is so stable, substantial amounts of heat or energy must be delivered to dissociate the base paired, hydrogen bonded strands, an in vitro process described as DNA melting. In the cell, however, the duplex DNA structure must be dissociated biochemically, not thermally, into single DNA strands before enzymatic DNA replication, DNA repair, or DNA transcription can occur. Helicases are the class of enzymes responsible for this dissociation of duplex DNA into DNA single strands (equation 1) (*3*). Because of the tight binding and thermodynamic stability of duplex DNA, helicases require the energy of ATP or other nucleotide triphosphates to drive the duplex strands apart into single strands.

One important characteristic of DNA intercalating anthracycline antibiotics is that they bind the base paired strands of DNA together more tightly than normal (*1*). This reversible intercalative binding increases the melting temperature of the duplex DNA higher than the normal melting temperatures, indicating that the DNA strands are bound together with increased stability.

A second characteristic of such anthracyclines is the geometric changes they produce in the duplex DNA structure upon intercalation. These interactive forces of DNA binding anthracyclines distort, deform, kink, elongate, and stiffen the duplex DNA (*1, 4*).

A third characteristic of DNA intercalating anthracyclines is their steric effects on the structural profile of duplex DNA. These drugs project out into the spatial realms of both the narrow and the wide grooves of duplex DNA (*5, 6*). They occupy space in the DNA structure, and they probably obstruct access to the DNA.

Our hypothesis suggests that the duplex DNA-anthracycline complex is a modified substrate for helicases (equation 2). This modified substrate for DNA helicase is more stable, requires more energy to dissociate the DNA into single strands, is distorted, and the bound anthracycline physically blocks regions of the duplex. Because of these observed characteristics of the DNA-anthracycline complex, we propose that the process of separating DNA strands of the DNA-anthracycline complex is hindered (equation 3). This hindrance of the helicase action by anthracyclines may be crucial to their action against cancer cells. Because the anthracycline action may result from the blockade of specific DNA centered processes of DNA replication and RNA transcription, it is critically important to determine the degree of helicase blockade caused by different anthracyclines. The degree of helicase blockade by the anthracyclines may then be correlated with parameters such as their DNA binding constants, their effect on DNA melting, their base-sequence binding specificities, their geometric relationship to regions of the duplex DNA, and other determinants of their pharmacologic actions.

In the helicase-DNA-anthracycline relationships, the term *inhibition* is not used in the classical enzyme-inhibitor context where the inhibitor interacts with the enzyme directly to form a complex. With the DNA modifying anthracyclines, we are determining effects on the helicase reaction caused by the DNA-anthracycline complex, a modified substrate. It is possible that the anthracyclines may interact directly with helicases, but we have not detected a direct anthracycline-helicase interaction.

The specificity of a DNA binding anthracycline against either a DNA or RNA process may depend on the types of helicases principally affected. The blockade of DNA helicases involved in DNA replication would lead specifically to inhibition of DNA replication. Similarly, a blockade of DNA helicase activity for transcription would inhibit the synthesis of new RNA. DNA binding drugs show this type of differential inhibition. Certain anthracycline antibiotics demonstrate equal inhibition of DNA and RNA synthesis (*7, 8*) whereas other anthracyclines show a preference for the inhibition of RNA synthesis (*8*). It is possible that this inhibition specificity occurs at the helicase level even though both types of anthracyclines intercalate into duplex DNA.

Eukaryotic helicases are a rich collection of enzymes that have varying substrate specificities, cofactor specificities, mechanistic differences, and structural differences (3). In view of this diversity, it is quite reasonable to propose that a cell's sensitivity to DNA interactive drugs may reside at the helicase level. Helicases have substrate specificities for DNA-DNA, DNA-RNA, or RNA-RNA duplexes. In addition, these enzymes have characteristics of polarity or directionality, and they move along polynucleotide strands in a specific direction, either 3' to 5' or 5' to 3'. This directional specificity may be reflected mechanistically as a specificity for DNA-drug blockade of helicase action. In the case of cell sensitivity to anthracyclines, we must question whether the helicases in drug-sensitive cell lines differ from those helicases of insensitive cells and whether sensitivity to the DNA binding drugs occurs through the modification of helicase activities. This wide variation in the specificity and directionality of the helicase properties suggests that these enzymes are very plausible sites for manipulation both by the cell and by the anthracyclines for the modulation of DNA activity. It is at this level that the action of anthracyclines may occur and that the development of new agents may be promising. Consequently, we have conducted research into the characteristics of helicase processing of duplex DNA and the structure-activity nature of anthracycline antibiotics for the modulation of the helicase activities.

Recently, several papers have appeared describing the blockade of helicase action by DNA binding drugs. We reported our findings about a series of anthracycline antibiotic analogs and their highly potent blockade of SV40 T antigen and eukaryotic helicases from human (HeLa) and murine (FM3A) cells (9) as well as helicase blockade by other DNA binding substances (10). Others report studies of the antibiotic CC1065 and its analogs and their interference with *Escherichia coli* and T4 phage helicases (11, 12). Another study of *E. coli* helicases shows the anti-helicase characteristics of a series of DNA intercalators including the anthracycline antibiotic nogalamycin (13).

Although anthracycline antibiotics have several well described actions, such as the inhibition of DNA polymerases, RNA polymerases, topoisomerase II, and repair enzymes, these activities have not been satisfactorily associated with the anticancer actions of the anthracycline molecules (14). Stemming from our original descriptions of a ternary complex formed by anthracycline antibiotic, DNA, and DNA polymerase (15, 16), and from the more recent descriptions of protein associated DNA breaks related to topoisomerase II (17), we investigated the ability of anthracyclines to block the DNA-helicase complexes.

In this report, we have extended our previous studies to include additional anthracycline analogs. Our data show that acylation of the amine of daunosamine decreases or eliminates the ability of the anthracycline to block helicase action. In contrast, alkylation of the sugar amino group has much less effect on the anti-helicase action. We also propose an hypothesis for the guanosine-cytidine (G/C) binding preference of anthracyclines specifically and for intercalating antibiotics in general.

Methods

Procedures. We used SV40 large T antigen, a well documented and studied helicase (18), to evaluate our hypothesis that the helicase process is affected by anthracycline antibiotics. We purified SV40 large T antigen according to the method of Simanis and Lane (*19*) as previously described (*9*). To assay the helicase activity, we synthesized a substrate for the enzyme consisting of an M13 carrier, single strand (ss) circular DNA annealed to a complementary 17mer probe primer (*20*). The significance of our selection of this specific 17mer sequence for the helicase substrate will be discussed later. This primer is a synthetic oligodeoxyribonucleotide that hybridizes to the 5' side of the multiple cloning site of M13mp19(+) DNA. We labeled the 5' end of the 17mer probe primer with ^{32}P as a signal atom, annealed the ^{32}P labelled 17mer probe primer to the M13 circular DNA, and used this molecular DNA duplex as a substrate for the helicases. The helicase reaction separates the two DNA species (Figure 1). The resultant DNA single and double strand products were isolated on polyacrylamide gel electrophoresis, visualized by autoradiography, and quantified by densitometry or radioanalysis. Details of our assay system were reported previously (*9*).

The T antigen helicase dissociated our synthetic DNA duplex substrate in a concentration and time dependent manner, and the reaction was readily quantifiable by measuring the amount of 17 base P primer that was dissociated from the DNA duplex pair separated in polyacrylamide gel electrophoresis and measured through the marker ^{32}P end label. By adjusting the enzyme concentration, we developed a standard reaction mixture that gave complete duplex DNA dissociation by 30 min.

Anthracycline antibiotics and analogs of the native anthracyclines were obtained from various sources. Doxorubicin and 4'-epidoxorubicin were provided by Farmitalia (Milan, Italy). Adria Laboratories (Columbus, Ohio) supplied 4-demethoxydaunorubicin. The other anthracycline analogs were supplied by the Drug Synthesis and Chemistry Branch, Developmental Therapeutic Program, National Cancer Institute. The anthracycline antibiotics were preincubated with M13-17mer DNA duplex helicase substrate for up to 2 h to assure completion of the reversible binding (equation 2) prior to the addition of the T antigen helicase and the start of the helicase reaction. The reaction mixture components were added, and the helicase assay was started by the addition of T antigen helicase and assayed as described previously (*9*).

Results and Discussion

SV40 Large T Antigen Helicase. In our appraisal of the anthracycline antibiotics, we selected compounds that had specific modifications of the aglycone part and of the sugar moiety. We tested a range of concentrations for each anthracycline against the standard duplex DNA dissociation reaction.

Starting with anthracyclines that have modifications in the aglycone system, we used daunorubicin as our standard anthracycline and compared structural changes of the anthracycline analogs with their binding effects on double stranded DNA and with their effects on SV40 T antigen helicase action. The first change of the anthraquinone chromophore that we compared was the absence of the 4-methoxy group in the analog 4-demethoxydaunorubicin (Ida-rubicin). This anthracycline, which is quite potent as an anticancer drug, has enhanced cellular uptake because of lipophilicity and develops a higher ∆ Tm

$$\text{duplex DNA} + n\text{NTP} \xrightarrow{\text{helicase}} \begin{array}{c}\text{2 complementary single DNA strands}\\ + n\text{NDP} + n\text{PO}_4\end{array} \qquad (1)$$

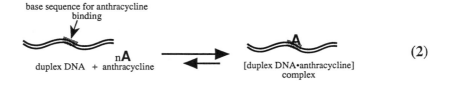

$$\begin{array}{c}\text{base sequence for anthracycline}\\ \text{binding}\end{array}$$

$$\text{duplex DNA} + \overset{n\mathbf{A}}{\text{anthracycline}} \rightleftharpoons \begin{array}{c}\text{[duplex DNA·anthracycline]}\\ \text{complex}\end{array} \qquad (2)$$

$$\text{[duplex DNA·anthracycline]} + n\text{NTP} \xrightarrow{\text{helicase}} \text{modified or no reaction} \qquad (3)$$

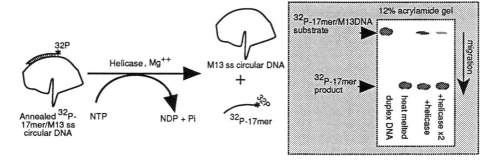

Figure 1.   Helicase assay reaction showing dissociation of duplex DNA substrate to product single strand M13 DNA and ^{32}P-17mer. Diagrammatic polyacrylamide gel electrophoresis depicts separation of reaction products.

with DNA than daunorubicin (*21, 22*). In its blockade of the T antigen helicase, however, 4-demethoxydaunorubicin is a less effective agent than daunorubicin (Figure 2) (Table I).

Comparing doxorubicin and daunorubicin, which have one difference at the C-14 position of ring A, we found that these two agents have identical IC_{50} values for blockade (Table I, Figures 2 & 3) of T antigen helicase. Although there are reported differences between these two compounds in their binding constants to double stranded DNA (*23*), the T antigen helicase showed equally potent blockade by both anthracyclines.

Modifications of the sugar portion of the anthracycline molecule comprise the second major class of analogs. Doxorubicin and 4'-epidoxorubicin provide one interesting comparison of modest modification of doxorubicin (Table I, Figure 3). The configurational inversion of the 4'-hydroxy group yields an analog that has slightly reduced anti-helicase activity. Therefore, the change at the 4' position of this agent suggests that the 4'-hydroxy affects binding to double stranded DNA and anti-helicase action.

Because of the importance of the amino group of daunosamine to anthracycline activity, we selected several 3'-*N*- substituted analogs to evaluate. The first series, *N*-acylated anthracyclines, have low cytotoxic activity and show a decreased binding affinity to double stranded DNA (*14, 25*). We found these compounds to be poor or ineffective blockers of T antigen helicase at concentrations up to 40 μM. This is seen in the series of *N*-formyl, *N*-acetyl, *N*-proprionyl, and *N*-butyryl daunorubicins (Table I, Figure 4). Presumably, the abolition of the basic property of the amino group by acylation decreases the DNA binding affinity of these analogs. However, a second group of *N*-acylated doxorubicin analogs, AD32 (*N*-trifluoroacetyl-14-*O*-valerate-doxorubicin) and AD41 (*N*-trifluoroacetyldoxorubicin), did not block the T antigen helicase at concentrations up to 50 μM (Table I). These compounds are cytotoxic and active as anticancer agents (*14*); but, these compounds do not bind effectively to DNA and do not block T antigen helicase.

The second series of amino substituted analogs, the *N*-alkyl anthracyclines, present quite a contrast to the *N*-acyl congeners. As a group, the *N*-alkyl daunorubicins and doxorubicins are quite active as blockers of T antigen helicase (Table I, Figure 4). This corresponds to their cytotoxic activity *in vitro* and to their excellent DNA binding characteristics (*14, 26*). Because the *N*-alkyl analogs retain the basic amino function, they retain DNA binding properties and the resultant anti-helicase action. The size, shape, and electronic properties of the alkyl groups show effects on their anti-helicase activities. As the alkyl groups enlarge (*N*-benzyl, *N,N*-dibenzyl, *N,N*-didecyl), the analogs lose anti-helicase efficacy. This correlates to the decrease of Δ Tm for these three analogs (Δ Tm = 10.2, 1.4, and 0.1, respectively). As we have shown previously, the correlation of anti-helicase action and Δ Tm for the anthracycline family indicates that other factors are affecting anti-helicase action. Perhaps for a limited series such as the *N*-alkyl series of one fundamental anthracycline, we may find that the correlation to Δ Tm holds as we had predicted previously (*9*).

The *N,N*-cyclic anthracycline derivatives like *N*-morpholinodaunorubicin retains a high anti-helicase activity, similar to its high anticancer action (*24*).

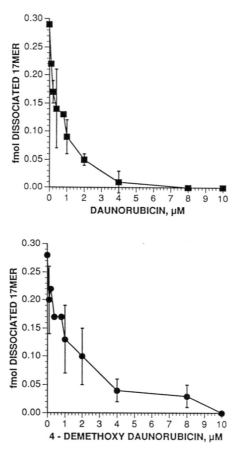

Figure 2.   Effects of daunorubicin and 4-demethoxydaunorubicin on SV40 large T antigen helicase activity.

Table I. **Anthracycline Blockade of SV40 T Antigen Helicase Activity**

Free Amino Anthracyclines	IC_{50} μM
Daunorubicin	0.4 ± 0.09
4-Demethoxydaunorubicin	1.8 ± 0.8
Doxorubicin	0.4 ± 0.2
4'-Epidoxorubicin	2.0 ± 2.0
N-Acyl Anthracyclines	
N-Formyldaunorubicin	>40
N-Acetyldaunorubicin	>40
N-Proprionyldaunorubicin	>40
N-Butyryldaunorubicin	>40
N-Trifluoroacetyldoxorubicin-14-*O*-valerate (AD-32)	>50
N-Trifluoroacetyldoxorubicin (AD-41)	>50
N-alkylated anthracyclines	
N-Propyldaunorubicin	1.8 ± 0.35
N,N-Dimethyldaunorubicin	0.9 ± 0.1
N,N-Diethyldaunorubicin	1.8 ± 0.4
N,N-Diethyldoxorubicin	1.5 ± 0.1
N,-Benzyldaunorubicin	1.5 ± 0.8
N,N-Dibenzyldaunorubicin	4.0 ± 3.0
N,N-Didecyldaunorubicin	15.5 ± 3.5
N-Morpholinodaunorubicin	0.80 ± 0.3
N,N,N-Trimethyl (N^+ Cl^-) daunorubicin	0.85 ± 0.2
Unclassified Anthracyclines	
Aclacinomycin	4.0 ± 2.8
Nogalamycin	0.2 ± 0.14
7-*O*-Methylnogarol	6.0 ± 3.0

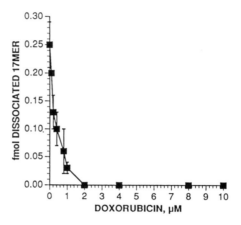

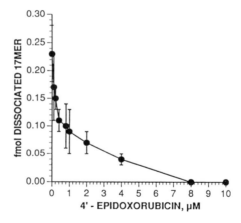

Figure 3.   Effects of doxorubicin and 4'-epidoxorubicin on SV40 large T
            antigen helicase.

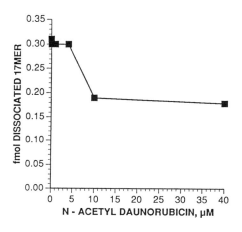

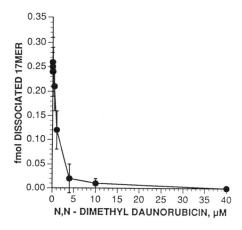

Figure 4.   Effects of *N*-acetyldaunorubicin and *N,N*-dimethyldaunorubicin on SV40 large T antigen helicase.

Surprisingly, the quaternary amine compound, N,N,N-trimethyl (N^+ Cl^-) daunorubicin also is very active, suggesting that the permanent charge of the amino nitrogen retains anti-helicase action.

We evaluated other anthracyclines that have complex structural modifications for their anti-helicase activities. The first anthracycline, aclacinomycin, which has numerous structural differences from our standard daunorubicin is considered to be a much more active inhibitor of RNA synthesis, especially of the nucleolar RNA synthesis (27). Although aclacinomycin had an IC_{50} of 4.0 μM for T antigen helicase, it is not as potent as daunorubicin or doxorubicin.

The second group of complex modified anthracyclines we studied included nogalamycin and 7-O-methylnogarol. Nogalamycin is the most active helicase blocker with an IC_{50} of 0.2 μM whereas 7-O-methylnogarol blocks with an IC_{50} of 6 μM. 7-O-Methylnogarol was previously not considered to intercalate into DNA; but recent studies indicated that the compound binds to double stranded DNA (28). Our studies support DNA binding since 7-O-methylnogarol is a good blocker of T antigen helicase. Both nogalamycin and 7-O-methylnogarol have the bridge ring sugar on the D ring of the chromophore. With nogalamycin, this bridge ring sugar and part of the D ring intercalate and protrude into the wide groove of duplex DNA (6). This structure complex may add an obstruction to helicase, which other compounds such as the 4-demethoxydaunorubicin do not possess.

Because the T antigen helicase is a virally induced enzyme, because other endogenous modulators that may affect the helicase reaction are not tested in our system, and because of complex cellular pharmacodynamics, attempts to correlate the cytotoxic effect of these anthracycline agents with their IC_{50} values on T antigen helicase would be seriously questionable. For example, one comparison of anthracycline inhibition of L1210 cell growth gives μM-ID_{50} values of 0.11, 0.31, and 0.48, respectively, to doxorubicin, 7-O-methylnogarol and nogalamycin (29). These ID_{50} values do not correspond directly with our T antigen helicase μM IC_{50} values of 0.4, 6.0, and 0.2 for the same three anthracyclines (Table I). These differences can readily be attributed to the reasons stated above. To assess helicase effects relative to drug cytotoxicity, we must be able to evaluate the relevant eukaryotic helicases and their interactions with duplex DNA-drug complexes.

The utility of our model helicase assay system is in providing important information describing the dynamic effects of DNA binding substances on the biomolecular processing of DNA. With this model system, we can obtain an understanding of the molecular and physical-chemical interactions of DNA binding substances and duplex DNA. Through the use of molecular modeling structure-activity relationships and complementary research data regarding antibiotic-DNA interactions, we will develop clearer pictures of how these anticancer drugs affect DNA biochemistry. However, to understand the differential effects of these anticancer drugs, we must investigate the helicases of the target cells for more accurate models of these mechanisms.

Eukaryotic Helicases From HeLa and FM3A Cells. Since the SV40 T antigen is a virally induced helicase, we felt it would be important to evaluate constitutive helicases from a human malignant cell line to determine if the action of DNA binding drugs extends to a human helicase. Following a published procedure, we purified a helicase activity from HeLa cells (*30*). The helicase assay system, the helicase substrate, and the inhibitor assessment procedure were described previously (*9*). For another eukaryotic helicase analysis, we purified a murine helicase from FM3A cells and assayed it for activity as described for T antigen. Both partially purified helicases are active in our system with our 17mer-M13 substrate.

Both eukaryotic helicases are blocked by doxorubicin in our standard assay and show concentration-dependent blockade (Figure 5). The HeLa enzyme with an estimated doxorubicin IC_{50} of 4×10^7 M is more sensitive to doxorubicin than the murine FM3A helicase, which has a doxorubicin IC_{50} of 9×10^7 M. These different sensitivities of eukaryotic helicases to a DNA binding drug agree with observations made on prokaryotic helicases (*11-13*). The prokaryotic helicases show very wide differences in sensitivities to DNA binding agents.

Characteristics of Anti-helicase Action of Anthracyclines. The activities, specificities, and other characteristics of helicases vary (*3*). It is noteworthy that the rates of unwinding and estimates of activities for the three helicases we have examined suggest different kinetic characteristics for these enzymes as well as different sensitivities to intercalating drugs (*9*). Such differences could contribute to different sensitivities to DNA binding drugs. Our findings of the different sensitivities of eukaryotic and viral helicases to drugs are in accord with recent observations concerning drug inhibition of prokaryotic helicases (*11-13*).

In addition to increasing stability, anthracycline intercalation also increases DNA helix rigidity and deforms, lengthens, and unwinds the DNA helix (*1*). Blockade of helicase activity could result from any of these effects or a combination of them. Other factors such as base sequence specificity, the duplex DNA base to drug ratio, and structural characteristics of the intercalating agents themselves must also be considered. Since the structural differences among the anthracycline analogs are varied and involve modifications and substitutions of the chromophore system as well as sugar modifications, we feel that a thorough examination of the kinetics of this inhibitory action with regard to structure-activity relationships is necessary.

Daunorubicin binding to duplex DNA favors G/C regions and particularly the triplets 5' A/T CG or 5' A/T GC (*31, 32*). Our selected 17mer-M13 duplex DNA substrate contains one AGC triplet region, which we presume is the preferential site for daunorubicin binding, if not for all the intercalating anthracyclines (Figure 6). A second triplet region, TGG, may also provide preferential binding for anthracycline intercalation, based on known preferences of intercalating antibiotics for C/G rich regions. We expect that the base sequence of our 17mer-M13 substrate affects the binding of the different anthracyclines and their anti-helicase actions, and we are conducting studies aimed at resolving these questions.

We also do not know the effect of binding multiple drug molecules into the 17mer-M13. The anti-helicase action at the lowest drug concentrations

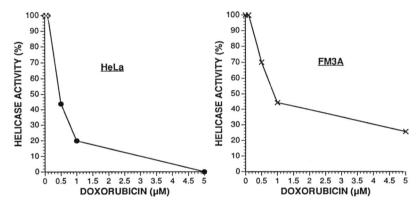

Figure 5.   Anti-helicase activity of doxorubicin on human (HeLa) and murine (FM3A) helicases.

Figure 6.   Substrate 17mer base sequence.

probably involves a single drug molecule, but the anti-helicase action of a drug may be much greater for multiple intercalated drugs, especially if they are in proximity. For example, cooperative DNA binding of anthracyclines may also participate in this action (*33*). When an anthracycline molecule intercalates into duplex DNA, not only does the drug alter the shape and topology of the DNA but the structural forces of the DNA modifies the drug shape (*5*). This mutual topologic distortion must be evaluated to understand this helicase blockade.

Several factors must be considered in our analysis of the anti-helicase drug action. We have already considered the Δ Tm effects of drug on DNA, and we have not found this factor to correlate well with the anti-helicase action of widely different groups of anthracyclines (*9*). We must also consider other factors such as the DNA-drug binding constant, DNA-drug residency, effect of drug on DNA helicity (unwinding angle), and steric factors of the drug blocking the wide and narrow groove. Most anthracyclines have A-ring sugars which protrude into the narrow groove which may interfere with helicase movement along the DNA chain. Similarly, some anthracyclines have sugars attached to the D ring which may block access to the wide groove of duplex DNA. These anthracycline analogs could be useful in the study of the mechanism of helicase action.

Comparing any of the similar anthracyclines, daunorubicin with 4-demethoxydaunorubicin, doxorubicin with 4'-epidoxorubicin, or even doxorubicin with daunorubicin, we can see that evaluating cytotoxicity is a complex issue, which we doubt will be directly relatable to blockade of SV40 T antigen helicase action. Since cytotoxicity depends on the cellular pharmacodynamics and cellular pharmacokinetics of these agents, as well as the DNA binding capacity and lipophilicity of the anthracyclines (*25*), it is unlikely that this comparison with the T antigen helicase will relate directly. Nevertheless, our model system of the T antigen helicase provides us with a new tool and options for gaining insight into important characteristics of drugs that bind to the double stranded DNA and for determining what effects occur due to this binding in a biochemical and molecular assessment. Other characteristics of the helicase blockade by anthracylines are being evaluated for a more comprehensive evaluation of structure-activity relationship.

Binding in the DNA narrow groove is a major component of the anthracycline DNA interaction since daunosamine, nogalose, or other sugars attached to the aglycone A ring are considered to lie in the narrow groove and affect drug binding to DNA. This narrow groove binding component must be of significant specificity and importance since *N*-acylation of daunosamine decreases binding affinity of these analogs and eliminates the anti-helicase activity of these *N*-acylated compounds. With *N*-alkylation and the retention of the basicity of the amino group, DNA binding appears less affected and the *N*-alkylated analogs retain potent anti-helicase activity. As the size and lipophilic character of the *N*-alkyl groups increases, the helicase blockade decreases.

General Hypothesis Relating Intercalation Base Pair Specificity and Anti-helicase Action. A correlation appears to exist between the G/C specific DNA binding preference of intercalating anthracycline antibiotics and the anti-helicase action of these antibiotics. Anthracycline antibiotics, actinomycins, echinomycin, and elsamicin, show preferential, if not specific, binding to G/C base paired sites of double stranded DNA. We have shown that these same antibiotics display anti-helicase activities (*10, 34*). In the enzymatic helicase process, this enzyme must dissociate electronically complemented A-T bonds and G-C bonds. G-C base pairs are bound by -16.79 kcal total binding energy, whereas A-T base pairs are bound by -7.00 kcal total binding energy (*35*). As shown in DNA melting, the G-C bonds are the most stable and the most difficult to dissociate. Therefore, we assume that the maximal catalytic power needed by any DNA helicase is the power needed to separate G-C base pairing. DNA helicases would have evolved to accomplish G-C separation as the maximal catalytic power needed, and G-C pairs probably are the limit for the dissociative enzymatic power of helicase. If the base paired or strand paired binding energy were greater than the normal G-C binding energy, then helicase might not be able to overcome this increased binding energy to separate the strands. When an anthracycline or other antibiotic molecule is intercalated next to a G-C position, the increase in base pair binding energy imparted by the intercalated molecule may exceed the maximal catalytic power of helicase. The anthracycline modified duplex DNA is then no longer a suitable substrate, and helicase blockade occurs. This correlation of inherently high G-C binding energy and the binding specificity of intercalating anthracyclines to the G-C base pair may be a basic mechanistic characteristic of the intercalating anthracyclines.

It is tempting to propose that the evolutionary selection of DNA intercalating antibiotics for G-C binding may have been partially determined by this relationship to the highest total binding energy. If this is so, the anti-helicase action of these antibiotics would be one of their most fundamental actions.

In evaluating the possible mechanisms that have been described for anthracycline antibiotics, it is only natural to compare these mechanisms according to experimental observations. Both topoisomerases and DNA helicases are affected by anthracycline binding to DNA, but the characteristics of these interactions differ substantially. Theoretically, helicases may be a more sensitive target for DNA-binding anthracyclines than topoisomerases. According to experimental data (*17*), topoisomerases bind to DNA at a site and catalyze topological interchange of duplex DNA strands. Accordingly, if an anthracycline molecule is bound to duplex DNA at a site X, the topoisomerase may attach to a different site Y, and not come in contact with the anthracycline. Then, topoisomerase can perform its topological interchange unhindered. Helicases, in contrast, bind to DNA and move along a DNA strand according to their enzyme's processivity characteristics (*3*). The helicases by virtue of their enzymatic mechanism must move through or by every base pair site of the duplex DNA (Figure 7). Because helicases move past every base pair site of the duplex DNA strands, helicases must encounter any anthracycline bound to the DNA at any site and should be affected by the anthracycline-duplex DNA complex. Helicases, therefore, should be susceptible to every anthracycline molecule bound to DNA.

Another significant difference between the action of anthracyclines on helicases and on topoisomerase II, is the effect of the antibiotics at different concentrations. Although topoisomerase II forms "cleavable complexes" (*17*) with anthracyclines at lower drug concentrations, at higher anthracycline concentrations no "cleavable complexes" or DNA strand breaks occur (*36*). Helicase blockade by anthracyclines is quite different. First, helicases are very sensitive to submicromolar anthracycline concentration. Second, as the anthracycline concentration is increased, the anti-helicase action increases. There is no decrease of helicase blockade at high anthracycline concentrations. In our observations, we found that the helicases are directly and stoichiometrically sensitive to anthracycline binding to the DNA substrate.

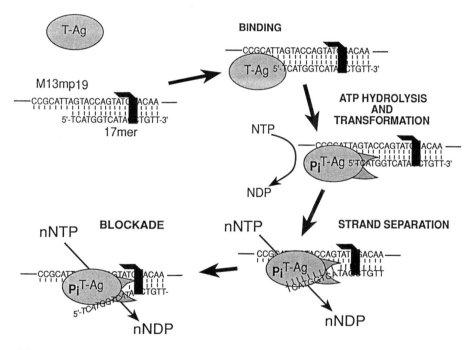

Figure 7.   Anti-helicase reaction model.   The free T-antigen helicase approaches duplex DNA containing intercalated anthracycline antibiotic.   The helicase binds to the single strand DNA. Reacting with nucleotide triphosphate, the helicase is transformed and proceeds along the DNA strand 3' to 5'.   The helicase continues consuming nucleotide triphosphate and separates the ds DNA to single strands. When the helicase comes in proximity to the intercalated anthracycline, the helicase procession is blocked.

We stress that our T antigen helicase system is an excellent model for examining drug structure-activity relationships and for determining drug-DNA interactions. However, in order to relate anti-helicase action to anticancer action and to evaluate anticancer activities to anthracycline structure, cancer cell helicases must be tested.

Acknowledgments

The authors appreciate the comments and information provided by Dr. Edward M. Acton and Dr. J.B. Chaires. This research was supported by the University of Maryland DRIF and Bressler Awards and American Cancer Society Maryland Division Awards.

Literature Cited

1. DiMarco, A.; Arcamone, F. *Arzneim. Forsch.* **1975**, 25, 368-375.
2. Reisner, D.; Römer (1973) *Thermodynamics and kinetics of conformational transitions in oligonucleotides and tRNA.* Academic Press, New York, **1973**, Vol. 1, pp. 455-484.
3. Matson, J.H.; Kaiser-Rogers, K.A. *Ann. Rev. Biochem.* **1990**, 59, 289.
4. Waring, M.J. *Ann. Rev. Biochem.* **1981**, 50, 159-192.
5. Gabbay, E.J.; Grier, D.; Fingerle, R.E.; Reimer, R.; Levy, R.; Pierce, S.W.; Wilson, W.P. *Biochemistry* **1976**, 15, 2062-2070.
6. Egli, M.; Williams, L.D.; Frederick, C.A.; Rich, A. *Biochemistry* **1991**, 30, 1364-1372.
7. Meriwether, D.W; Bachur, N.R. *Cancer Res.* **1972**, 32, 1137-1142.
8. DuVernay, V.H.; Essery, J.M.; Doyle, T.W.; Bradner, W.T.; Crooke, S.T. *Mol. Pharmacol.* **1979**, 15, 341-356.
9. Bachur, N.R.; Yu, F.; Johnson, R.; Hickey, R.; Wu, Y.; Malkas, L. *Mol. Pharmacol.* **1992**, 41, 993-998.
10. Bachur, N.R.; Johnson, R., Yu, F.; Hickey, R.; Applegren, N.; Malkas, L. *Mol. Pharmacol.* **1993**, 44, 1064-1069.
11. Maine, I.P.; Sun, D.; Hurley, L.H.; Kodadek, T. *Biochem.* **1992**, 31, 3968-3975.
12. Sun, D.; Hurley, L.H. *J. Med. Chem.* **1992**, 35, 1773-1782.
13. George, J.W.; Ghate, S.; Matson, S.W.; Besterman, J.M. *J. Biol. Chem.* **1992**, 267, 10683-10689.
14. Bodley, A.; Lin, L.F.; Israel, M.; Ramakrishman, S.; Yoshihiro, K.; Guiliani, F.C.; Kirchenbaum, S.; Silber, R.; Potmesil, M. *Cancer Res.* **1989**, 49, 5969-5978.
15. Goodman, M.F.; Bessman, M.J.; Bachur, N.R. *Proc. Natl. Acad. Sci. U.S.A.* **1974**, 71, 1193-1196.
16. Goodman, M.F.; Lee, G.M.; Bachur, N.R. *J. Biol. Chem.* **1977**, 252, 2670-2674.
17. Liu, L.F. *Ann. Rev. Biochem.* **1989**, 58, 351-375.
18. Stahl, H.; Droge, P.; Knippers, R. *EMBO J.* **1986**, 5; 1939-1944.

19. Simanis, V.; Lane, D.P. *Virology* **1985**, 144, 88-100.
20. Tuteja, N.; Tuteja, R.; Rahman, K.; Kang, L.; Falaschi, A. *Nucleic Acids Res.* **1991**, 18, 6785-6792.
21. Zunino, F.; Gambetta, R.; DiMarco, A.; Luoni, G.; Zaccara A. *Biochem. Biophys. Res. Comm.* **1976**, 69, 744-750.
22. Supino, R.; Necco, A.; Dosdia, T.; Casazza, A.M.; DeMarco, A. *Cancer Res.* **1977**, 37, 4523-4528.
23. Arcamone, F. (1981) In *Doxorubicin, Anticancer Antibiotics*; Stevens, G., Ed.; Medicinal Chemistry Series; Academic Press, New York, 1981, Vol. 17.
24. Naff, M.B.; Plowman, J.; Narayanan, V.L. In *Anthracyclines in the National Cancer Institute Program. Anthracycline Antibiotics*; Khadem, E.L., Ed., Academic Press, NY, 1982, pp. 1-57.
25. Hoffman, D.; Berscheid, H.G.; Böttger, D.; Hermentin, P.; Sellacek, H.H.; Kraemer, H.P. *J. Med. Chem.* **1990**, 33, 166-171.
26. Acton, E.M. In *N-Alkylation of Anthracyclines. Anthracyclines: Current Status and New Developments*; Crook S.T.; Reich, S.D., Eds. Academic Press, NY, 1980, pp 15-24.
27. Crooke, S.T.; DuVernay, V.H.; Galvan, L.; Prestayko, A.W. *Mol. Pharmacol.* **1978**, 14, 290-298.
28. Wierzba, K.; Sugimoto, Y.; Matsuo, K.; Toko, T.; Takeda, S.; Yamada, Y.; Tsukagoshi, S. Jpn J. *Cancer Res.* **1990**, 81, 842-849.
29. Bhuyan, B.K.; Neil, G.L.; Li, L.H.; McGovren, J.P.; Wiley, P.F. In *Chemistry and biological activity of 7-con-O-methylnogarol (7-con-OMEN). Anthracyclines: Current Status and New Developments* Crooke, S.T.; Reich, S.D., Eds., Academic Press, NY, 1980, pp. 365-395.
30. Malkas, L.; Hickey, R.; Li, C.; Pederson, N.; Baril, E.F. *Biochemistry* **1990**, 29, 6263-6274.
31. Chaires, J.B.; Fox, K.R.; Herrera, J.E.; Britt, M.; Waring, M.J. *Biochemistry* **1987**, 26, 8227-8236.
32. Chaires, J.B.; Herrera, J.E.; Waring, M.J. *Biochemistry* **1990**, 29, 6145-6153.
33. Graves, D.E.; Krugh, T.R. *Biochemistry* **1983**, 22, 3941-3947.
34. Bachur, N.R.; Johnson, R.; Yu, F.; Hickey, R.; Malkas, L. *Proc. Amer. Assoc. Cancer Res.*, **1993**, abst. 2097, p. 352.
35. Kudritshaya, Z.G.; Danilov, V.I. *J. Theor. Bio.* **1976**, 59, 303-318.
36. Wasserman, K.; Markovets, J.; Jaxel, C.; Capranicio, G.; Kohn, K.W.; Pommier, Y. *Mol. Pharmacol.* **1990**, 38, 38-45.

RECEIVED June 3, 1994

Chapter 14

Membrane Biophysical Parameters Influencing Anthracycline Action

Ratna Mehta and Thomas G. Burke[1]

Division of Pharmaceutics, College of Pharmacy, and The Comprehensive
Cancer Center, The Ohio State University, Columbus, OH 43210–1291

Membrane interactions of the anthracyclines are thought to play important roles in the three major clinical findings concerning these agents, namely antitumor activity, cardiotoxicity, and multidrug resistance (MDR). The anthracycline drugs are thought to gain access to their intracellular targets in cancer cells by a passive diffusion mechanism. Drug binding to the lipid bilayer domains of the plasma membrane precedes the transport event, and anthracycline interactions with lipid bilayers are known to be very sensitive to modifications in both drug structure as well as lipid composition of the membrane. In cardiotoxicity, positively charged doxorubicin's affinity for negatively charged cardiolipin, a lipid abundant in heart tissue, is thought to be involved in drug localization in the heart tissue. Through modification in doxorubicin's structure, its selective binding for negatively charged lipid has now been attenuated, with a concomitant reduction in cardiotoxic potential. An alternative approach to circumventing doxorubicin cardiotoxicity is liposomal delivery of the medication, where the biodistribution of the drug is altered. In MDR, a plasma membrane-based drug efflux pump known as P-glycoprotein (PGP) is thought to play a role in reducing drug levels in resistant cells to sublethal levels. PGP is thought to be capable of removing drugs directly from the lipid bilayer domains of the plasma membranes of MDR cells. Interestingly, certain simple modifications to doxorubicin's structure, modifications which alter lipid bilayer interactions of the drug but retain potency at the topoisomerase II target, have resulted in agents displaying high levels of retention and activity in MDR cells. The central role of membrane interactions in the three major clinical findings concerning the anthracyclines provides strong justification for pursuing a more complete and thorough elucidation of the membrane biophysical parameters of this important class of anticancer drugs.

Antitumor Activity

The anthracycline compounds used in cancer chemotherapy are antibiotics isolated from the Streptomyces species (1). The most prominent of these broad spectrum

---

[1]Corresponding author

anti-cancer agents is doxorubicin, which is widely employed in treating several solid tumors including carcinomas and soft tissue sarcomas. The close structural analogue daunomycin is found to be more effective in the treatment of various leukemias. These compounds consist of a more or less planar (2) anthraquinone nucleus attached to an aminosugar via a glycosidic linkage. The planar aglycone portion of these drugs effectively intercalates between the DNA base pairs, and it appears that the aminosugar imparts stability to this binding through its interaction with the sugar phosphate backbone of DNA (1). The primary mechanism of antitumor activity of these drugs in rapidly proliferating tumor cells is considered to be interference (2-5) with DNA topoisomerase II activity. This enzyme is critical to the DNA replication process (6). Interference with topoisomerase II results in the inhibition of mitotic activity, cell growth, and proliferation. It has also been demonstrated that the anthracyclines can damage cancer cells through the generation of free radicals upon interaction with membrane-based redox enzymes (7,8). The free radicals generated cause lipid peroxidation and other deleterious effects leading to membrane lesions and leakage of cytoplasmic contents. Some research groups have indicated that this process may be a possible mechanism by which doxorubicin exerts its antitumor activity in human breast cancer cells in vitro (9,10). Also, polymer-immobilized doxorubicin has been shown to be cytotoxic without entering cells (11,12), suggesting that interactions of the drug at the membrane level can prove to be lethal. Yet another school of thought has indicated that anthracycline-induced cytotoxicity may possibly be caused by interference of the drug with the cellular signal transduction system (13-15).

Thus, anthracyclines may kill tumor cells by either one or more mechanisms depending upon the biochemistry of the cells. In the case of intracellular targets such as the DNA-topoisomerase II mediated mechanism, membrane interactions are necessary for the drug to reach the cell nucleus. It is generally accepted that intracellular uptake of these compounds occurs predominantly via passive transport (16-20), and no indication exists so far of a specific carrier mechanism for the uptake of these drugs. The rate and extent of uptake and intracellular retention of anthracyclines is dependent upon a number of factors (18-20) including the polarity of the compound, the pK_a of ionizable functional groups in the molecule and the pH of the cell, the extent of binding of the drug to cellular constituents, and temperature. Doxorubicin, daunomycin, and their several analogues bearing the aminosugar exhibit a pK_a of 7.6 - 8.2 depending upon the ionic strength (18,21). Thus, at physiological pH, these compounds exist as both the charged and uncharged species. It is predominantly the uncharged form (18-20,22) that is able to penetrate the membrane barrier via passive diffusion, the driving force being the activity gradient created by the difference in extra- and intracellular concentrations of the unionized free drug. Once inside the cell, depending upon the intracellular pH, some of the drug is thought to convert to the ionized form, which can bind significantly to various macromolecular components, reducing the intracellular amounts of the drug in its free form (16,17,23). Hence, the rate and extent of the drug's transport is considered to be influenced by the ability of the anthracycline analogues to interact with and traverse the lipid bilayer domains of cell membrane. It is therefore of interest to understand in full anthracycline interactions with the phospholipid components of biological membranes.

Towards this goal, exploitation of the intrinsic fluorescence properties of the anthracyclines has allowed for the structural basis of drug binding to lipid bilayers to be characterized in some detail. Fluorescence anisotropy titration has been employed as a sensitive technique to determine the equilibrium binding affinities of several anthracyclines for sonicated phospholipid model membranes (25-27) and biological membranes (28). The overall association constants (K) of various anthracyclines for electroneutral phospholipid vesicles like dimyristoylphosphatidylcholine (DMPC) under near physiological conditions have been found to range from about 200 M^{-1} for doxorubicin (26) to about 70,000 M^{-1} for the new experimental compound 3'-O-benzyldoxorubicin-14-valerate (29). This greater than 300-fold difference in binding affinities achieved through analogue development is impressive. Undoubtedly the differential membrane interactions of highly lipophilic anthracycline agents impact on cellular accumulation, subcellular localization, and the biological activities of these agents relative to doxorubicin (29).

Some understanding of the influence of the structural features in the aglycon and the aminosugar residues of the anthracyclines has been made possible through systematic modifications in the drug molecule. Binding affinities for neutral membranes have been compared for a series of doxorubicin and daunorubicin analogues bearing modifications only in the aglycon portion of the molecule, leaving the aminosugar portion unchanged. Substitutions that increase the hydrophobicity of the anthracyclines have led to higher binding affinities for both fluid-phase and solid-phase neutral bilayers. Within both the doxorubicin and daunomycin series, binding affinities for lipid bilayers have been shown to correlate reasonably well with drug uptake in various cell lines. In general, additional correlations between membrane binding, cellular uptake, and cytotoxic potency were also observed (see Table I).

Although the aminosugar is not directly implicated in anthracycline cytotoxicity, it does seem to play an important role in cellular transport of these drugs and their ability to bind to various macromolecular sites. But unlike aglycon modifications, substitution of the primary amine in the sugar residue of the compound affects not only the ability of the compound to interact with the cellular membranes, but also changes the pK_a of the amine which in turn influences the degree of ionization of the drug at physiological pH (30). The altered pK_a of the drug is also thought to affect many other pharmacological considerations such as the rate and extent of uptake into the cells as well as its ability to bind to various intracellular macromolecular sites including DNA-binding sites, thereby altering the capability of the drug to induce cellular damage. Thus, the aminosugar modification impacts not only on membrane binding but other factors as well. As a result, the effect of aminosugar modification on membrane binding and cellular uptake processes has been difficult to correlate and predict.

In general, it appears that for the anthracyclines to exert their favorable antitumor activities, these drugs must interact with and traverse cellular membranes and accumulate in intracellular target sites such as the nucleus in sufficient quantities to elicit cytotoxic effects. Certain simple modifications in the aglycon region of doxorubicin and daunorubicin can be employed to promote drug uptake and, in turn, enhance antitumor activity, although in each case the effect of drug modification on host toxicity needs to be considered. While consequences of aglycone modification

TABLE I: **Relationship Between Membrane Affinity, Cellular Uptake, and Cytotoxicity for Aglycone-Modified Anthracyclines**

Anthracyclines in the order of decreasing hydrophobicity and decreasing affinity for fluid-phase bilayers (*25*).

1.	Carminomycin	5.	5-Iminodoxorubicin
2.	4-Demethoxydaunomycin	6.	Doxorubicin
3.	5-Iminodaunomycin	7.	Daunomycinol
4.	Daunomycin	8.	Doxorubicinol

Cell Line	Drug Uptake[a]	Reference
L1210	4 > 6 , 7 > 8	(*31*)
L1210	1 > 4 > 6	(*32*)
Ehrlich ascites	1 > 4 > 6	(*33*)
HeLa	2 > 4	(*34*)

Cell Line	In Vitro Cytotoxicity[a]	Reference
HeLa	2 > 4	(*35*)
HeLa	4 > 6 > 7	(*36*)
CCRF-CEM	1 > 2 > 4 > 6	(*37*)
HL-60	1 > 4 > 7	(*38*)
HL-60	4 > 6	(*30*)

a   Numbers correspond to numbers of compounds in this table.

on tumor cell uptake are relatively straightforward to predict by knowing the equilibrium lipid bilayer association constants or some less sophisticated approximation thereof (e.g. octanol:buffer partition coefficients), the consequences of aminosugar modification on tumor cell uptake and cytotoxicity are much more difficult to predict.

Cardiotoxicity

Manifestation and Dose Dependency. A characteristic limitation of chemotherapy employing doxorubicin or daunorubicin is drug-induced cardiotoxicity (*39*). The less common acute cardiotoxic effects may manifest in the form of arrhythmias associated with myocardial ischemia within hours of drug administration, possibly due to adrenergic stimulation and vasospasm (*40-42*). However, more common is the cumulative, dose-related chronic myocardial damage that develops during the course of chemotherapy in patients irrespective of their myocardial history. This toxicity limits the total or cumulative amount of the drug that can be administered. Although rarely observed at cumulative dose levels below 450 mg/m^2 body surface area, the incidence of doxorubicin related cardiomyopathy increases progressively (*43-44*) to about 15% at levels of 600 mg/m^2 and to as high as 30-40% at 700 mg/m^2, often leading to congestive heart failure.

Cardiac Tissue Composition and Anthracycline Drug Localization. Mitochondria, the primary site for cell respiratory functions, are present in abundance in cardiac tissue. The diphosphatidylglycerols, more commonly known as cardiolipins, constitute a major component of these mitochondrial membranes (45). The cardiolipin phospholipids are concentrated in the inner membrane of the mitochondria. Biophysical studies on cardiolipin-containing model membranes (17,46-51) indicate that there exist strong electrostatic interactions between the highly anionic cardiolipins and the primary amine of the anthracycline drug, which is positively charged at the physiological pH. The strong drug cardiolipin interactions are believed to facilitate localization of the anthracyclines in the cardiac tissue. In fact, a good correlation between drug-cardiolipin association constant and drug-induced cardiotoxicity has been reported (51), and the doxorubicin-cardiolipin complex is believed to alter the membrane environment such that it results in the inhibition of some membrane-based redox enzymes. This can lead to membrane damage and cell death.

Histological changes observed in cardiac tissue due to the administration of doxorubicin include a disarray of myofilaments, vacuolization of the sarcoplasmic reticulum, degeneration of the mitochondria with increased electron density and generation of myelin-like figures. The myelin-like figures are generally related to the formation of a complex between the drug and the phospholipids (47). The existence of a negative electrochemical gradient inside the phosphorylating mitochondria is believed to accelerate the uptake of the cationic drug into this organelle.

The presence of doxorubicin in heart tissue has been implicated in several membrane-based redox activities that result in impairment of mitochondrial functions such as oxygen consumption and ATP synthesis (52-54). The reduction in rhythmic contractions of the myocardiac cells correlate with a decrease in ATP and phosphocreatine concentrations (55,56) in the heart cells. This effect is seen to be proportional to the doxorubicin concentration (57-60) in the cells.

Drug Interactions with Redox Enzymes and Generation of Reactive Species. The cardiotoxic action of doxorubicin and several related anthracyclines is attributed to the generation of free radical species (61-63) such as superoxide anion, hydrogen peroxide, and hydroxyl radical, the last being the most reactive and destructive of these (64). The mechanism of free radical formation is believed to be a multistep, multifaceted process occurring within the mitochondrial membranes and sarcosomes in the cardiac cells. Doxorubicin, daunorubicin, and many other anthracycline derivatives possessing an anthraquinone nucleus, promote electron flow from NADH or NADPH to molecular oxygen (61-63) through enzymatic processes. The drug interacts with the NADH-dehydrogenase in the mitochondrial membranes. The anthracyclines also interact to some extent with microsomal NADPH cytochrome P-450 reductase, an event resulting in the conversion of drug to the semiquinone free radical form. Transfer of a single electron from the drug free radical to molecular oxygen gives rise to the superoxide anion. Thus, the reversible reduction of the anthracycline drug enables it to shuttle electrons to oxygen without effectively undergoing any change. The superoxide anion is known to undergo dismutation to produce hydrogen peroxide. According to currently prevailing hypotheses, although the superoxide anion and the hydrogen peroxide by themselves do not initiate significant oxidative damage to the cellular components, the presence of transition metals like iron or copper in the cell can catalyze formation of the more deleterious hydroxyl radical by the mechanisms commonly known as the Fenton reaction or the iron catalyzed Haber-Weiss reaction (*63,64*). Various findings suggest yet another potential mechanism of oxyradical generation that involves anthracycline chelation of iron mobilized from ferritin, its intracellular storage site, to form a very stable 3:1 complex of drug with the ferric ion (*65*). This complex is considered to be capable of hydroxyl radical generation (*66,67*). Thus, the presence of ferric ion in the cell is considered to be critical to the initiation and promotion of reactions generating hydroxyl radical.   The highly reactive hydroxyl radical is implicated in the peroxidation of the membrane lipids leading to extensive damage to the mitochondrial membranes, endoplasmic reticulum, and nucleic acids (*68,69*).   Catalase, a free radical scavenger in cells, is either present in very low amounts or is completely absent in cardiac cells (*70*), leaving them particularly susceptible to free radical damage. Other cellular defenses like glutathione peroxidase and superoxide dismutase are either depressed in the cells exposed to adriamycin, or they are overwhelmed in their ability to scavenge the generated free radicals. The generation of free radicals is thought to be attenuated when drug is bound to a nucleic acid (*71,72*).

Approaches to Overcoming Cardiotoxicity. Various strategies have been employed to overcome anthracycline-induced cardiotoxicity. These take into account either the reduction of drug localization in the cardiac tissue or the manipulation of the capacity of the drug or cellular species to generate free radicals and to cause lipid peroxidation leading to cardiotoxicity.

Cardioprotective Agents. Since damaging effects to the cellular membranes occur as a result of the generation of free radicals and reactive species, compounds capable of scavenging free radicals including alpha-tocopherol (*62,73*) and N-acetylcysteine (*74*) have been tried as protective agents. As the participation of free iron is considered to be essential for the generation of hydroxyl radicals, an iron-chelating agent can also offer protection against cardiotoxicity. Among a wide variety of compounds tested, the one compound that seems to hold significant promise is (+)-1,2-bis(3,5-dioxopiperazinyl-1-yl)propane (also known as ICRF-187) (*73-75*). Although the exact mechanism by which ICRF-187 exerts its cardioprotective action remains to be determined, it is known that the agent forms several decomposition products in aqueous solution (*76-79*) capable of chelating metal ions (*79*).   ICRF-187 is thought to be capable of crossing cell membranes and hydrolyzing within cells (*76*), including cardiac cells (*80*). The hydrolysis products are open-ring compounds such as ICRF-198 with a structure similar to EDTA (*76*). The open-ring product acts as a strong metal-chelating agent. Results of some clinical trials (*81*) have indicated that ICRF-187 provides protection against adriamycin-induced cardiotoxicity without affecting its antitumor activity. More recently in preclinical studies the agent PZ-51 or Ebselen has been shown to be protective against anthracycline-induced cardiac damage (*82*).

Structural Modification of the Anthracyclines. It is generally accepted that the anthracycline drugs bearing the cationic primary amine selectively localize in the heart tissue due to the abundance of cardiolipins, and that the cardiotoxicity induced by these drugs is dose-related. Also, the mechanism of antitumor activity of these drugs and the mechanism of anthracycline-induced cardiotoxicity are thought to occur by separate pathways, the former being primarily related to the topoisomerase II-DNA-drug interactions whereas the latter being related to free radical generation through membrane-based interactions. As discussed below it has been shown to be possible to modify the cardiac behavior of the drug without significantly affecting its cytotoxic activity.

One approach in this direction has been related to the design of adriamycin and daunomycin analogues that result in either decreased accumulation of the drug in the heart or reduced capacity of the drug to participate in the generation of free radicals in the cardiac cells. An indication of the effectiveness of structural modification of the anthracycline in reducing drug accumulation in cardiac tissue may be obtained from its relative affinities for various types of phospholipid membranes. Doxorubicin analogues bearing a 14-valerate side chain substituent and various amino group substituents have been evaluated for their comparative affinity for negatively charged lipid membranes including cardiolipin (49,83). It has been observed that incorporation of the hydrophobic valerate group in the parent anthracycline promoted binding of the compound to both neutral as well as negatively charged bilayers to the same high levels such that the drug no longer exhibited a binding preference for the anionic membranes. The selectivity factor, computed as the ratio of the overall association constants (K) for negatively charged DMPG to that for electroneutral DMPC (i.e., K_{DMPG}/K_{DMPC}), was found to be 17 for adriamycin and 0.7 for adriamycin-14-valerate. The amino-substituted derivatives of adriamycin also followed the same trend upon valerate group incorporation indicating that the presence of the 14-valerate side chain alone results in markedly reduced selectivity for the negatively charged membranes. Results of the study were in agreement with the hypothesis that the membrane selectivity of doxorubicin was related to the electrostatic attraction between the ionized amine on the drug residue and the negative charge on the phospholipid membranes. N-trifluoroacetyladriamycin, which is uncharged at pH 7.4, displayed the lowest selectivity factor in the series where the other compounds were positively charged and did exhibit high levels of selective binding to negatively charged lipids when the 14-valerate side chain was not present. N-trifluoroacetyladriamycin-14-valerate (AD32), which contains both structural modifications (*i.e.* nonbasic amino group and 14-valerate side chain) that abrogate selective binding to negatively-charged membranes, has been shown to be less cardiotoxic (84) and therapeutically superior to adriamycin in animal model studies (85,86). But the poor water solubility (87) of AD32 limits its administration to a long-duration, continuous intravenous infusion at high dilution in a surfactant-containing vehicle. The surfactant-containing vehicle used for AD32 administration was found to be associated with a chest pain-inducing side effect (84).

Besides the case of valerate-containing anthracyclines such as AD-32 and AD-198, other close structural analogues of doxorubicin have proven to be less cardiotoxic than the parent in experimental models. The 3'-deamino-3'-(3-cyano-4-morpholinyl)doxorubicin (MRA-CN) congener of doxorubicin has been shown to possess not only significantly higher antitumor potency compared with doxorubicin, but more important it has been shown to be noncardiotoxic at therapeutic dose levels (*88*). Although unique activity of this drug at the DNA site (*89*) is postulated for the favorable activity of MRA-CN, data generated using model membrane systems may provide some explanation for the reduced cardiotoxicity of this compound. In a comparative study of anthracycline associations with electroneutral and negatively charged membranes under physiological conditions, the nonbasic cyanomorpholino derivative of both doxorubicin and daunomycin exhibited much lower binding selectivity for negatively charged model membranes compared with the parent compounds (*26*). The reduced affinity of 3'-deamino-3'-(3-cyano-4-morpholinyl)doxorubicin for negatively charged lipids (e.g. cardiolipin) offers a potential explanation for the reduced cardiotoxicity observed in the case of this analogue.

3'-Hydroxy-3'-deaminodoxorubicin or hydroxyrubicin (*90*), a synthetic analogue of adriamycin in which the amine is replaced by a hydroxyl residue, is yet another example of how a structural modification to doxorubicin's amino group results in an agent with attenuated cardiotoxicity but equal or superior antitumor activity (depending on cancer cell line and phenotype). Hydroxyrubicin has been shown to induce fewer cardiac lesions in animal models as compared with doxorubicin, suggesting it is a less potent cardiotoxic agent (*91*). While the association constant of doxorubicin for DMPC has been shown to increase over five fold upon incorporation of small amounts of cardiolipin (DMPC:CL ratio of 3:1), no significant increase is observed with hydroxyrubicin, indicative of an absence of selective interactions of this drug with cardiolipins (*91*). An added advantage of hydroxyrubicin is that the agent has been found to possess not only superior or equivalent cytotoxicity in different tumor cell lines, but also improved activity in multidrug resistant cells (*91*).

Modification of the anthracycline aglycone by imino substitution of the quinone ring oxygen has been shown to result in lower cardiac toxicity (*92*), probably due to the abrogation of the capacity of the compound to generate free radicals through participation in the one electron reduction process. It has been demonstrated that 5-iminodaunomycin fails to produce significant quantities of the superoxide anion (*61*) in both the sarcosomes and in the mitochondria when compared with the parent drugs and several other quinone-bearing anthracyclines. The imino derivative has also demonstrated an increased capacity to induce DNA strand breaks (*93*) in some tumor cell lines, but the transient character of these strand breaks renders this compound a less potent cytotoxic agent than doxorubicin.

Liposomal Formulations. Besides the potential to enhance antitumor activity, liposomal delivery has been shown to also reduce the doxorubicin-induced cardiotoxicity in several animal models (*94-100*). The decrease in cardiotoxicity is considered a result of altered biodistribution and reduced cardiac uptake of the drug due to encapsulation in liposomes. Liposomes exhibit preferential affinity for tissues with a sinusoidal capillary system and for the reticuloendothelial tissues such as in the liver or spleen, whereas the uptake of liposomes is much lower in extravascular compartment of tissues with continuous capillaries such as the skeletal and cardiac muscles and the nervous tissue possibly due to the relatively much lower endocytosis capability of these tissues. In this way, the modified biodistribution of liposome-encapsulated doxorubicin may account for the reduced cardiotoxicity of this drug.

Although several phospholipid compositions of liposomes, including ones containing neutral and positively charged phospholipids (94) have demonstrated effectiveness in lowering cardiotoxicity, doxorubicin encapsulated in negatively charged liposomes seems to offer an added advantage since the net negative charge on the drug-bearing liposomal bilayer may potentially result in a decreased interaction with the negatively charged surface of the cardiolipin-rich heart tissue (97-98).

Multidrug Resistance

A serious limitation of many natural product antitumor agents including doxorubicin and daunomycin is the development of acquired or pleiotropic multidrug resistance or MDR (101-103). Evidence to date indicates that tumor cells are capable of developing resistance not only to the drug to which they are exposed, but also to other structurally unrelated cytotoxic compounds. This cross-resistance poses a major problem in the treatment of many cancers; hence, the understanding of MDR and approaches to overcoming it has been a much pursued area of cancer research (101-155). Sensitive and resistant cell-lines of several tumors have been employed in the study of MDR. In the most common form of MDR, which involves the participation of a 170-kDa plasma membrane glycoprotein, reduced intracellular levels of doxorubicin and daunomycin have been observed. In anthracycline-resistant sublines of P388 leukemia (104), reduced levels of cellular drug accumulation were found to correlate with resistance to the cytotoxic effects of these agents. Similar changes have also been seen in sublines of Ehrlich carcinoma (105-107) and Chinese hamster ovary cells (108). The decreased drug levels in these cells were found to be due to enhanced efflux of these drugs by an active outward transport mechanism (109-112). The evidence of an efflux pump was provided by an increase in intracellular drug levels similar to those in sensitive cells upon incorporation of inhibitors of oxidative phosphorylation such as sodium azide or 2,4-dinitrophenol in glucose-free medium, and rapid efflux of the drugs upon addition of glucose to the medium.

The most striking alteration in MDR cell membranes has been the overexpression of a 170kDa molecular weight glycoprotein (113,114,115), referred to as P-glycoprotein (PGP) for 'permeability' glycoprotein. This membrane glycoprotein has been isolated from the resistant phenotypes of a variety of cell lines (116) and correlation could be detected between the degree of drug resistance and the amount of PGP in the membrane. The 1280 amino acid PGP expressed by the mdr1 gene (117,118) consists of 12 transmembrane regions and two domains that appear to be ATP-binding sites (119,120). It is thought that the two domains come together in the plasma membrane to form a channel for the energy-driven efflux of various drugs from the cell.

Some drugs involved in MDR were seen to bind to isolated membranes of resistant cells, whereas little or no specific binding by the same drug was seen in the isolated membranes of the sensitive or revertant phenotypes (121). By contrast, since the drug retention levels in whole cells of the resistant type are seen to be lower than in sensitive cells, it appears that these drugs may first associate with PGP and are then effluxed by the pump, through its ATPase-like activity (*122*). It is also believed that the drug-binding sites and the ATP-binding sites are probably different (*123*).

Substrates for the drug-binding sites of PGP include a variety of cytotoxic agents of natural origin such as doxorubicin, daunomycin, the vinca alkaloids, actinomycin D, mitoxantrone, and taxol. The only features common to several of these drugs are their amphipathic nature and a tendency to be positively charged at the neutral pH (*124*). Thus, PGP seems to exhibit low binding specificities. Some preliminary investigations seem to indicate a three-dimensional similarity between several of these compounds, which may also share some common volume element and similar molar refractivities (*124*).

Other Alterations Related to MDR. Apart from the overexpression of PGP, several other alterations have been reportedly linked to MDR. Differences in the distribution of daunomycin in the plasma membranes of drug-sensitive and drug-resistant P388 leukemia cells have been indicated based on fluorescence energy transfer and iodide quenching studies (*125*). The relatively deeper location of the drug in resistant cell membranes is suggested to be effected partly due to a decrease in the proportion of phosphatidylserine and an increase in cholesterol proportion in the cell membranes. The presence of phosphatidylserine increases both the affinity and stoichiometry of daunomycin binding to model lipid membranes containing phosphatidylcholine, whereas cholesterol causes the opposite effect (*28*). In some cases resistance appears to be induced by the enhancement of superoxide dismutase (*126*) or glutathione peroxidase (*126,127*) in tumor cells. Other morphological alterations and lesions induced by MDR in several cell lines include an increase in plasma membrane fluidity (*128,129*), increased cellular fragility to osmotic shock (*130*), and higher intramembrane particle density observed upon freeze-fracture microscopy (*131*).

Other membrane-based differences between sensitive and resistant cell types reported in literature include changes in the composition of oligosaccharides from a high mannose type of N-linked oligosaccharides to the complex type (*132*), disturbed glycosylation patterns of glycoconjugates (*102,133*), and alterations in lipid composition of the membranes. Some resistant-cell bilayers (*134*) have shown higher triglyceride content and a lower phosphatidylcholine:sphingomyelin ratio than sensitive cells, whereas some others (*135*) have exhibited decreased phosphatidylserine and slightly higher cholesterol content than sensitive cell membranes.

An "atypical" form of MDR, in which neither overexpression of PGP nor decreased drug levels are observed, is exhibited by some tumors treated with drugs such as epipodophyllins (*136*), mitoxantrone (*137*), and m-AMSA (*138*). This form of MDR is attributed to altered topoisomerase II activity by the mutagenic effects of these drugs and appears not to be linked with membrane alterations.

Approaches to the Circumvention of MDR

MDR Modulators. A broad class of relatively nontoxic compounds including some calcium channel blockers (verapamil, nifedipine) and others like quinidine, cyclosporine, and more, have been found to modulate or reverse PGP-based MDR (*139,140,141*). Among these, verapamil has been most widely employed in MDR studies in several experimental and clinical tumors. Although the precise mechanism of MDR modulation by these compounds is not very clear, it has been proposed that the MDR modulators bind competitively to the PGP at the drug-binding site and inhibit the activity of the multidrug transporter, thereby enabling the antitumor drugs like doxorubicin to remain within the cell and exert their cytotoxic action. Other investigators postulate that both direct interactions with PGP and indirect lipid interactions of modulators should be considered as mechanisms by which modulators can reverse MDR (142).

Photoaffinity labelling studies with verapamil have indicated this compound itself to be a good substrate for PGP drug-binding site (143,144). Verapamil is also known to increase the cytosolic pH and ATP consumption in PGP-expressing resistant tumor cells (145,146). Verapamil was also shown to restore the morphology of plasma membranes of P388 cell lines resistant to daunomycin (147). The high density of intramembrane particles associated with the MDR in these cell lines were observed to revert to a density pattern similar to those displayed by drug-sensitive cells.

Preliminary results of some clinical trials suggest that verapamil can effectively reverse MDR (148); however, the inherent pharmacological properties of these drugs at the concentrations employed for MDR modulating effect renders these compounds as potentially toxic. The cardiotoxic potential of verapamil has necessitated the search for somewhat better candidates for clinical purposes. The optically active isomer, R-verapamil, and a few other derivatives of verapamil tested in transgenic mice expressing the human *mdr1* gene, have exhibited encouraging results with respect to circumventing MDR without significant displays of toxic side effects (127). An approach has been the use of two or more modulators of MDR, each at lower dose levels, in a combination to achieve a high level of drug retention in cancer cells with minimal side-effects (126).

Liposomal Formulations in MDR. Recently, liposomal formulations of doxorubicin have demonstrated improved activity against MDR compared with the free drug. Early studies were conducted in Chinese hamster LZ cells, which were 3,000-fold resistant to doxorubicin compared to the parental V-79 cells (149). In these, the liposomal doxorubicin was shown to produce sensitivity seven times higher than that achieved with the free drug.

Since a relatively high incidence of MDR is observed during chemotherapy for gastrointestinal cancers, the influence of liposomal drug delivery on MDR has been also studied in some human colon cancer cells (150). The resistant cell lines, including both the nonselected and the one selected for resistance to doxorubicin, exhibited improved sensitivity upon treatment with liposome-encapsulated doxorubicin than with the free drug. Since no difference was observed between DNA strand breakage caused by the free drug and that induced by liposome-encapsulated doxorubicin in any of the cell lines, the enhanced cytotoxicity of liposomal doxorubicin in these resistant cells is attributed to some non-DNA based effect. In mouse UV-2237M fibrosarcoma cell studies, doxorubicin in multilamellar vesicles of phosphatidylcholine(PC)-phosphatidylserine(PS) were found to produce higher levels of cytotoxicity in both doxorubicin-sensitive and doxorubicin-resistant cells compared to the free drug (151). The phospholipid composition was seen to influence the outcome of the drug-mediated cytostasis. The levels of doxorubicin-mediated cytotoxicity in both types of cells increased with the increasing proportion of PS in the liposomes. Since no appreciable changes in the intracellular doxorubicin concentration or in the doxorubicin-induced DNA cleavage was found upon use of liposomal drug in these cells, the enhanced sensitivity of these tumor cells is believed to occur due to localized damage to the PGP-containing plasma membrane through PS-mediated release of the drug (*151*).

To gain further insight into the role of liposomes in improving sensitivity of MDR tumor cells to anthracyclines like doxorubicin, the effects of empty liposomes of cardiolipin, PC, and cholesterol in tumor cells have been recently investigated (*152*). The empty liposomes were found to exert greater cytotoxicity on MDR cells than to sensitive ones, leading to the hypothesis that the cardiolipin-containing liposomes could be, in some manner, altering the function of PGP and impairing its drug efflux activity. Currently, although very little is clear regarding the application of liposomal drug delivery to overcome MDR in the clinical environment, investigations in transgenic animals (*153,154*) have shown encouraging results and active research is ongoing in this area. Also, liposomes are being investigated as carrier systems for delivery of antisense oligonucleotides in the effort to overcome MDR (*153*).

Structural Considerations of Anthracyclines in PGP-Mediated MDR. The generally accepted model of the plasma membrane-based PGP is that of an energy dependent "pump" which binds to broadly specific substrates at the cytosolic side and actively transports these out of the cell through a porous channel formed by the transmembrane segments of this protein (*155*). An alternative model of a "flippase" has been recently proposed (*156*), which suggests that the cytotoxic drug substrates gain access to this membrane-based transporter directly through the inner leaflet of the bilayer. The glycoprotein utilizes ATP and flips over to the outer leaflet of the bilayer along with the drug, which then exits through a passive diffusion process. The flippase model suggests the ability of the drug to intercalate into the lipid bilayer, as a primary determinant of its PGP-based transport, whereas the pump model emphasizes the importance of the substrate's binding affinity to the transport protein in determining the specificity of the efflux mechanism.

Considering either or both of these mechanisms to be relevant to the active efflux of the various cytotoxic drugs by PGP, two features of the drug substrates seem to be of major importance: (1) the lipophilicity of the drug determines its ability to partition into the lipid bilayer, and its relative location within the bilayer would depend upon, besides other factors, the degree of lipophilicity of the compound; (2) the charge on the drug substrate also seems to be a major determinant of the PGP-mediated transport. Recently, lipophilic cations that share some structural characteristics of typical PGP substrates have exhibited the ability to confer MDR in vitro, which resulted in decreased intracellular accumulation and reversal by verapamil, the competitive inhibitor of PGP (*157*). Thus, the data suggests that a cationic functional group on moderately lipophilic cytotoxic agents may be possible factors facilitating drug-PGP interactions.

Comparison of cytotoxic activity of some anthracyclines, including doxorubicin as well as the morpholinyl and cyanomorpholinyl derivatives of doxorubicin and its imino analogue in sensitive and resistant P388 cell lines (*158*) has indicated some interesting trends. The imino derivatives were found to be less cytotoxic than their parent analogues, possibly due to their decreased capacity to generate free radicals and cause membrane lesions. The morpholinyl and cyanomorpholinyl analogs of doxorubicin, in which the primary amine is substituted by tertiary amino groups, exhibited improved potency in both the sensitive and resistant cells. More important, these compounds significantly lowered the resistance index (i.e., IC_{50} in resistant cells/ IC_{50} in sensitive cells, where IC_{50} is the drug concentration producing 50% inhibition of cell growth). Basicity of the compound appeared to be an important factor in determining cytotoxicity as well as the ability to induce MDR, since among the two derivatives, the less basic cyanomorpholino compound was more potent, inducing least resistance. The PGP inhibitor, verapamil, which substantially reversed the resistance to doxorubicin, had no appreciable effect on the morpholinyl and cyanomorpholinyl derivatives, suggesting that the less basic compounds are less influenced by the efflux protein. Similar behavior, with respect to cytotoxicity in sensitive and resistant cell lines and the influence of verapamil, has also been observed with another analogue, 3'-deamino-3'-morpholino-13-deoxo-10-hydroxy-carminomycin (*159*).

The role of the primary amine in the sugar portion of the anthracyclines in MDR has been further substantiated with studies using hydroxyrubicin (*91*). Apart from other favorable effects, hydroxyrubicin has demonstrated the ability to overcome, to some extent, the *mdr1*-mediated resistance both in vitro and in vivo in selected tumors.

In a similar manner, the deaminated annamycin congener of doxorubicin has also exhibited superior in vitro cytotoxicity compared with doxorubicin in resistant phenotypes (*160*). Also uptake and intracellular accumulation of annamycin was higher than doxorubicin in the resistant cells. But of special significance were the drug efflux studies that suggest the possibility that different mechanisms of efflux exist for annamycin versus doxorubicin. While doxorubicin is retained over sufficiently long periods in sensitive cells, it is rapidly effluxed from the PGP-containing resistant cells. On the other hand, the efflux profile of annamycin in both sensitive and resistant cells was more or less identical, indicating that the transport of annamycin in resistant cells may not be mediated by PGP. This difference in cellular efflux mechanism may, to some extent, account for the superior activity of annamycin in resistant cells. Biophysical investigations are now underway aimed at elucidating, on a molecular level, differences between the interactions of annamycin and doxorubicin with the plasma membranes of wild-type versus drug-resistant cells. Hopefully the completion of these research efforts will provide additional insight into the details of anticancer drug interactions with the *mdr1* protein.

ACKNOWLEDGEMENTS. The support of the National Institutes of Health (CA 55320, CA 50270, CA 16058) and the American Heart Association-Ohio Affiliate is gratefully acknowledged.

Literature Cited

1.   Chabner B.A.; Myers C.E. In *Cancer: Principles and Practice of Oncology;* DeVita V.T.,Jr., Hellman S., Rosenberg S.A., Eds.; J.B. Lippincott Co.: Philadelphia, PA, 1989, Vol. 1; 349-395.

2. Tewey K.M.; Chen G.L.; Nelson E.M.; Liu L.F. *J. Biol. Chem.* **1984**, *259*, 9182.

3. Potmesil M.; Kirshenbaum S.; Israel M.; Levin M.; Khetarpal V.K.; Silber R. *Cancer Res.* **1983**, *43*, 3528.

4. Bodley A.; Liu L.F.; Israel M.; Seshadri R.; Koseki Y.; Giuliani F.C.; Kirschenbaum S.; Silber R.; Potmesil M. *Cancer Res.* **1989**, *49*, 5969.

5. Neidle S.; Pearl L.H.; Skelly J.V. *Biochem J.* **1987**, *243*, 1.

6. Liu L.F.; Rowe T.C.; Yang L.; Tewey K.M.; Chen G.l. *J. Biol. Chem.* **1983**, *258*, 15365.

7. Goormaghtigh E.; Pollakis G.; Ruysschaert J.M. *Biochem. Pharmacol.* **1983**, *32*, 889.

8. Sun I.L.; Crane F.L. In *Oxireduction at the plasma membrane: Relation to Growth and Transport*; CRC Press: Boca Raton, FL, 1990; Vol. 1, 257-280.

9. Doroshow J.H. *Biochem. Biophys. Res. Commun.* **1986**, *135*, 330.

10. Sinha B.K.; Katki A.G.; Batist G.; Cowan K.H.; Myers C.E. *Biochemistry* **1987**, *26*, 3776.

11. Tritton T.R.; Yee G. *Science* **1982**, *217*, 248.

12. Tokes Z.A.; Rogers K.E.; Rembaum A. *Proc. Natl. Acad. Sci. USA* **1982**, *79*, 2026.

13. Zhao F.K.; Chuang L.F.; Israel M.; Chuang R.Y. *Anticancer Res.* **1989**, *9(1)*, 225.

14. Abraham I.; Hunter R.J.; Sampson K.E.; Smith S.; Gottesman, M.M.; Mayo J.K. *Mol. Cell. Biol.* **1987**, *7*, 3098.

15. Tritton T.R.; Hickman J.A. In *Experimental and Clinical Progress in Cancer Chemotherapy*; Muggia, F.M., Ed.; Martinus Nijhoff, The Hague, 1985; 81-131.

16. Siegfried J.M.; Burke T.G.; Tritton T.R. *Biochem. Pharmacol.* **1985**, *34(5)*, 593.

17. Goldman R.; Facchinetti T.; Bach D.; Raz A.; Shinitzky M. *Biochim. Biophys. Acta* **1978**, *512*, 254.

18. Dalmark M.; Storm H.H. *J. Gen. Physiol.* **1981**, *78*, 349.

19. Dalmark M.; Johansen P. *Mol. Pharmacol.* **1982**, *22*, 158.

20. Dalmark M. *Scand. J. Clin. Lab. Invest.* **1981**, *41*, 633.

21. Arcamone F.; Cassinelli G.; Franceschi G.; Penco S.; Pol C.; Radaelli S.; Selva A. In *International Symposium on Adriamycin*; Carter S.K.; DiMarco A.; Ghione M.; Krakoff I.H.; Mathe G., Eds.; Springer-Verlag, Berlin, 1972; 9-22.

22. de Duve C.; de Barsy T.; Poole B.; Trouet A.; Tulkens P.; van Hoof F. *Biochem. Pharmacol.* **1974**, *23*, 2495.

23. Pilgram W.J.; Fuller W.; Hamilton L.D. *Nature: New. Biol.* **1972**, *235*, 17.

24. Neidle S.; Taylor G. *Biochim. Biophys. Acta* **1977**, *479*, 450.

25. Burke T.G.; Tritton T.R. *Biochemistry* **1985**, *24*, 1768.

26. Burke T.G.; Sartorelli A.C.; Tritton T.R. *Cancer Chemother. Pharmacol.* **1988**, *21*, 274.

27. Burke, T.G.; Israel, M.; Seshadri, R.; Doroshow, J.H. *Biochim. Biophys. Acta* **1989**, *982*, 123.
28. Escriba P.V.; Ferrer-Montiel A.V.; Ferragut J.A.; Gonzalez-Ros J.M. *Biochemistry* **1990**, *29*, 7275.
29. Burke T.G.; Mehta R.; Perez-Soler R.; Priebe W. *Proc. Am. Assoc. Cancer Res.* **1993**, *34*, 331.
30. Burke T.G.; Morin M.J.; Sartorelli A.C.; Lane P.E.; Tritton T.R. *Mol. Pharmacol.* **1987**, *31*, 552.
31. Bachur N.R.; Steele M.; Meriwether W.D.; Hildebrand R.C. *J. Med. Chem.* **1976**, *19*, 651.
32. Kessel D. *Biochem. Pharmacol.* **1979**, *28*, 3028.
33. Seeber S.; Lith H.; Crooke S.T. *J. Cancer Res. Clin. Oncol.* **1980**, *98*, 109.
34. Casazza A.M.; Barbieri A.; Funagalli A.; Geroni M.C. *Proc. Am. Assoc. Cancer Res.* **1983**, *24*, 25.
35. Supino R.; Necco A.; Dasdia T.; Casazza A.M.; DiMarco A. *Cancer Res.* **1977**, *37*, 4523.
36. DiMarco A., Casazza A.M.; Dasdia T.; Giuliani F.; Lenaz L.; Necco A.; Soranzo C.; *Cancer Chemother. Rep.* **1973**, *57(1)*, 269.
37. Schwartz H.S.; Kanter P.M. *Cancer Treat. Rep.* **1979**, *63*, 821.
38. Zwelling L.A.; Kerrigan D; Michaels S. *Cancer Res.* **1982**, *42*, 2687.
39. Gianni L.; Corden B.J.; Myers C.E. *Reviews in Biochemical Toxicology* **1983**, *5*, 1.
40. Bristow M.R.; Mason J.W.; Billingham M.E.; Daniels J.R. *Ann. Intern. Med.* **1978**, *88*, 168.
41. Ippoliti G.; Casirola G.; Marini G.; Invernizzi R. *Lancet* **1976**, *1*, 430.
42. Mancuso L.; Marchi S.; Canonico A. *Cancer Treat. Rep.* **1985**, *69*, 241.
43. Praga C.; Beretta G.; Vigo P.L., et al. *Cancer Treat. Rep.* **1979**, *63*, 827.
44. Von Hoff D.D.; Layard M.W.; Basa P.; Davis H.L.; Von Hoff A.L.; Rozencweig M; Muggia F.M. *Ann. Intern. Med.* **1979**, *91*, 710.
45. White D.A. In *Form and Function of Phospholipids*; Ansell G.B.; Hawthorne J.N.; Dawson R.M.C., Eds.; B.B.A. Library Vol. 3; Elsevier Scientific Pub. Co., New York, NY, 1973, 441-482.
46. Goormaghtigh E.; Chatelain P.; Caspers J.; Ruysschaert J.M. *Biochim. Biophys. Acta* **1980**, *597*, 1.
47. Duarte-Karim M.; Ruysschaert J.M.; Hildebrand J. *Biochem. Biophys. Res. Commun.* **1976**, *71*, 658.
48. Henry N.; Fantine E.O.; Bolard J.; Garnier-Suillerot A.; *Biochemistry* **1985**, *24*: 7085-7092.
49. Burke T.G.; Israel M.; Seshadri, R.; and Doroshow J.H.; *Cancer Biochem. Biophys.* **1990**, *11*, 177.
50. Goormaghtigh E.; Vandenbranden M.; Ruysschaert J.-M.; Kruijff B. *Biochim. Biophys. Acta* **1982**, *685*, 137.
51. Goormaghtigh E.; Chatelain P.; Caspers J.; Ruysschaert J.-M. *Biochem. Pharmacol.* **1980**, *29*, 3003.
52. Bachmann E.; Weber E.; Zbinden G. *Agents and Actions* **1975**, *5*, 383.
53. Bachmann E.; Zbinden G. *Toxicol. Lett.* **1979**, *3*, 29.

14. MEHTA & BURKE   *Membrane Biophysical Parameters*   **237**

54. Arena E.; Arico M.; Biondo F.; D'Allessandro N. et al In *EORTC Int'l Symposium Adriamycin Review*; Staquet M.; Tagnon H., Eds.; European Press Medikon, Ghent, Belgium, 1975, 160-172.
55. Seraydarian M.W.; Artza L.; Goodman M.F. *J. Mol. Cell Cardiol.* **1977**, *9*, 375.
56. Seraydarian M.W.; and Artza L. *Cancer Res.* **1979**, *39*, 2940.
57. Breed J.G.; Zimmermann A.N.; Meyler F.L.; Pinedo H.M. *Cancer Treat. Rep.* **1979**, *63*, 869.
58. Lowe M.C.; Smallwood J.I. *Cancer Chemother. Pharmacol.* **1980**, *5*, 61.
59. Lampidis T.J.; Henderson I.C.; Israel M.; Canellos G.P. *Cancer Res.* **1980**, *40*, 3901.
60. Necco A.; Dasdia T.; Di Francesco D.; Ferroni A. *Pharmacol. Res. Commun.* **1976**, *8*, 105.
61. Gervasi P.G.; Agrillo M.R.; Lippi A.; Bernardini N.; Danesi R.; Del Tacca M. *Res. Commun. Chem. Pathol. Pharmacol.* **1990**, *67*, 101.
62. Thomas C.E.; Aust, S.D.; *J. Free Radicals Biol. Med.* **1985**, *1*, 293.
63. Doroshow J.H. *Cancer Res.* **1983**, *43*, 4543.
64. Halliwell B.; Gutteridge J.M.C. *Free Radicals in Biology and Medicine*; Clarendon Press, Oxford, 1988, 188-276.
65. May P.M.; Williams G.K.; Williams D.R. *Inorg. Chim. Acta* **1980**, *46*, 24.
66. Gutteridge J.M.C. *Biochem. Pharmacol.* **1984**, *33*, 1725.
67. Myers C.E.; Gianna L.; Simone C.B.; Klecker R.; Greene R. *Biochemistry* **1982**, *21*, 1707.
68. Bachur N.R.; Gordon S.L.; Gee M.V. *Cancer Res.* **1978**, *38*, 1745.
69. Summerfield F.W.; Tappel A.L. *Anal. Biochem.* **1981**, *11*, 77.
70. Doroshow J.H.; Reeves J. *Proc. Am. Assoc. Cancer Res.* **1980**, *21*, 266.
71. Sato S.; Iwaizuni M.; Handa K.; Tamura Y. *Gann* **1977**, *68*, 603.
72. Calendi E.; Di Marco A.; Reggiani M.; Scarpinato B.; Valentini L. *Biochim. Biophys. Acta* **1965**, *103*, 25.
73. Herman E.H. *Lab. Invest.* **1983**, *49*, 69.
74. Herman E.H.; Ferrans V.J.; Myers C.E.; Van Vleet J.F. *Cancer Res.* **1985**, *45*, 276.
75. Eastland E. *Drugs of the Future* **1987**, *12*, 1017.
76. Dawson K.M. *Biochem. Pharmacol.* **1975**, *24*, 2249.
77. Hasinoff B.B. *Drug Metab. Dispos.* **1991**, *19*, 74.
78. Burke T.G.; Lee T.D.; van-Balgooy J.; Doroshow J.H. *J. Pharm. Sci.* **1991**, *80*, 338.
79. Huang, Z.-X.; May P.M.; Quinlan K.M.; Williams D.R.; Creighton A.M. *Agents and Actions* **1982**, *12*, 536.
80. Doroshow J.D.; Burke T.G.; van Balgooy C.; Akman S.; Verhoef V. *Proc. Am. Assoc. Cancer Res.* **1991**, *32*, 332.
81. Speyer J.L.; Green M.D.; Kramer E.; Rey M.; Sanger J.; Ward C.; Dubin N.; Ferrans V.; Stecy P.; Zeleniuch-Jaquotte A.; Wernz J.; Feit F.; Slater W.; Blum R.; Muggia F. *N. Engl. J. Med.* **1988**, *319*, 745.
82. Pritsos, C.A.; Sokoloff, M.; Gustafson, D.L. *Biochem. Pharmacol.* **1992**, *44*, 839.
83. Karczmar, G.S. and Tritton, T.R. *Biochim. Biophys. Acta.*, **1979**, *557*, 306.

84. Garnick M.B.; Griffin J.D.; Sack M.J.; Blum R.H.; Israel M.; Frei E.III In *Anthracycline Antibiotics in Cancer Therapy;* Muggia F.M.; Young C.W.; Carter S.K., Eds.; Martinus Nijhoff Publ., The Hague, 1984; 541-548.
85. Israel M.; Modest E. J.; Frei E.III *Cancer Res.* **1975**, *35*, 1365.
86. Parker L.M.; hirst M.; Israel M. *Cancer Treat. Rep.* **1978**, *62*, 119.
87. Israel M.; Modest E.J. *U.S. Patent No.4,035,566,* **1977**.
88. Sikic B.I.; Ehsan M.N.; Harker W.G.; Friend N.F.; Brown B.W.; Newman R.A.; Hacker M.P.; Acton E.M. *Science* **1985**, *228*, 1544.
89. Beigleiter A.; Jesson M.I.; Johnston J.B. *Cancer Commun.* **1990**, *2(1)*, 7.
90. Horton D.; Priebe W.; Varela O. *J. Antibiot.* **1984**, *37*, 853.
91. Priebe W.; Van N.T.; Burke T.G.; Perez-Soler R. *Anti-Cancer Drugs* **1993**, *4*, 37.
92. Myers C.E.; Muindi R.F.; Zweier J.; Sinha B.K. *J. Biol. Chem.* **1987**, *262*, 11571.
93. Zwelling L.A.; Kerrigan D.; Michaels S. *Cancer Res.* **1982**, *42*, 2687.
94. Rahman A.; Kessler A.; More N.; Sikic B.; Rowden G.; Woolley P.; Schein P.S. *Cancer Res.* **1980**, *40*, 1532.
95. Gabizon A.; Dagan A.; Goren D.; Barenholz Y.; Fuks Z. *Cancer Res.* **1982**, *42*, 4734.
96. Herman E.H.; Rahman A.; Ferrans V.J.; Vick J.A.; Schein P.S. *Cancer Res.* **1983**, *43*, 5427.
97. Forssen E.A.; Tokes Z.A. *Cancer Res.* **1983**, *43*, 546.
98. Forssen E.A.; Tokes Z.A. *Proc. Natl. Acad. Sci. U.S.A.* **1981**, *78(3)*, 1873.
99. Cowans J.W.; Creaven P.J.; Greco W.R.; Brenner D.E.; Tung Y.; Ostro M.: Pilkiewicz F.; Ginsberg R.; Petrelli N.; *Cancer Res.* **1993**, *53*, 2796.
100. Gabizon A.; Catane R.; Uziely B.; Kaufman B.; Safra T.; Cohen R.; Martin F.; Huang A.; Barenholz Y. *Cancer Res.* **1994**, *54*, 987.
101. Ling V. In *Molecular Cell Genetics.*; Gottesman M.M., Ed., John Wiley, New York, NY, 1985; 773-787.
102. Biedler J.; Peterson R.H.F. In *Molecular Actions and Targets for Cancer Chemotherapeutic Agents.*; Sartorelli A.C.; Bertino J.R.; Lazo J.S.; Eds., Academic Press, New York, NY, 1981; 453-482.
103. Johnson R.K.; Chitnis M.P.; Embrey W.M.; Gregory E.B. *Cancer Treat. Rep.* **1978**, *62*, 1535.
104. Inaba M.; Johnson R.K. *Biochem. Pharmacol.* **1978**, *27*, 2123.
105. Dano K. *Cancer Chemother. Rep.* **1972**, *56*, 321.
106. Dano K. *Cancer Chemother. Rep.* **1972**, *56*, 701.
107. Skovsgaard T. *Cancer Res.* **1978**, *38*, 1785.
108. Riehm H.; Biedler J.L. *Cancer Res.* **1971**, *31*, 409.
109. Dano K. *Biochim. Biophys. Acta* **1973**, *323*, 466.
110. Skovsgaard T. *Biochem. Pharmacol.* **1977**, *26*, 215.
111. Skovsgaard T. *Cancer Res.* **1978**, *38*, 1785.
112. Inaba M.; Kobayashi H.; Sakurai Y.; Johnson R.K. *Cancer Res.* **1979**, *39*, 2200.
113. Riordan J.R.; Ling V. *Pharmacol. Ther.* **1985**, *28*, 51.
114. Gottesman M.M.; Pastan I. *Trends Pharmacol.Sci.* **1988**, *9*, 54.
115. Moscow J.A.; Cowan K.H. *J. Natl. Cancer Inst.* **1988**, *80*, 14.
116. Kartner N.; Riordan J.R.; Ling V. *Science* **1983**, *221*, 1285.

117. Fojo A.T.; Whang-Pang J.; Gottesman M.M.; Pastan I. *Proc. Natl. Acad. Sci. U.S.A.* **1985**, *82*, 7661.
118. Ruiz J.C.; Choi K.; Von Hoff D.D.; Roninson I.B.; Wahl G.M. *Mol. Cell. Biol.* **1989**, *9*, 109.
119. Chen C.J.; Chin J.E.; Ueda K.; Clark D.P.; Pastan I.; Gottesman M.M.; Roninson I.B. *Cell* **1986**, *47*, 381.
120. Cornwell M.M.; Tsuruo T.; Gottesman M.M.; Pastan I. *FASEB J.* **1987**, *1*, 51.
121. Cornwell M.M.; Gottesman M.M.; Pastan I. *J. Biol. Chem.* **1986**, *261*, 7921.
122. Hamada H.; Tsuoro T. *J. Biol.Chem.* **1988**, *263*, 1454.
123. Naito M.; Hamada H.; Tsuoro T. *J. Biol. Chem.* **1988**, *263*, 11887.
124. Zamora J.M.; Pearce H.L.; Beck W.T. *Mol. Pharmacol.* **1988**, *33*, 454.
125. Ferrer-Montiel A.V.; Gonzalez-Ros J.M.; Ferragut J.A. *Biochim. Biophys. Acta* **1992**, *1104*, 111.
126. Lehnert M.; Dalton W.S.; Roe D.; Emerson S.; Salmon S.E. *Blood* **1991**, *77(2)*, 348.
127. Mickisch G.H.; Merlino G.T.; Aiken P.M.; Gottesman M.M.; Pastan I. *J. Urol.* **1991**, *146*, 447.
128. Wheeler C.; Rader R.; Kessel D. *Biochem. Pharmacol.* **1982**, *31*, 2691.
129. Siegfried J.A.; Kennedy K.A.; Sartorelli A.C.; Tritton T.R. *J. Biol. Chem.* **1983**, *258*, 339.
130. Riordan J.R.; ling V. *J. Biol. Chem.* **1979**, *254*, 12701.
131. Arsenault A.L.; Ling V.; Kartner N. *Biochim. Biophys. Acta* **1988**, *938*, 315.
132. Narasimhan S.; Wilson J.R.; Martin E.; Schacter H. *Can. J. Biochem.* **1979**, *57*, 83.
133. Kessel D.; Beck W.T.; Kukuruga D.; Schulz V. *Cancer Res.* **1991**, *51*, 4665.
134. Ramu A.; Glaubiger D. Weintraub H. *Cancer Treat. Rep.* **1984**, *68*, 637.
135. Kaplan O.; Jarowzewski J.W.; Clarke R.; Fairchild C.R.; Schoenlein P.; Goldenberg S.; Gottesman M.M.; Cohen J.S. *Cancer Res.* **1991**, *51*, 1638.
136. Glisson B.; Gupta R.; Smallwood-Kentro S.; Ross W. *Cancer Res.* **1986**, *46*, 1934.
137. Harker W.G.; Slade D.L.; Dalton W.S.; Meltzer P.S.; Trent J.M. *Cancer Res.* **1989**, *49*, 4542.
138. Per S.R.; Mattern M.R.; Mirabelli C.K.; Drake F.H.; Johnson R.K.; Crooke S.T. *Mol. Pharmacol.* **1987**, *32*, 17.
139. Dalton W.S.; Grogan T.M.; Meltzer P.S.;Scheper R.J.; Durie B.G.M.; Taylor C.W.; Miller T.P.; Salmon S.E. *J. Clin. Oncol.* **1989**, *7*, 415.
140. Hindenburg A.A.; Baker M.A.; Gleyzer E.; Stewart V.J.; Case N.; Taub R.N. *Cancer Res.* **1987**, *47*, 1421.
141. Kanamaru H.; Kakehi Y.; Yoshida O.; Nakanishi S.; Pastan I.; Gottesman M.M. *J. Natl. Cancer Inst.* **1989**, *81*, 844.
142. Wadkins R.M.; Houghton P.J. *Biochim. Biophys. Acta* **1993**, *1153*, 225.
143. Qian X.-d; Beck W.T. *Cancer Res.* **1990**, *50*, 1132.
144. Safa A.R. *Proc. Natl. Acad. Sci. U.S.A.* **1988**, *85*, 7178.
145. Broxterman H.J.; Pinedo H.M.; Kuiper C.M.; Kaptein L.C.M.; Schuurhuis G.J.; Lankelma J. *FASEB J.* **1988**, *2*, 2278.
146. Keizer H.G.; Joenje H. *J. Natl. Cancer Inst.* **1989**, *81*, 706.
147. Garcia-Segura L.M.; Soto F.; Planells-Cases R.; Gonzalez-Ros J.M.; Ferragut J.A. *FEBS Lett.* **1992**, *314(3)*, 404.
148. Dalton W.S. *Proc. Am. Assoc. Cancer Res.* **1991**, *31*, 520.

149. Thierry A.R.; Jorgensen J.T.; Forst D.; Belli J.A.; Dritschilo A.; Rahman A. *Cancer Commun.* **1989** 1: 311.

150. Oudard S.; Thierry A.R.; Jorgensen T.J.; Rahman A. *Cancer Chemother. Pharmacol.* **1991**, *28*, 259.

151. Fan D.; Bucana C.D.; O'Brian C.A.; Zwelling L.A.; Seid C.; Fidler I.J. *Cancer Res.* **1990**, *50*, 3619.

152. Thierry A.R.; Dritschilo A.; Rahman A. *Biochem. Biophys. Res. Commun.* **1992**, *187(2)*, 1098.

153. Thierry A.R.; Rahman A.; Dritschilo A. *Biochem. Biophys. Res. Commun.* **1993**, *190(3)*, 952.

154. Mickisch G.H.; Rahman A.; Pastan I.; Gottesman M.M. *J. Natl. Cancer Inst.* **1992**, *84(10)*, 804.

155. Gottesman, M.M. *Cancer Res.* **1993**, *53*, 747.

156. Higgins C.F.; Gottesman M.M. *TIBS* **1992**, *17*, 18.

157. Gros P.; Talbot F.; Tang-Wai D.; Bibi E.; Kaback H.R. *Biochemistry* **1992**, *31*, 1992.

158. Streeter D.G.; Taylor D.L.; Acton E.M.; Peters J.H. *Cancer Chemother. Pharmacol.* **1985**, *14*, 160.

159. Watanabe M.; Komeshima N.; Naito M.; Isoe T.; Otake N.; Tsuruo T. *Cancer Res.* **1991**, *51*, 157.

160. Ling Y.-H.; Priebe W.; Yang L.-Y.; Burke T.G.; Pommier Y.; Perez-Soler R. *Cancer Res.* **1993**, *53*, 1583.

RECEIVED July 18, 1994

Chapter 15

Signal Transduction Systems in Doxorubicin Hydrochloride Mechanism of Action

Paul Vichi, James Song, Jean Hess, and Thomas R. Tritton

Department of Pharmacology and Vermont Cancer Center, University of Vermont College of Medicine, Burlington, VT 05405

One of the hallmarks of cancer is loss of cellular growth regulation. Now that it has been established that many oncogenes code for proteins involved in signalling and growth control, there has developed an increased recognition that an understanding of these basic processes offers new prospects for controlling neoplasia. In this presentation we discuss two ideas stemming from our laboratory's work on this concept: (1) the ability of adriamycin to kill susceptible cells appears to involve a series of events initiated at the plasma membrane, and proceeding through the protein kinase C signal transduction pathway to ultimate damage to DNA in the nucleus; and (2) the drug brefeldin A can inhibit adriamycin's cytotoxic action, presumably by blocking the flow of information between the Golgi network and the nucleus.

Because adriamycin (ADR) is such a useful and important drug in the treatment of cancer, there has been an extensive amount of work aimed at exploiting and explaining its actions. In fact, this symposium exemplifies the fact that the anthracycline field is lively and flourishing, even after three decades of active research following the discovery of daunomycin by Arcamone in 1961. A number of the contributions seek to improve on the natural product by the time-honored approach of congener synthesis. Other workers have focussed on the effects and side effects of these agents using the tools of pharmacology and of cellular and molecular biology. Finally, a group of investigators has explored powerful biophysical methods to characterize the details of drug binding to its principal cellular loci.

One difficulty faced by all these investigators lies in defining exactly where in a cell an anthracycline is expected to bind and carry out its biologic function. Without accurate knowledge of the critical site (or sites) of interaction one cannot rationally design either new structures or experiments testing mechanistic hypotheses. The two types of binding sites for anthracyclines that have received the most attention are DNA and membranes. DNA in particular has been prominently discussed throughout the history of development of these drugs because most members of the group bind to DNA with reasonably high affinity (*1*). Membranes are also a high capacity site, however, and the anthracyclines are remarkably potent at modulating the structural and functional properties of biological membranes (*2*). The question thus becomes--is there any relationship between the actions of the anthracyclines at these two loci, or is it merely superfluous that the drugs happen to bind at these two places? In this contribution we will discuss evidence from our laboratory suggesting that there is a functional linkage between the cell surface and DNA in the nucleus and that there are at least two pathways by which information may be passed.

Relationship Between DNA and Cell Surface Actions Revealed by Temperature Studies

We will summarize here a series of experiments that exploited temperature to study adriamycin's mechanism of action (*3-5*). The original hypothesis guiding these experiments was that adriamycin would not be taken into cells at low temperature. This proved to be incorrect: while drug uptake declines as the temperature is lowered, there is still appreciable intracellular accumulation even at 0°C. Interestingly, however, cytotoxicity is lost at low temperature, and no adriamycin dose or exposure time causes cell death as long as the temperature is kept below a critical point of about 20°C. Loss of cellular accumulation cannot explain this result because the drug is still internalized at low temperature, and we have also eliminated altered subcellular distribution, metabolism, and redox chemistry as explanatory factors for the cytotoxicity temperature dependence. A key finding, however, is that topoisomerase II-mediated DNA damage does show a temperature profile parallel to cytotoxicity. Thus, above 20°C adriamycin provokes DNA damage and cell death, whereas below this temperature neither of these occurs. It would therefore seem to be true that DNA damage is functionally required for cytotoxicity to ensue.

The temperature response also offers an approach to loading cells with adriamycin to test whether the presence of intracellular drug--with none on the outside--is a necessary and sufficient condition to cause cytotoxicity. The experiment is conducted as follows (Table I): First, the cells are loaded with drug by exposure at 0°C. Because the temperature is low there are no cytotoxic sequelae even though there can be considerable drug uptake. Following equilibration of uptake at 0°C, the extracellular drug is washed away, leaving a cellular population with considerable drug accumulation but little or none available to interact with the external plasma membrane. Next, the cells are shifted to a permissive temperature (usually, 37°C) and survival measured by cloning in soft agar. The major result is that no cytotoxicity ensues unless drug is available for interaction with the cell surface no matter how mush drug is located intracellularly. Thus, the nuclear DNA can be exposed to virtually unlimited amounts of adriamycin, but there is no untoward consequence if there is no drug available to bind the plasma membrane.

TABLE I. Low Temperature Cellular Loading With Adriamycin

1. Expose at 0°C - cell has intracellular and extracellular drug
2. Wash at 0°C - cell has intracellular drug
3. Shift to 37°C - toxicity only ensues if free drug is present in medium

At first look the results described above seem to present a paradox. The temperature studies lead to two seemingly contradictory conclusions about adriamycin: (1) DNA damage is required for cytotoxicity and (2) plasma membrane interaction is required for cytotoxicity. The dilemma can be resolved, however, by postulating that information or signals are transmitted between the cell surface and the nucleus, and that these signals control the ability of the cell to sustain DNA damage and to ultimately expire. We will next describe two pathways by which such signals could be sent (summarized in Table II).

TABLE II. Physiologic Differences of Temperature

High Temperature (>20°C)	Low Temperature (<20°C)
ADR cytotoxic	ADR not cytotoxic
ADR induces DNA damage	ADR does not induce DNA damage
Plasma membrane fluid	Plasma membrane solid
Golgi mediated transport	Golgi transport inhibited

Pathways for Signal Transduction Between the Cell Surface and The Nucleus

Pathway 1: Protein Kinase C. There are a large number of signal transduction pathways used by cells to control growth by cell surface interactions. These include the activation of receptors for growth factors, hormones, and mitogens, and associated pathways of G factors, cyclic nucleotides, kinases, ion movements, and phosphoinositide metabolism (6). We began our investigations into such matters by recognizing that adriamycin has an affinity for binding phospholipids (7), and thus that phospholipid metabolism may be a reasonable target for disruption by the drug (8). We have found that S180 cells exposed to adriamycin show a rapid and sustained increase in the production of diacylglycerol resulting from the breakdown of both phosphatidylinositol and phosphatidylcholine. Thus, adriamycin activates phospholipase C, probably by disruption of bilayer structure rather than by a direct action on this phospholipid hydrolyzing enzyme.

Production of diacylglycerol is not an innocuous event since this lipid species is an intracellular activator of protein kinase C (PKC) (9). One would predict, therefore, that if adriamycin stimulates the production of diacylglycerol, it should also lead to an increase in the activity of PKC. This prediction is in fact borne out--exposure of S180 cells to adriamycin causes a dose-dependent increase in the activity of PKC in the cytosol, although with no change in the activity of the membrane PKC fraction. We have observed this drug-dependent activation of PKC in all cell types examined, but we note that in some systems (*e.g.*, CEM, KB) the PKC response occurs without any measurable increase in diacylglycerol production. Thus, there must be an alternate pathway for PKC activation that we have not yet identified or characterized.

Once we recognized that adriamycin caused activation of a kinase, the next question became—what, if anything, is the crucial substrate for this kinase that might regulate the cellular response to drug? An obvious candidate for such a substrate is topoisomerase II (topo II). It is well known that adriamycin causes a stabilization of the so-called cleavable complex of topo II/DNA, and that this event is involved in the cytotoxic action of the drug (*10*). Thus, phosphorylation of topo II by PKC could be a regulatory event that links the cell surface actions of adriamycin to the nuclear DNA damage, both of which are required for cytotoxicity. We have shown using an in vitro system (*11*) that addition of PKC and its required cofactors to nuclear extracts causes a large increase in the ability of topo II to decatenate kinetoplast DNA (Table III). Thus, phosphorylation by PKC is capable of activating the DNA damaging ability of topo II.

TABLE III. Activation of Topoisomerase II by Protein Kinase C

Condition	Relative Decatenation of Kinetoplast DNA
Nuclear extract	low
Nuclear extract, Ca^{2+}, Phosphatidyl serine	low
Nuclear extract, PKC	low
All	high

Although one is gratified to be able to show in an in vitro system that PKC can modulate topo II activity, one wonders whether this can also occur in an intact living cell. To address this question we took advantage of the fact that the phorbol ester tumor promoter 12-tetradecanoyl phorbol-13-acetate (TPA) can both activate and down regulate PKC, depending on the length of exposure (*12*). When cells are exposed to TPA for 30 min, PKC is activated (Table IV). If this PKC activation is accomplished in the presence of adriamycin, the cells accumulate more topoisomerase-mediated DNA damage and there is consequently more cell kill. Conversely, if PKC is effectively inhibited by down regulation, adriamycin causes less accumulation of DNA damage and less associated cell death. Thus, there appears to be a functional correspondence between the activity of PKC, the induction of DNA damage by topo II, and cytotoxicity in cells exposed to the anthracycline.

Figure 1 shows the adriamycin controlled information flow between the cell surface and the nucleus via the PKC pathway. This is undoubtedly an incomplete picture, and many details remain to be revealed by further experimentation, but this pathway offers a useful pathway for thinking about how adriamycin carries out its cytotoxic mission.

TABLE IV. Effect of TPA on ADR-Induced DNA Damage and Cytotoxicity

Treatment	PKC activity	DNA breaks	Cytotoxicity of ADR
TPA, 0.5 hr ADR, 2 hr	increase	increase	increase
TPA, 24 hr ADR, 2 hr	decrease	decrease	decrease

Plasma membrane--> binding	Phospholipid--> binding	PKC --> activation	TopoII/DNA --> stabilization	DNA --> damage	Cell death

Figure 1. Involvement of PKC in ADR-induced cytotoxicity.

Pathway 2: The Golgi Connection. It has been known for many years that the cellular processes of endocytosis and excocytosis show a temperature dependence reminiscent of the adriamycin cytotoxicity profile, namely loss of function at temperatures below about 20°C. This occurs because the vesicular fusions required for intracellular trafficking of proteins and lipids are inhibited at low temperature (*13*). With the recent discovery that the natural product Brefeldin A (BFA) blocks these reactions as well (*14,15*), we realized that this reagent might be analogous to the temperature variable in regulating adriamycin's action. Our reasoning was that if intracellular trafficking through the Golgi network was involved in mediating the signals or modifying proteins necessary to transduce adriamycin's action, then BFA could protect against adriamycin cytotoxicity just as temperature does. In fact, we have found that this is the case (*16*): pretreatment of cells with BFA prior to exposure to adriamycin causes a several fold increase in the IC_{50} for the anticancer agent (Table V). The response to BFA is different from the response to temperature, however, in three important respects: (1) the protection by BFA is not absolute since high doses of adriamycin can overcome the

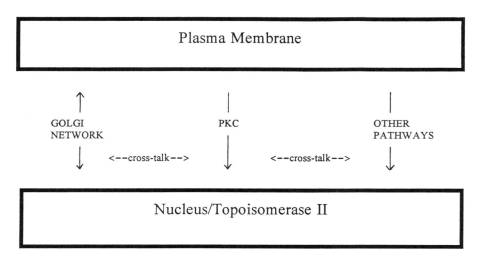

Figure 2. Pathways that may mediate signals between the cell surface
and the nucleus in adriamycin-treated cells.

effect; (2) BFA does not alter the uptake or accumulation of BFA into cells; and (3)
BFA pretreatment does not reduce the ability of adriamycin to cause DNA damage,
even though it does protect the cells from cytotoxicity, and thus effects a functional
decoupling of DNA damage from cell death.

TABLE V. Effect of BFA on ADR Cytotoxicity in L1210 Cells

Condition	ADR IC$_{50}$
ADR alone, 2 hr exposure	30 nM
BFA pretreatment, 2 hr Followed by ADR, 2 hr	100 nM

Taken together, these results support the proposal that vesicular traffic through the Golgi
network can contribute to or serve as an intermediary signal in promoting adriamycin's
cytotoxic action. Several key questions remain to be answered, however, including
where along the sequence of events the Golgi apparatus is positioned, what is the nature
of the "traffic" through the Golgi that carries or promotes the cytotoxic signal, and why
this pathway can be overcome or avoided by raising the drug dose, since this is not the
case for temperature.

Conclusion

The scheme in Figure 2 illustrates the ideas presented in this paper. We have discussed two pathways that may mediate signals between the cell surface and the nucleus when cells are exposed to the cytotoxic anticancer agent adriamycin. This figure should not be interpreted too literally: it is offered as a framework for designing experiments to test the importance and relevance of these conncections. We show pathways 1 and 2 as parallel, but in fact they may be sequential or connected by "cross-talk" reactions. Also, there may be additional or alternate pathways whereby adriamycin translocates signals to the nucleus, and these may also be linked in signalling networks. Clearly, additional details will need to emerge before we can confidently state that we understand adriamycin's ability to be a cytotoxic agent, but the scheme shown offers a unified hypothesis that links the various cell surface and nuclear actions of the drug into a self-consistent picture.

Literature Cited

1. Neidle, S. *Prog. Med. Chem.* **1979**, *16*, 151-220.
2. Tritton, T.R. *Pharmacol. Therap.* **1991**, *49*, 293-309.
3. Vichi, P., Robison, S., and Tritton, T.R. *Cancer Res.* **1989**, *49*, 5575-5580.
4. Lane, P., Vichi, P., Bain, D.L., and Tritton, T.R. *Cancer Res.* **1987**, *47*, 4038-4042.
5. Vichi, P., and Tritton, T.R. *Cancer Res.* **1992**, *52*, 4135-4138.
6. Nigg, E. *Adv. Cancer Res.* **1990,** *55*, 271-310.
7. Murphree, S.A., Tritton, T.R., Smith, D.L., and Sartorelli, A.C. *Biochim. Biophys. Acta* **1981**, *649*, 317-324.
8. Siegfried, J.M., Kennedy, K.A., Sartorelli, A.C., and Tritton, T.R. *J. Biol. Chem.* **1983**, *258*, 339-343.
9. Kishimoto, A., Takai, Y., Mor, T., Kikkawa, U. and Nishizuka, Y. *J. Biol. Chem.* **1980**, *255*, 2273-2276.
10. Tewey, K.M., Rowe, T.C., Yang, L., Halligan, B.D., and Liu, L.F. *Science* **1984**, *226*, 466-468.
11. Posada, J., Vichi, P. and Tritton, T.R. *Cancer Res.* **1989**, *49*, 6634-6639.
12. Song, J.S., Bhushan, A., and Tritton, T.R. *Proceedings of AACR* **1992**, *33*, 436.
13. Kuismanen, E. and Saraste, J. In *Methods in Cell Biology*; Tartakoff, A.M., Ed., Academic Press, San Diego, CA, 1989, 32; pp. 257-274 .
14. Klausner, R.D., Donaldson, J.B., and Lippincott-Schwartz, J. *J. Cell Bio.* **1992**, *116*, 1071-1080.
15. Misumi, Y., Misumi, Y., Miki, K., Takatsuki, A., Tamura, B., and Ikehara, Y. *J. Biol. Chem.* **1986**, *261*, 11398-11403.
16. Vichi, P.J., and Tritton, T.R. *Cancer Res.* **1993,** *53*, 5237-5243.

RECEIVED June 3, 1994

Chapter 16

Analysis of Multidrug Transporter in Living Cells

Application to Uptake and Release of Anthracycline Derivatives by Drug-Resistant K562 Cells

Arlette Garnier-Suillerot, Frédéric Frézard, Marie-Nicole Borrel, Elene Pereira, Jolanta Tarasiuk, and Marina Fiallo

Laboratoire de Chimie Bioinorganique (LPCB UA CNRS 198), Université Paris Nord, 74 rue Marcel Cachin, Bobigny 93012, France

Among the 1 million deaths per year due to cancer in Europe, it has been estimated that about 90% of them were influenced by the problem of multidrug resistance (MDR) to chemotherapeutic agents. In vitro, the cellular pharmacological basis for MDR appears to be: a reduced steady state accumulation of drugs caused by the overexpression of the mdr1 gene which encodes for a plasma glycoprotein (P-gp) of 170-180kDa. P-gp is thought to function as an energy-dependent drug efflux pump.

To analyze the function of the multidrug transporter in living cells we have developed a new method with which it is possible to follow the uptake of fluorescent drug by cells, continuously, as incubation of the drug with the cells proceeds. This technique has been used to study the mechanism of anthracycline accumulation in drug-sensitive and drug-resistant K562 cells in the absence and in the presence of various inhibitors.

The anthracycline antibiotic adriamycin (ADR) is one of the most potent anticancer drugs in clinical use (1). Other anthracycline derivatives such as daunorubicin (DNR), 4'-O-tetrahydropyranyladriamycin (THP-ADR), carminomycin (CAR) and aclacinomycin (ACM) also have outstanding antitumor activity (Scheme 1).

One of the major obstacles to chemotherapy is that, after repeated treatments, cellular resistance to drug appears, and it has been estimated that among the millions of deaths per year due to cancer world-wide, 90% of them are influenced by the problem of drug resistance. The problem is that the tumor cells become resistant not only to the drug that has been used during the treatment but also to other drugs that are structurally and functionally unrelated. This is termed "multidrug resistance" (MDR).

MDR is now a well-characterized phenomenon (2-5). However, the mechanisms of MDR are far from elucidated and, in fact, modifications of several cellular functions have been observed. In molecular terms, the best known and most widely found change associated with MDR is the overexpression of P-glycoprotein (P-gp), a plasma membrane protein that acts as an ATP-dependent efflux pump able to eliminate drugs and lipophilic substances from the cells (6-10).

This general concept, however, does not explain the exact mechanism of drug removal by P-gp. This protein, encoded by the mdr1 gene, belongs to a superfamily of ABC-transport proteins (ABC for ATP binding cassette) that have six transmembrane

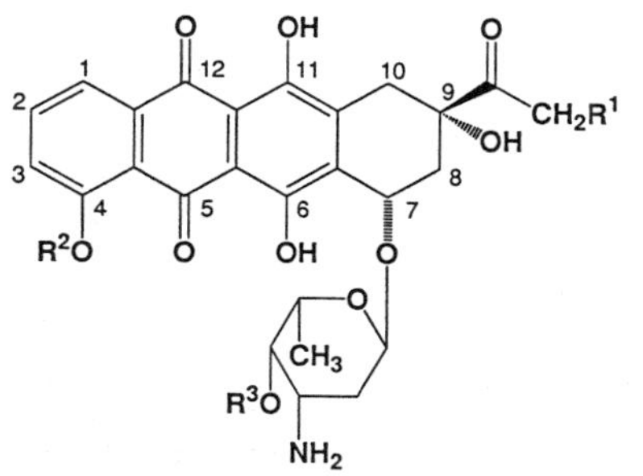

	R^1	R^2	R^3
Adriamycin	OH	CH$_3$	H
Daunorubicin	H	CH$_3$	H
Carminomycin	H	H	H
THP-Adriamycin	OH	CH$_3$	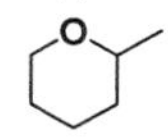

Scheme 1

spanning regions followed by a nucleotide binding site (11). Its members are implicated in the ATP-dependent transport of various substrates across cell membranes. The human cystic fibrosis transmembrane regulator (CFTR) also belongs to this family (12). The mechanism of the transport process is not known for any of the transporters of the ABC superfamily, including one reconstituted in artificial vesicles (13).

Several models are being considered to explain how various lipophilic drugs are removed from the cells in an ATP-dependent manner. For instance, (i) drug enters the cell by passive diffusion through the lipid bilayer and binds to P-gp on the cytoplasmic side of the membrane or (ii) drug binding to P-gp occurs not within the cell but rather inside the bilayer (14).

Recently, additional functions have been found for P-gp: it has been shown (i) that P-gp has a chloride channel function (15) and (ii) that P-gp in plasma membranes serves as a channel for steady ATP release from the cell (16). It has also been suggested that several properties of P-gp may be explained if it acts as a "flippase" to transport drugs from the inner leaflet of the lipid bilayer to the outer or to the external medium (17).

A number of molecules are able to modulate MDR and to enhance the cytotoxicity of anticancer drugs in MDR cells. Tsuruo et al. were the first to demonstrate that verapamil could enhance the cytotoxicity of vincristine and vinblastine in murine MDR cells (18). Since then, many other membrane-active agents with this property have been found (19). Indeed, these membrane-active agents appear to antagonize the "binding" of anticancer drug to P-gp. In studies using membrane isolated from MDR cells, it has been shown that photoaffinity analogs of anticancer drugs that are substrates for P-gp (e.g., vinblastine (20) and DNR (21)), or compounds that block P-gp function (e.g., verapamil (22)) appear to bind to P-gp. However, it is not known if they bind to the same site on the protein. Whether P-gp serves as a "strict" ATP-dependent drug efflux pump with a common binding site for all substrates has been questioned because of the difficulties in showing competitive interactions among related drugs in intact cells.

To analyze the function of the multidrug transporter in living cells, we have developed a method that uses the changes in the fluorescent characteristics of the transported compound when it moves between the intra and extracellular medium to follow (i) the uptake of fluorescent drugs by cells, continuously, as incubation of the drug with the cells proceeds and (ii) the functionality of P-gp without isolating this enzyme. This means that we can measure directly within the cells the kinetics of the reaction catalyzed by P-gp (i.e., the active efflux of the drug) and the substrate concentration (i.e., the free drug concentration in the cytosol) (23-28). This method does not perturb the drug distribution in the various cell compartments and is a simple and rapid technique for measuring membrane transport rates or changes in membrane transport rates.

We have used this method to study the mechanism of anthracycline accumulation in drug-sensitive and drug-resistant erythroleukemia K562 cells.

The Spectrofluorometric Method

Up until now, the methods used to determine the drug concentration in the different cell compartments have involved fractionating drug-loaded cells, a process that can lead to redistribution of drug in the various cellular components during the fractionation procedure. The spectrofluorometric method that we have developed is based on the following three observations: (i) The major binding sites for most anthracycline derivatives inside the cell are provided by the nucleus. (ii) Anthracyclines are fluorescent molecules and their fluorescence is quenched when they intercalate between the base pairs in the DNA only; their interactions with other cellular components such as membrane do not yield fluorescence quenching. (iii) The passage of the drug through the plasmic membrane is the limiting step for its kinetics of uptake by the cell. Under these conditions, the monitoring of the decay of fluorescence of anthracycline in the presence of cells is a means of measuring the kinetics as well as gaining insight into the mechanisms of drug transport through the plasmic membrane (Figure 1).

Initial Rate of Uptake of Anthracycline Derivatives by K562 Cells

We have studied the uptake of ADR, DNR, THP-ADR, CAR, and ACM by drug-resistant and -sensitive cells and have clearly shown that, in both cases, the influx occurs by passive diffusion of the neutral form of the drug through the lipid bilayers (26). Once inside the cells, the drugs intercalate between the base pairs of DNA in the nuclei, which are strong binding sites for these molecules. Thus, the anthracyclines gain access to cells by passive diffusion of the neutral form of the drug under the action of a driven force provided by DNA in the nucleus. This was corroborated by the study of the uptake of anthracycline derivatives into large unilamellar vesicles (LUV) in response to a driven force provided by DNA encapsulated inside the LUV (27).

We have compared the initial rate of uptake (V_+) of the five anthracycline derivatives. For these measurements, 10^6 cells/ml were incubated with 1 μM drug at pH 7.25 and at 37°C. Our main conclusions were as follows:

(i) the initial rate of uptake is the same in both drug-sensitive and -resistant cells.

(ii) the initial rate of uptake is correlated with the pK_a of deprotonation of the drug, and the kinetics of uptake increases as the pK_a of deprotonation decreases. This is in agreement with the fact that the uptake occurs by free permeation of the neutral form of the drug. Also, we observed a correlation between the initial rate of uptake and the degree of resistance ρ: ρ decreases as the initial rate of uptake V_+ increases. This means that the MDR efflux system having a finite speed (29), highly lipophilic compounds can overhelm the efflux system since drug uptake occurs too rapidly for efflux to effectively compete.

(iii) The outer membrane fluidity of sensitive and resistant cells has been determined as a function of temperature (28) and we have observed a thermotropic transition at 20°C in agreement with literature data (30). However, the initial rates of uptake determined for THP-ADR and ACM at temperatures ranging from 5°C to 40°C do not exhibit this thermotropic transition, and when the rates are depicted in an Arrhenius plot, straight lines with only one slope are found. The activation energies thus obtained are 45±9 kJ mol^{-1}. This suggests that the passage of the unprotonated forms of the anthracycline molecules occurs in lipid domains whose transition temperature is lower than 5°C (28).

Determination of the Free Drug Concentration in the Cytosol

The pH inside the cytosol, 7.25±0.05, is the same in both drug-sensitive and -resistant K562 cells. In sensitive cells, at the steady state, the equilibrium transmembrane concentration verified the relation $[DH^+]_i/[DH^+]_e = [H^+]_i/[H^+]_e$, where $[DH^+]_i$ and $[DH^+]_e$ are the concentration of the protonated form of the free drug inside and outside the cells, respectively. In other words, the concentration of drug in the neutral form is the same inside the cytosol and in the extracellular medium $([D^0]_i = [D^0]_e)$. Inside the cell, the drug free in the cytosol is in thermodynamic equilibrium with the drug bound to the nucleus. Under these conditions it is very easy to determine the free drug concentration in the cytosol of sensitive cells, which depends on the extracellular pH and the temperature only (Scheme 2).

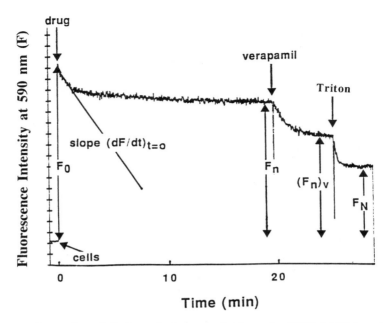

Figure 1 Uptake of 4'-O-tetrahydropyranyladriamycin (THP-ADR) by drug-resistant K562 cells. F, the fluorescence intensity at 590 nm (λex=480 nm) is recorded as a function of time. Cells (2x106) are suspended in a cuvette filled with 2 ml buffer at pH_e 7.25 with vigorous stirring. At t = 0, 20 μl of a 100 μM stock (THP-ADR) solution is added to the cells, yielding a C_T=1 μM THP-ADR solution. The fluorescence intensity is then F_0. The slope of the tangent to the curve F=f(t) at t=0 is $(dF/dt)_{t=0}$, and the initial rate of uptake $V_+ = (dF/dt)_{t=0}$ (C_T/F_0). Once the steady state is reached, the fluorescence is F_n and the concentration of drug intercalated between the base pairs in the nucleus is C_n=$C_T(F_0-F_n)/F_0$. When the steady state is reached, verapamil is added. At the new steady state, the fluorescence intensity is $F_n{}^V$. The addition of 0.05% Triton X-100 yields the equilibrium state. The fluorescence intensity is then F_N.

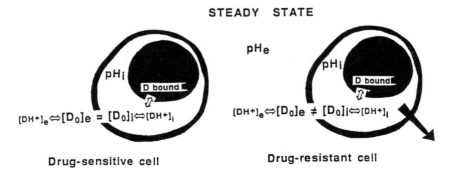

Scheme 2

In the case of drug-resistant cells, due to the presence of the P-gp, the transmembrane equilibrium does not exist, and the free drug concentration in the cytosol is lower than in the extracellular medium ($[D^0]_i < [D^0]_e$). However, the thermodynamic equilibrium between the free drug in the cytosol and the drug bound to the nucleus is the same as in sensitive cells. Under that condition and from the knowledge of the amount of drug bound to the nucleus, we can determine the amount of free drug in the cytosol (27). This point was clearly demonstrated by the observation that the amount of drug bound to the nucleus at the steady state, measured at different temperatures (5°C-40°C), increased as the temperature decreased. We have determined the enthalpies of formation of the drug-nucleus complex inside the living cells at the steady state as well as those of the drug-DNA complex. In both cases the values compare and are characteristic of an exothermic process (28).

Determination of the Kinetics of Active Efflux V_a

In the following section, we will focus mainly to the data obtained with THP-ADR because this drug enters the cells very rapidly and the steady state is reached within about 20 minutes at 37°C. The rate of P-gp-mediated efflux of THP-ADR can be determined at the steady state, taking into accounts the following points: (i) the passive influx and efflux of the drug occurs through free permeation of the neutral form of the drug; (ii) whatever the type of cells, i.e., either drug-sensitive or drug-resistant, the rate of drug influx at the steady state is equal to that of drug efflux; (iii) for drug-resistant cells, the efflux should be composed of two terms, namely, a passive efflux of the neutral form of the drug and a P-gp-mediated efflux of the drug (26).

It follows that for sensitive cells $(V_+)_s = (V_-)_s$ and for resistant cells $(V_+)_s = (V_-)_s + (V_a)_s$, where $(V_+)_s$ and $(V_-)_s$ are the passive influx and efflux of the neutral form of the drug at the steady state, respectively, and $(V_a)_s$ is the P-gp-mediated active efflux at the steady state (Scheme 3). Thus,

$$(V_a)_s = (V_+)_s - (V_-)_s$$

$$(V_a)_s = P_+^0 \, n \, S \, ([D^0]_e - [D^0]_i)_s$$

where P_+^0 is the permeability constant for the neutral form of the drug, S the membrane exchange area per cell, n the number of cells in 1 cm^3, and $([D^0]_e)_s$ and $([D^0]_i)_s$ the neutral drug concentration at the steady state in the extracellular medium and in the cytosol, respectively. Here, we make the reasonable assumption that the permeability coefficients are the same for the passive influx and efflux ($P_+^0 = P_-^0$) and are the same at t=0 and at the steady state. The permeability coefficient can thus be determined from the initial rate of uptake as follows $(V_+)_{t=0} = P_+^0 \, n \, S \, ([D^0]_e)_{t=0}$, where $([D^0]_e)_{t=0}$ is the extracellular neutral drug concentration at t=0. It follows that

$$(V_a)_s = (V_+ / [D^0]_e)_{t=0} \, ([D^0]_e - [D^0]_i)_s$$

as under our experimental conditions, the extracellular pH$_e$ is equal to the cytosolic pH$_i$

STEADY   STATE

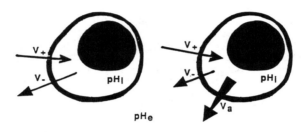

Drug-sensitive cell   Drug-resistant cell

$$V_+ = V_-$$ $$V_+ = V_- + V_a$$

V_+ kinetics for the passive influx of the neutral form of the drug

V_- kinetics for the passive efflux of the neutral form of the drug

V_a kinetics for the active P-glycoprotein-mediated efflux of the drug.

Scheme 3

$([D^0]_e-[D^0]_i)_s/([D^0]_e)_{t=0} = (C_e-C_i)_s/(C_e)_{t=0}$, where C_e and C_i are the extracellular and intracellular drug concentrations respectively. It follows that

$$(V_a)_s = (V_+/C_e)_{t=0}\cdot(C_e-C_i)_s$$

Direct determination of the P-gp-mediated efflux of THP-ADR can be made when cells are incubated in the presence of N_3^- and in the absence of glucose (30 min, $[N_3^-]$ = 10 mM). In these conditions, there is no ATP synthesis and the P-gp-mediated active efflux of THP-adriamycin is blocked (*31*). The incorporation of THP-ADR in these energy-depleted resistant cells thus compares and the free drug concentration is the same in the cytosol and in the extracellular milieu. The subsequent addition of glucose (5 mM) gives rise to the ATP synthesis via the glycolysis pathway and, after about 30 s, one observes an increase in the fluorescence signal due to the release of drug from the cells (Figure 2).
The rate of active efflux V_a is then equal to

$$V_a=(dF/dt)_{glu}\ C_T/F_0$$

where $(dF/dt)_{glu}$ is the slope of the tangent to the curve $F = f(t)$ at the time corresponding to the glucose addition.
Whatever the method used, we can simultaneously determined V_a and C_i. The plot of V_a as a function of C_i clearly shows a saturation of the active efflux (Figure 3). A double reciprocal plot of $1/V_a$ as a function of $1/C_i$ (Lineweaver-Burk plot) yields the Michaelis constant $K_M = 0.5\pm0.3$ µM (Figure 4) (*29*). This value is in good agreement with those reported in the literature using membrane isolated from resistant cells.
Similar experiments were performed in the presence of verapamil, and our data strongly suggest that the inhibition of P-gp-mediated efflux of THP-ADR by verapamil is a noncompetitive process (Figure 4). This is at variance with the data reported in the literature, which leads to the conclusion that this inhibition was competitive.
We have also determined the kinetics of active efflux of THP-ADR as a function of temperature. The activation thus obtained is very low, 25 ± 7 kJmol^{-1}, and we can tentatively suggest that P-gp may facilitate the transbilayer flux of drug by forming a low resistance nonspecific shunt between protein and lipid.
Using the same method, we have compared the P-gp-mediated efflux of THP-ADR in the different phases of the drug-resistant cell cycle. For this purpose, centrifugal elutriation, which allows the isolation of cell subpopulation according to specific cell-cycle phase in a short time and with minimal metabolic perturbation, was used. The value of $V_a / n\ S\ C_i$ thus calculated is nearly constant whatever the cycle phase. This means that the number of molecules of THP-ADR that are actively effluxed per cell and per second is constant and does not depend on the cycle phase. This shows that the amount of P-gp that is functionally active at the beginning of the cell cycle and that pumps out THP-ADR, is not modified during cell cycle. To account for this observation, the following two mechanisms can be evoked: (i) like the vast majority of the many different proteins and RNA molecules present in a cell, P-gp is synthesized continuously throughout interphase, but the newly synthesized P-gp is not functional before mitosis, (ii) P-gp is synthesized late in G2 (*32*). Experiments have been undertaken to quantify the amount of P-gp present in the different phases.
In conclusion, using the fluorometric method that we have developed, it is possible to study the functionality of P-gp directly in living cells.

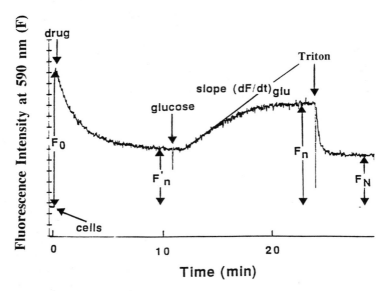

Figure 2 Direct determination of the active kinetics of efflux V_a of 4'-*O*-tetrahydropyranyladriamycin (THP-ADR) under the effect of P-glycoprotein. Drug-resistant K562 cells (10^6/ml) are incubated with 1µM THP-ADR in the presence of 10 mM N_3^- and in the absence of glucose. F, the fluorescence intensity at 590 nm (λ_{ex}=480nm), is recorded as a function of time. At the steady state, the fluorescence intensity is F_n', and the concentration of drug intercalated between the base pairs equal to $C_n' = C_T(F_0-F_n')/F_0$. At t = t_{glu}, 5 mM glucose is added, yielding the release of THP-ADR. The slope of the tangent to the curve is $(dF/dt)_{t_{glu}}$, and the kinetics of release of THP-ADR $V_a = (dF/dt)t_{glu}$ (C_T/F_0). At the new steady state the fluorescence intensity was F_n. The addition of 0.05% Triton X-100 yielded the equilibrium state. The fluorescence intensity was then F_N.

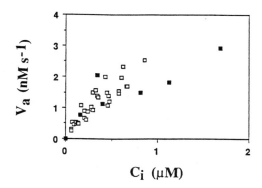

Figure 3 Variation of the kinetics of P-glycoprotein-mediated efflux of THP-adriamycin (THP-ADR) as a function of C_i, the cytosol free-drug concentration.

Resistant cells (10^6/ml) are incubated in the presence of various concentrations of THP-ADR ranging from 0.25 to 3 μM. The incubation medium is either the glucose containing Hepes buffer at pH_e 7.25 ($\square$) or the glucose depleted-Hepes buffer in the presence of 10 mM N_3^- ($\blacksquare$).

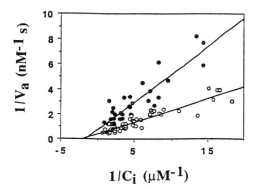

Figure 4 Non competitive inhibition of THP-adriamycin (THP-ADR) transport by verapamil. The kinetics of active efflux V_a and the cytosol drug concentration C_i have been determined in the absence ($\circ$) and in the presence ($\bullet$) of 2 μM verapamil. The figure shows the double reciprocal plot (Lineweaver Burk plot) of the data. The two fitted curves were calculated by linear regression. The correlation coefficients are 0.94 (in the absence of verapamil) and 0.91 (in the presence of 2 μM verapamil).

Acknowledgments: This investigation was supported by: l'Association pour la Recherche sur la Cancer (ARC), l' Universite Paris Nord et le Centre National de la Recherche Scientifique.

Literature Cited.

1. Arcamone, F. *Med. Chem.* **1980**, *16*, 1-41.
2. Pastan, I.H.; Gottesman, M.M. *N. Engl. J. Med.***1987**, *316*, 1388-1393.
3. Bradley, G.; Juranka, P.F.; Ling, V. *Biochim. Biophys. Acta* **1988**, *948*, 87-128.
4. Beck, W.T.; Danks, M.R. In *Molecular and Cellular Biology of Multidrug Resistance in umor Cells*; Roninson I.B., Eds; Plenum Press: NY, 1991. pp 3-46.
5. Gottesman, M. M.; Pastan, I. *Annu. Rev. Biochem.* **1993**, *62*, 385-427.
6. Dano, K. *Biochim. Biophys. Acta* **1973**, *323*, 466-483.
7. Riordan, J.R.; Ling, V. *Pharmacol. Ther.* **1985**, *28*, 51-76.
8. Skovsgaard, T. *Cancer Res.***1978**, *38*, 1785-1791.
9. Juliano, R.L.; Ling, V. *Biochim. Bioiphys. Acta* **1976**, *455*, 152-162.
10. Beck, W.T.; Mueller, T.J.; Tanzer, L.R. *Cancer Res.***1979**, *39*, 2070-2076
11. Roninson, I.B. In *Molecular and Cellular Biology of Multidrug Resistance in Tumor Cells*; Roninson I.B., Eds; Plenum Press: NY, 1991. pp 189-209.
12. Hyde, S.C.; Emsley, P.; Hartshorn, M.J.; Mimmack, M.M.; Gileadi, U.; Pearce, Gallgher, S.R.; Gill, M.P.; Hubbard, R.E.; Higgins, C.F. *Nature* **1990**, *346*, 362-365.
13. Bishop, L.; Agbayani, R.; Ambudkar, S.V.; Maloney, P.C.; Ames, G.F. *Proc. Natl. Acad. Sci. U.S.A.* **1986**, *86*, 6953-6957.
14. Gros, P.; Croop, J.; Housman, D.E. *Cell* **1986**, *47*, 371-380.
15. Gill, D.R.; Hyde, S.C.; Higgins, C.F.; Valverde, M.A.; Mintenig, G.M.; Sepulveda, F. V. *Cell* **1992**, *71*, 23-32.
16. Abraham, E.H.; Prat, A.G., Gerwec; L., Seneveratne, T.; Arceci, R.; Kramer, R.; Guidotti, G.; Cantiello, H.F. *Proc. Natl. Acad. Sci. U.S.A.* **1993**, *90*, 312-316.
17. Higgins, C.F.; Gottesman, M.M. *TIBS* **1992**, *17*, 18-21.
18. Tsuruo, T.; Iida, H; Tsukagoshi, S.; Sakurai, Y. *Cancer Res.* **1981**, *41*, 1967-1972.
19. Pearce, H.L.; Safa, A.R.; Bach, N.J.; Winter, M.A.; Cirtain, M.C.; Beck, W.T. *Proc. Natl. Acad. Sci. U.S.A.* **1989**, *86*, 5128-5132.
20. Safa, A.R.; Glover, C.J.; Meyers, M.B.; Biedler, J.L.; Felsted, R.L. *J. Biol. Chem.* **1986**, *261*, 6137-6140.
21. Busche, R.; Tummler, B.; Riordan; J.R.; Cano-Gauci, D.F. *Mol. Pharmacol.***1989**, *30*, 388- 397.
22. Safa, A.R. *Proc. Natl. Acad. Sci. U.S.A.***1988**, *85*, 7178-7191.
23. Tarasiuk, J; Frezard, F.; Garnier-Suillerot, A.; Gattegno, L. *Biochim. Biophys. Acta* **1989**, *1013* , 109-117.
24. Frézard, F.; Garnier-Suillerot, A. *Biochim. Biophys. Acta* **1990**, 1036, 121-127.
25. Frézard, F.; Garnier-Suillerot, A. *Biochim. Biophys. Acta* **1991**, *1091*, 29-35.
26. Frézard, F ; Garnier-Suillerot, A. *Eur. J. Biochem.* **1991**, *196*, 483-491.
27. Frézard, F ; Garnier-Suillerot, A. *Biochemistry* **1991**, *30*, 5038-5043.
28. Tarasiuk, J.; Garnier-Suillerot, A *Eur. J. Biochem.* **1992**, *204*, 693-698.
29. Pereira, E.; Borrel, M.N.; Fiallo, M.; Garnier-Suillerot, A. *Biochim. Biophys. Acta* **1994**, *1225*, 208-216.
30. Deliconstantinos, G.; Kopeikina-Tsiboukidou, L.; Villiotou, V. *Biochem. Pharmacol.***1987**, *36*, 1153-1161.
31. Beck, W.T. *Adv. Enzyme Regul.***1984**, *22*, 207-227.
32. Tarasiuk, J.; Foucrier, J.; Garnier-Suillerot, A.*Biochem. Pharmacol.***1993**, *45*, 1801-1808.

RECEIVED July 13, 1994

Chapter 17

Role of Reactive-Oxygen Metabolism in Cardiac Toxicity of Anthracycline Antibiotics

James H. Doroshow

Department of Medical Oncology and Therapeutics Research, City of Hope National Medical Center, Duarte, CA 91010

Anthracycline antibiotics enhance reactive oxygen radical formation in adult rat heart myocytes. Reactive oxygen metabolites are formed in essentially all intracellular compartments of the myocyte and play an important role in the disruption of critical cardiac homeostatic functions including mitochondrial energy production and calcium sequestration in the sarcotubular system. To date, the only agent known to decrease anthracycline cardiac toxicity effectively in the clinic is the EDTA analog ICRF-187, which has been shown in cell-free systems to chelate iron avidly. The studies reported here demonstrate that ICRF-187 is able to bind and efflux free iron from adult rat heart myocytes. Since iron plays a critical role in catalyzing the formation of strong oxidants, such as the hydroxyl radical, after anthracycline redox cycling, these experiments support a role for reactive oxygen species in the cardiac damage produced by the anthracycline antibiotics.

The anthracycline antibiotics (doxorubicin and daunorubicin) play a critical role in the treatment of both hematologic malignancies and cancers of the breast, lung, and ovary (1). Unfortunately, they produce a chronic cardiomyopathy that limits their clinical usefulness. Since the introduction of the anthracyclines over 20 ago, a wide variety of hypotheses have been suggested to explain their cardiac toxicity. In general, data supporting these hypotheses can be divided into those studies suggesting a direct effect of the antibiotics on one or more biochemical processes in the myocyte and those that suggest a role for anthracycline-enhanced reactive oxygen radical formation in the etiology of anthracycline cardiac toxicity. Outlines of the experimental results supporting these alternate hypotheses are shown in Tables I and II.

0097–6156/95/0574–0259$08.00/0
© 1995 American Chemical Society

Table I. Non-oxidative Mechanisms of Anthracycline Cardiac Toxicity[a]

Inhibition of mitochondrial electron transport enzymes: succinate dehydrogenase and oxidase; cytochrome c oxidase; requires > 0.5 mM doxorubicin

Direct membrane binding: cardiolipin; facilitates mitochondrial injury

Direct interaction with ryanodine receptor enhancing calcium release from cardiac sarcoplasmic reticulum; inhibition of sarcolemmal calcium ATPase

Down regulation of cardiac beta adrenergic receptor density

Inhibition of cardiac metmyoglobin reductase

Altered cardiac glucose metabolism

Inhibition of specific cardiac mRNA's: c-actin, α-actin, troponin I, myosin light chain 2; but not β-actin

[a]Summarized from refs. 2, 18-25.

Many of the direct effects of the anthracycline antibiotics involve alterations in critical membrane functions, including inhibition of mitochondrial electron transport, plasma membrane or sarcotubular ion transport, and receptor function. The affinity of doxorubicin for cardiolipin may play an important role in producing toxic effects on cardiolipin-rich cardiac mitochondria. Unfortunately, most of the studies that have demonstrated impaired function of cardiac mitochondria or the calcium pump of the sarcoplasmic reticulum have employed anthracycline concentrations far in excess of those ever achieved in vivo; thus, the ultimate importance of such effects is unclear. However, recent studies demonstrating specific inhibition of cardiac (rather than skeletal muscle) mRNAs which code for critical myofibrillar proteins are particularly noteworthy (2). Since cardiac myofibrils are one major site of injury after exposure to anthracycline antibiotics that has been consistently demonstrated in both animal model systems and in man, these studies are of substantial potential importance despite the fact that the mechanism involved in the downregulation of their RNAs remains to be determined.

The ability of various flavoproteins to reduce the anthracycline quinone has been known for over 15 years (3). However, the demonstration of an active cycle of reduction and oxidation of the anthracycline quinone moiety in vivo leading to the generation of reactive oxygen species in the heart occurred much more recently (4). Reduction of the anthracycline quinone occurs at multiple sites within the heart, including: (i) complex I of the mitochondrial electron transport chain where acceptance of an electron by the quinone moiety occurs between NADH dehydrogenase and an iron-sulfur center, (ii) the sarcoplasmic reticular membrane

Table II. Oxidative Mechanisms of Anthracycline Cardiac Toxicity[a]

Inhibition of calcium sequestration by cardiac sarcoplasmic reticulum; decrease IC_{50} for doxorubicin 10-20 fold by enzymatic drug activation

Inhibition of NADH dehydrogenase between its flavin and iron-sulfur center N-1: oxygen dependent; occurs at the site of anthracycline reduction; requires low micromolar anthracycline concentrations

Generalized membrane lipid peroxidation

Oxidation of oxymyoglobin: potential for the production of "ferryl" myoglobin

Iron "delocalization"

[a]Summarized from refs. *5, 8, 10, 14, 26,* and *27.*

where reduction is a consequence of electron transfer from the NADPH cytochrome P-450 reductase present at that site, and (iii) the cytoplasmic compartment where the oxidation of oxymyoglobin to metmyoglobin by the anthracycline leads to reduction of the quinone (*5-7*). In an aerobic environment, molecular oxygen is rapidly reduced by the anthracycline semiquinone forming the superoxide anion, and subsequently, hydrogen peroxide. These reactions are associated with "site specific" inhibition of mitochondrial NADH dehydrogenase and sarcoplasmic calcium ATPase and occur at the low micromolar levels of anthracycline that are available after drug treatment in man (*8,9*). It is likely that the final common end product of the reductive metabolism of the anthracycline antibiotics catalyzed by these flavin dehydrogenases is the generation (through a metal-enhanced Haber-Weiss reaction) of species with the chemical reactivity of the hydroxyl radical; it is clear, however, that this may be either the free hydroxyl radical or an oxo-metal compound of elevated oxidation state, such as "perferryl" iron (*5,10-12*).

Support for the hypothesis that metal-dependent free radical species contribute to the cardiac toxicity of the anthracyclines comes not only from the demonstration of the production of such radicals in subcellular compartments, but also from spin-trapping studies in the intact heart where hydroxyl radical-like intermediates have been identified after treatment with doxorubicin in vivo (*4*). Furthermore, clinical studies in man have shown that the administration of ICRF-187--an iron chelating derivative of EDTA--to patients receiving doxorubicin significantly diminishes the dose-dependent decrease in cardiac ejection fraction associated with anthracycline therapy (*13*).

However, these results do not provide the basis for a complete understanding of the pathophysiology of the cardiac injury produced by doxorubicin. For example, while it has recently been demonstrated that anthracycline-stimulated free radical formation leads to the release of protein-bound iron from ferritin in a cell-free system, no such data exist for intact cells, including cardiac myocytes (*14*). Thus, the source of the intracellular iron that must be available to catalyze strong oxidant formation is unknown. Furthermore, while it is clear that ICRF-187 both chelates iron in vitro and protects the heart from doxorubicin treatment in the clinical setting, very little information is currently available to link experiments done in the absence of cells with the results of the clinical trials of ICRF-187. Therefore, in the experiments reported here, we have begun to examine the question of whether intracellular iron chelation could explain the cardioprotective mechanism of action of ICRF-187.

Materials and Methods

Doxorubicin was obtained from Adria Laboratories, Columbus, OH. ICRF-187 and its ^{14}C-labeled derivative were obtained from the Drug Synthesis Branch, National Cancer Institute. Adult rat heart myocytes were prepared by collagenase perfusion of 200-250 g male Sprague-Dawley rat hearts by a previously described technique (*15*). The cytotoxic effect of doxorubicin on myocytes was determined by the loss of rodlike morphology and the ability to exclude 0.1% trypan blue dye. Cells were incubated in Tyrode's buffer at 37°C for 3 h in a shaking water bath with or without doxorubicin or ICRF-187 at the indicated concentrations. To measure uptake of ICRF-187, the labeled drug (400-1000 μM) was added to an equal volume of myocytes; cells were mixed, incubated for the indicated periods of time, and then 1-ml aliquots were overlaid on 0.4 ml of silicone oil and immediately centrifuged for 30 s at 16,000 X g at room temperature. After washing the pellet, 0.1 ml of trifluoroacetic acid was added and the cells were sonicated for 30 s; the sonicate was then centrifuged for 30 s and the supernatant taken for scintillation counting. For efflux experiments, after 30 min of drug uptake, cells were placed in fresh buffer; aliquots were assayed for cell-associated total radioactivity at specified time points after centrifugation through silicone oil. Iron uptake experiments were performed using the same number of myocytes treated with 50 μg/ml of ferric ammonium citrate. Intracellular iron stores in myocytes were assessed using established methods (*16*). Statistical significance was determined using the unpaired Student's t-test.

Results

Cardioprotection by ICRF-187. We found that the cytotoxic effect of doxorubicin in adult rat heart myocytes could be significantly decreased by exposure to ICRF-187. As shown in Table III, ICRF-187 decreased the toxic effect of doxorubicin after 3 h of drug incubation in a dose-dependent fashion. At a 10:1 molar ratio of the cardioprotective agent to doxorubicin, the membrane integrity and light-microscopic morphology of the myocytes were completely intact at the end of the doxorubicin exposure period.

Table III. Effect of ICRF-187 on Doxorubicin Toxicity (100 μM) for Myocytes

ICRF-187 Concentration (μM)	Myocyte Survival (% Control $\pm$ SE)
0	42 $\pm$ 3
50	53 $\pm$ 4
100	66 $\pm$ 3
250	82 $\pm$ 4[a]
1000	96 $\pm$ 4[a]

[a]$P < 0.05$

Cellular Pharmacology of ICRF-187. To assess the mechanism of ICRF-187 cardioprotection, it was necessary to examine the cellular pharmacokinetics of this drug. Using [^{14}C]-ICRF-187, we found that drug uptake was extraordinarily rapid. As demonstrated in Figure 1, maximum levels of myocyte-associated radioactivity were detectable within 60 s of drug exposure and did not increase with increasing exposure times. Furthermore, alterations in temperature or ATP status had no effect on the amount of cell-associated radioactivity (data not shown). Efflux of the myocyte-associated radioactivity was equally rapid and essentially complete within 1 min (Figure 2). Using the HPLC method that we had previously demonstrated to be capable of separating ICRF-187 from its major metabolite ICRF-198 (which accounts for the majority of the metal chelating capacity of the molecule)(*17*), we also found that conversion of the parent drug to its metabolite is complete inside heart cells within 60 s and involves approximately 25-30% of the molar equivalents of the parental compound (data not shown).

Intracellular Iron Binding by ICRF-187. The initial approach to evaluating the ability of ICRF-187 to bind intracellular iron in the heart involved studies with iron-loaded myocytes. As shown in Figure 3, we found that by using ferric ammonium citrate (as well as a series of other iron chelates) it was possible to increase intracellular iron stores 2 to 3-fold after 1-3 h of incubation. In myocytes that were iron-loaded, the toxicity of doxorubicin (50 μM) increased from a control level of 15 $\pm$ 3% to 52 $\pm$ 2%, $P < 0.05$. Perhaps of greatest importance, however, as shown in Figure 4, we found that treatment of iron-loaded myocytes with ICRF-187 led to the complete elimination of the intracellular iron that had been introduced into those cells compared with iron-loaded myocytes treated with buffer alone.

Discussion

In this chapter, we have reviewed some of the biochemical mechanisms that have

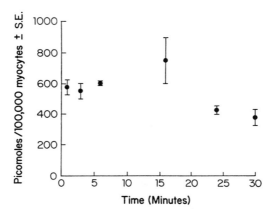

Figure 1.   Uptake of ICRF-187 by adult rat heart myocytes.

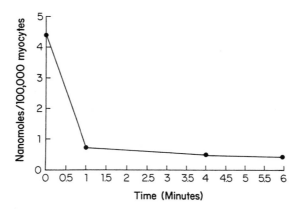

Figure 2.   Efflux of ICRF-187 from adult rat heart myocytes.

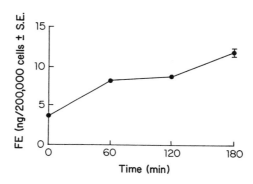

Figure 3.   Uptake of exogenous free iron by rat heart myocytes.

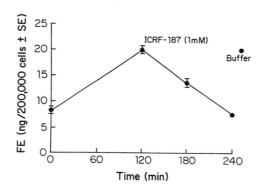

Figure 4.   Effect of ICRF-187 on intracellular iron content in iron-loaded myocytes.

been suggested to explain the cardiac toxicity of the anthracycline antibiotics. From a clinical perspective, however, the clear demonstration that an iron-chelating agent can significantly decrease the functional heart damage produced by doxorubicin provides a compelling rationale for the free radical hypothesis of cardiac injury. To the extent that strong oxidizing species with the chemical characteristics of the hydroxyl radical produced as a byproduct of cardiac anthracycline metabolism are responsible for some portion of the heart damage caused by this class of drugs, several features of the pathophysiology of anthracycline cardiac toxicity require further explanation. Principally, the role of myocardial iron or iron-proteins as critical catalysts of oxidant injury remains to be explained completely.

The experiments presented here, using a model system that employs beating adult rat heart myocytes provide an initial approach to this problem. We have shown that the clinically useful cardioprotective agent ICRF-187 protects rat heart myocytes from the toxic effects of doxorubicin in a dose-dependent manner. The drug enters and effluxes from cardiac myocytes by a process that is extraordinarily rapid and is not energy- or temperature-dependent, and thus, is likely to be diffusion-mediated. Furthermore, it is converted to its chelating metabolite almost immediately after uptake. Finally, ICRF-187 treatment of myocytes loaded with iron leads to the efflux of the entire pool of exogenously-added metal. This is the first demonstration in cells of the ability of ICRF-187 to chelate a transition metal in a form that allows rapid cellular export of potentially damaging (unbound) species of iron. Since iron-loaded myocytes have been shown here to possess enhanced sensitivity to the toxic effects of doxorubicin, it is tempting to speculate that the protective effect of ICRF-187 (and hence, one of the major mechanisms of toxicity of the anthracyclines) is, in fact, related to the complexation of unbound iron in the heart in a form that is not conducive to the Fenton reaction. It is important to point out, however, that no data currently exist demonstrating that treatment with doxorubicin actually leads to iron "delocalization" from protein-bound species in intact myocytes or that ICRF-187 forms unreactive complexes with such species. Current studies underway in our laboratory are directed toward providing the data to evaluate the precise role of intracellular protein-bound iron stores in the cardiac toxicity of the anthracycline antibiotics.

Acknowledgments

I wish to thank Cap van Balgooy for his excellent technical assistance.
This study was supported by grant CA33572 from the National Cancer Institute.

Literature Cited

1. Young, R.C.; Ozols, R.F.; and Myers, C.E. *N. Engl. J. Med.* **1981**, *305*, 139-153.
2. Ito, H.; Miller, S.C.; Billingham, M.E.; Akimoto, H.; Torti, S.V.; Wade, R.; Gahlmann, R.; Lyons, G.; Kedes, L.; and Torti, F.M. *Proc. Natl. Acad. Sci. U.S.A.* **1990**, *87*, 4275-4279.

3. Handa, K.; and Sato, S. *Gann* **1975**, *66*, 43-47.

4. Rajagopalan, S.; Politi, P.M.; Sinha, B.K.; and Myers, C.E. *Cancer Res.* **1988**, *48*, 4766-4769.

5. Doroshow, J.H.; and Davies, K.J. *J.Biol.Chem.* **1986**, *261*, 3068-3074.

6. Doroshow, J.H. Cancer Res. **1983**, *43*, 4543-4551.

7. Doroshow, J.H. In *Membrane Lipid Oxidation*; Vigo-Pelfrey, C., Ed.; CRC Press: Boca Raton, FL, 1990, Vol. 1; pp. 303-314.

8. Marcillat, O.; Zhang, Y.; and Davies, K.J.A. *Biochem.J.* **1989**, *259*, 181-189.

9. Harris, R.N.; and Doroshow, J.H. *Biochem.Biophys.Res.Commun.* **1985**, *130*, 739-745.

10. Puppo, A.; and Halliwell, B. *Free Rad.Res.Commun.* **1988**, *4*, 415-422.

11. Harel, S.; Salan, M.A.; and Kanner, J. *Free Rad.Res.Commun.* **1988**, *5*, 11-19.

12. Harel, S.; and Kanner, J. *Free Rad.Res.Commun.* **1988**, *5*, 21-33.

13. Green, M.D.; Alderton, P.; Gross, J.; Muggia, F.M.; and Speyer, J.L. *Pharmacol.Ther.* **1990**, *48*, 61-69.

14. Thomas, C.E.; and Aust, S.D. *Arch.Biochem.Biophys.* **1986**, *248*, 684-689.

15. Jacobson, S.L. In *Isolated Adult Cardiomyocytes*; Piper H.M., and Isenberg G., Eds.; CRC Press: Boca Raton, FL, 1989, Vol. 1 Structure and Metabolism; pp. 43-80.

16. Fish, W.W. *Methods Enzymol.* **1988**, *158*, 357-364.

17. Burke, T.G.; Lee, T.D.; van Balgooy, J.; and Doroshow, J.H. *J.Pharm.Sci.* **1991**, *80*, 338-340.

18. Muhammed, H.; Ramasarma, T.; and Ramakrishna Kurup, C.K. *Biochim.Biophys.Acta* **1982**, *722*, 43-50.

19. Goormaghtigh, E.; Brasseur, R.; Huart, P.; and Ruysschaert, J.M. *Biochemistry* **1987**, *26*, 1789-1794.

20. Pessah, I.N.; Durie, E.L.; Schiedt, M.J.; and Zimanyi, I. *Mol.Pharmacol.* **1990**, *37*, 503-514.

21. Cheneval, D.; Muller, M.; Toni, R.; Ruetz, S.; and Carafoli, E. *J.Biol.Chem.* **1985**, *260*, 13003-13007.

22. Bocherens-Gadient, S.A.; Quast, U.; Nussberger, J.; Brunner, H.R.; and Hof, R.P. *J.Cardiovasc.Pharmacol.* **1992**, *19*, 770-778.

23. Taylor, D.; and Hochstein, P. *Biochem.Pharmacol.* **1978**, *27*, 2079-2081.

24. Chatham, J.C.; Hutchins, G.M.; and Glickson, J.D. *Biochim.Biophys.Acta* **1992**, *1138*, 1-5.

25. Papoian, T.; and Lewis, W. *Am.J.Pathol.* **1992**, *141*, 1187-1195.

26. Doroshow, J. In *Progress in Chemotherapy*; Berkada B., and Kuemmerle H., Eds.; Ecomed: West Germany, 1988, Vol. 3; pp. 203-205.

27. Myers, C.E.; McGuire, W.P.; Liss, R.H.; Ifrim, I.; Grotzinger, K.; and Young, R.C. *Science* **1977**, *197*, 165-167.

RECEIVED June 3, 1994

Chapter 18

Amelioration of Anthracycline-Induced Cardiotoxicity by Organic Chemicals

Donald T. Witiak[1] and Eugene H. Herman[2]

[1]School of Pharmacy, University of Wisconsin—Madison,
Madison, WI 53706
[2]Division of Research and Testing, U.S. Food and Drug Administration,
Laurel, MD 20708

The amelioration of anthracycline-induced cardiotoxicity by numerous organic chemicals is discussed. These include ion regulators, receptor site antagonists, and inhibitors of mediator release, energy regulators, enzyme inhibitors, membrane stabilizers, anthracycline uptake inhibitors, and inactivators, antioxidants and chelating materials, and various miscellaneous substances. Of these the bis(2,6-dioxopiperazine)s remain as the potentially most useful drugs.

Recently published reviews (*1,2*) describe the various organic chemicals that serve to ameliorate acute and chronic anthracycline-induced toxicities at the cellular, animal, and clinical levels. In this review drugs are categorized according to the rationale proposed for their protective activity. Whereas major efforts have been directed towards decreasing myocardial concentrations of anthracyclines and their metabolites and the development of less cardiotoxic anthracyclines, this summary focuses upon the concurrent administration of substances that block anthracycline-induced toxicities.

As discussed in other chapters, anthracycline antitumor activity may be attributable to multiple mechanisms (*1-3*) such as (a) intercalation into DNA, (b) formation of ternary complexes with DNA topoisomerase II, and (c) radical-induced DNA strand breakage. The latter may involve anthracycline semiquinone radical-induced hydroxyl radical (HO•) formation, anthracycline semiquinone radical-induced superoxide anion radical (O_2^-)generation, and/or anthracycline-ferric ion complex catalyzed Fenton reactions (*3*).

Anthracycline-induced cardiotoxicity also is a function of multiple mechanistic possibilities (*1*). Acute effects such as hypotension are inhibited by antagonists of neurotransmitters (*4,5*). Subacute toxicity is rare, is unrelated to cumulative anthracycline dose, and manifests itself within 4 weeks as pericarditis-myocarditis with pericardial effusion and ventricular dysfunction (*6*). Nucleolar segregation also

0097–6156/95/0574–0268$10.16/0

occurs quickly after doxorubicin administration, but the significance of this reversible alteration is unknown (*1,7*). Dose-dependent, chronic cardiomyopathy following continued anthracycline therapy results in significant morbidity and mortality (*1*). Such toxicity has been attributed to anthracycline binding to (a) nuclear and mitochondrial DNA (*8*), (b) membrane phospholipids (*9*), and (c) contractile protein (*10*), as well as to anthracycline-induced (a) generation of reactive oxygen species (ROS) and lipid peroxides (*1,11-14*), (b) toxicity to specific enzymes (*1*), (c) modification of electrolyte levels and Ca^{+2} transport (*1,5*), and (d) release of endogenous chemical mediators (*1,5*).

Chemical ameliorators of anthracycline-induced cardiotoxicity are categorized according to their proposed mechanisms of action. Most of these agents have not been researched at the clinical level, and thus considerable work remains in order to determine their usefulness. In many cases, studies in animals suggest that some of these drugs have a low probability of success, but the rationales are of sufficient interest to warrant limited discussion. The classes we considered here include (a) ion regulators, (b) receptor site antagonists, (c) inhibitors of mediator release, (d) energy regulators, (e) enzyme inhibitors, (f) membrane stabilizers, (g) anthracycline uptake inhibitors, (h) anthracycline inactivators, (i) some miscellaneous entities, and (j) antioxidants that scavenge ROS or block ROS production via chelating mechanisms.

Ion Regulators

Compounds studied have included calcium channel blockers (CCBs) such as verapamil, diltiazem, prenylamine and nifedipine, and the calmodulin inhibitor trifluoperazine. Hearts of doxorubicin-treated rabbits have increased Ca^{+2} concentrations (*15*), and CCBs inhibit sarcolemmal Na^+/Ca^{+2} exchange, block Ca^{+2} slow channels, and alter mitochondrial Ca^{+2} transport (*1*). Verapamil protects isolated rat myocytes against doxorubicin-induced ATP depletion and morphological changes (*16*), and decreases doxorubicin-induced Ca^{+2} uptake (*17*). Verapamil also protects against rabbit cardiotoxicity in large doses (*18,19*), and in rats given low doses of doxorubicin the QaT interval shows less change when verapamil is administered (*20*). Prenylamine also exhibits some protective activity in rabbits and mice (*21-23*), but does not affect doxorubicin-enhanced hydroperoxide chemiluminescence in liver or myocardial tissue (*23*). Studies with diltiazem have been less successful (*16*), as have been a number of other studies using verapamil or nifedipine in mice (*24,25*). In some cases, CCBs actually stimulate cardiotoxicity (*24,26*) and decrease animal survival times (*27,28*). CCB-enhanced doxorubicin-induced cytotoxicity in tumor, heart, and normal cells may be due to increased anthracycline uptake (*1,29*). Further work, however, is required to determine the efficacy of CCB at the clinical level (*1*). In animal models the calmodulin inhibitor trifluoperazine is inactive or increases heart morphologic damage and lethality (*30*).

Receptor Site Antagonists and Inhibitors of Mediator Release

Adrenergic blockers phenoxybenzamine and propranolol ameliorate some acute cardiac effects, and in high doses certain combinations of histamine H_1 (chlor-

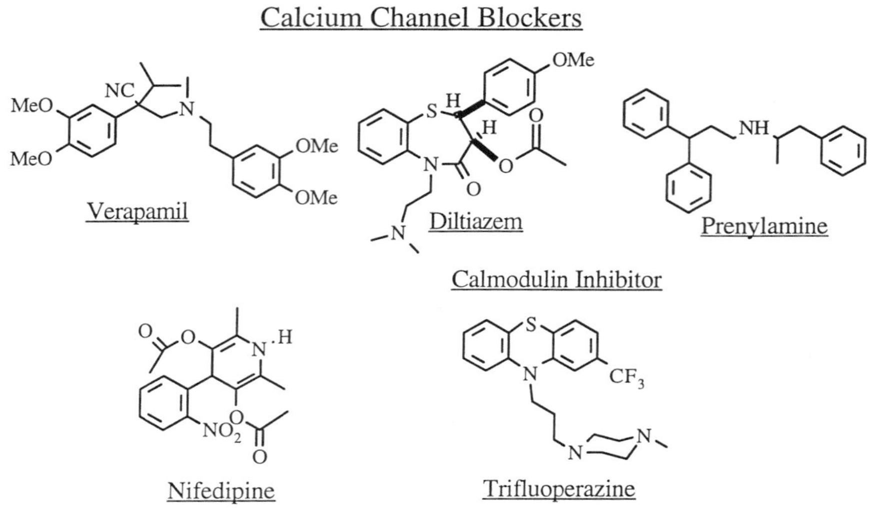

Calcium Channel Blockers

Verapamil   Diltiazem   Prenylamine

Calmodulin Inhibitor

Nifedipine   Trifluoperazine

pheniramine) and H_2 (cimetidine) antagonists with α-adrenergic (phentolamine) and ß-adrenergic (metoprolol) blockers inhibit anthracycline-induced myocardial lesions (*1*). Cromolyn inhibits release of the slowly reacting substance of anaphylaxis (SRS-A) and cromolyn and theophylline also alter acute anthracycline-induced toxicity (*1*). In mice, cromolyn (200 mg/kg) or theophylline (100 mg/kg) ameliorate doxorubicin-induced loss in body weight, lethality, and myocardial lesions (*31*).

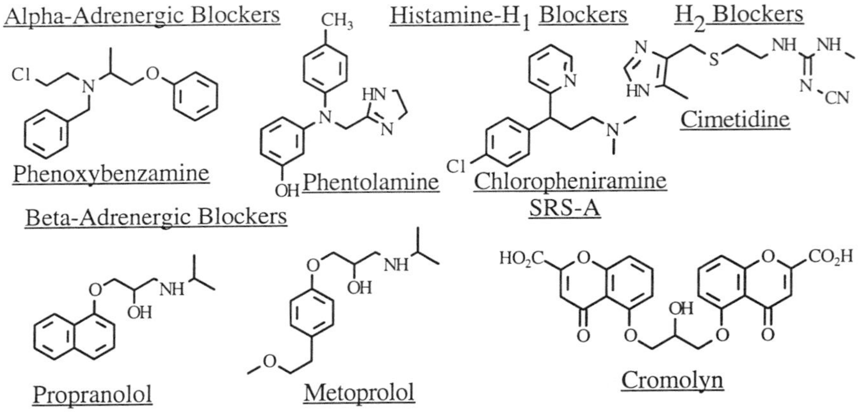

Alpha-Adrenergic Blockers   Histamine-H$_1$ Blockers   H$_2$ Blockers

Cimetidine

Phenoxybenzamine   Phentolamine   Chlorpheniramine

SRS-A

Beta-Adrenergic Blockers

Propranolol   Metoprolol   Cromolyn

However, it is not likely that histamine or catecholamines are important to the development of chronic anthracycline-induced cardiomyopathy; these neurotransmitters primarily induce focal necroses that differ from the generalized damage induced by the anthracyclines (*1,32-34*).

Energy Regulators

An energy regulator such as adenosine partially protects against anthracycline-induced loss of rhythmic and functional integrity in cultured myocytes (*35,36*) and increases the survival of mice treated with a single high-dose (17.5 mg/kg) of doxorubicin (*37*), but the mechanism of protection is unknown (*1*).   Similarly, inosine, an inotropic compound that stimulates production of high energy phosphate bonds in myocytes (*38,39*), only provides a very limited protection against anthracycline-induced myocardial electrical and morphological alterations in rats (*40*).   Pharmacological doses of the cellular metabolite, fructose-1,6-diphosphate, stimulate glycolysis and ATP synthesis (*41*) and, possibly for these reasons, ameliorates doxorubicin-induced increases in myocardial catalase and lipid peroxidation activities in rats (*42*). Although further work is necessary to define the degree of protection elicited by the monosaccharide derivative against the development of cardiomyopathy, this compound is known to attenuate doxorubicin-induced acute electrocardiogram (ECG), but not negative inotropic changes (*43*).

Highly polar L-carnitine is found in high concentrations in skeletal and heart muscle mitochondria and is required for fatty acid oxidation.   This compound, or its propanyl derivative, ameliorate doxorubicin-induced respiratory depression in rat heart slices (*44*) and attenuate acute changes in heart rate, coronary blood flow, and contractile force in isolated perfused rat or dog hearts (*45,46*).   L-Carnitine increases the life span of mice (*47-49*) and rabbits (*50*) and reduces histopathological changes in rats (*46*), rabbits (*50*), and monkeys (*51*) receiving doxorubicin.   Some protection is observed in humans.   In one uncontrolled study, individuals who were administered L-carnitine with daunorubicin or doxorubicin did not show significant increases in creatine-kinase MB isozyme levels or myocardial contractility (*52*).   Additional work, however, is necessary to confirm both the extent and mechanism of protection.

Adenosine   Inosine

L-Carnitine

Fructose-1,6-diphosphate

Enzyme Inhibitors

Four enzymatic targets that have received attention are cyclooxygenase, sodium-potassium ATPase (Na^+-K^+ ATPase), phosphodiesterase, and glutathione peroxidase.   S-Ibuprofen inhibits cyclooxygenase and in turn prostaglandin (PG) biosynthesis, whereas doxorubicin stimulates PG synthesis in vitro (*53*) and increases coronary blood PG levels in vivo (*18*). However, the racemic 2-arylpropanoic acid, the *R* enantiomer of which is converted to the active *S* antipode in vivo (*54*), does not protect against doxorubicin-induced weight loss or mortality in mice (*55*).

Both the cardiac glycosides and the anthracyclines interfere with membrane Na^+-K^+ ATPase (*56*). Thus, the cardiac glycosides were investigated for their protective activity. Strophanthin, the aglycone of which is strophanthidin, prevents doxorubicin-induced alterations in isolated rabbit or dog hearts (*57,58*), digoxin inhibits the anthracycline-induced negative inotropic effects in paced cat hearts (*58*), and ouabain blocks anthracycline-induced depressant activity in isolated myocytes (*59*). The mechanism or mechanisms by which these agents operate remain controversial (*1*), but in clinical studies very little protection is observed (60-63).

Phosphodiesterase (PDE) inhibitors of type II (sulmazole) and type III C (amrinone and milrinone), the latter of which block the cyclic AMP-specific, cyclic GMP-inhibited PDE isozyme (PDE III C) and are useful for the treatment of congestive heart failure (*64*) ameliorate anthracycline-induced negative inotropic effects observed in isolated guinea pig atrium (*65-68*). Amrinone also reduces lethality induced by high dose doxorubicin (18 mg/kg) (*65*), but clinical studies have not been carried out to determine the usefulness of this drug in preventing cardiomyopathy. The nonselective and weak PDE inhibitor theophylline, in combination with cromolyn, also ameliorates anthracycline-induced loss in body weight and lethality in mice (*31*), but more work is necessary to determine whether theophylline is truly useful in the prevention of cardiomyopathy.

Glutathione peroxidase (GP) detoxifies ROS; and selenium, as sodium selenite, is an essential nutritional element and cofactor for GP. Selenium deficiency potentiates doxorubicin-induced lethality in rats (*69*), and selenium administration delays death (*70*). A decrease in mortality and cardiomyopathy severity occurs when selenium and the antioxidant vitamin E are given to doxorubicin-treated weanling rabbits, (*71,72*) whereas large doses of vitamin E alone only partially protect hearts of animals against chronic anthracycline administration (*72*). Currently, selenium seems to provide the most protection when animals are treated prior to and during short-term anthracycline dosing (*73*). However, much more work is needed before the clinical efficacy of sodium selenite can be determined (*1,74*).

Membrane Stabilizers

Dextran sulfate, steroids such as methylprednisolone, and the ß-amino sulfonic acid taurine are among those substances possessing membrane-stabilizing properties. Since anthracyclines cause disruptions in the structure and function of cellular membranes, compounds having membrane-protective properties are of interest. Thus, whereas it is not known whether dextran sulfate has cardioprotective activity, pretreatment with the polymer does ameliorate depression of mitochondrial activity and produce small increases in survival times of doxorubicin-treated mice (*75*). Additionally, methylprednisolone exhibits limited protection for ten weeks against myocardial damage induced by doxorubicin (*76*). Like the adrenocorticoids, taurine has many pharmacological actions in addition to stabilizing membranes.

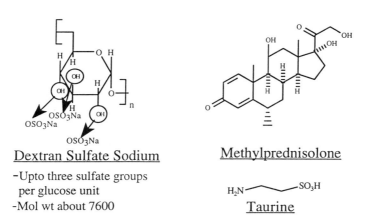

Taurine also modulates calcium movement across cell membranes and scavenges ROS (77). Concentrations of taurine found in the heart change under varying pathological conditions,   and the amino acid is known to protect chick hearts against anthracycline-induced decreases in high energy phosphate and contractility (78). Fewer mice administered a single dose (15 mg/kg) of doxorubicin die when 25 mg/kg of taurine is given for 6 consecutive days (77), but under such conditions it is not possible to determine whether this is true protection; death due to high doses   is a function of gastrointestinal toxicity, not cardiotoxicity (1).

Anthracycline Uptake Inhibitors

Doxycycline, a 6-deoxy-5-hydroxy analogue of tetracycline, enhances the survival of mice treated with a high dose (18 mg/kg) of daunorubicin (79,80). Such protection may be a function of changes in the tissue distribution of the structurally related anthracycline. However, when 2 mg/kg doxorubicin was given weekly to rats also administered 10 mg/kg of tetracycline, no evidence of protection was observed over an 8-week period (81). Like these potential anthracycline uptake inhibitors, use of a doxorubicin specific antibody for cardioprotection has been disappointing. Antibody protection against a single high dose of doxorubicin in mice produced <40% 30-day survivors, but myocardial alterations were somewhat decreased likely owing to decreased anthracycline tissue levels in the antibody-treated animals (82).

Tetracycline   Doxorubicin Specific
                                                      Antibody

Anthracycline Inactivators

Oxomorpholinyl radical dimers, introduced by Koch and his coworkers (83-87), are discussed elsewhere in this monograph. Though these interesting compounds ameliorate acute high dose doxorubicin toxicity and may be of use in controlling extravasation injury (86,87), the drugs require further study in order to determine their usefulness in chronic anthracycline-induced cardiotoxicity models and their effect on anthracycline antitumor efficacy (1).

3,5,5-Trimethyl-2-oxomorpholine-3-yl
Radical Dimers

Miscellaneous Drugs

Several compounds with differing pharmacological properties have been evaluated. The cytostatic agent, Damvar, enhances antitumor activity, increases survival of tumor-bearing mice, reduces mortality, and attenuates acute myocardial toxicity when given in conjunction with doxorubicin (88). However, posttreatment studies are too short and limited to allow for the assessment of the possible eventual development of cardiotoxicity. The pharmaceutical solvent polyethylene glycol 400 increases the survival rates and reduces myocardial lesions in mice administered high doses of doxorubicin, but it is not known which of the many biological activities of the polymer is attributable to the protective effect (*89*). Inorganic compounds, such as bismuth subnitrate, increase tissue levels of metallothionein. Subsequent scavenging of intracellular free radicals by increased concentrations of this cysteine-rich protein in heart and other tissue may be responsible for the increased survival of bismuth subnitrate-pretreated mice given a single high dose of doxorubicin (*90*). Interestingly, bismuth subnitrate pretreatment also ameliorates doxorubicin-induced gastrointestinal and bone marrow toxicities (*90*). Zinc and cadmium salts are protective too, perhaps for similar reasons (*91*).

Damvar (cytostatic)

Bismuth subnitrate

Zn and Cd Salts

Polyethylene glycol 400 (PG-400)

$$H(OCH_2CH_2)_nOH$$
$$n = 8.2-9.1$$

Antioxidants

Antioxidant compounds are by far the largest class of organic chemicals explored for their protective effects. Two endogenous enzymes, superoxide dismutase (SOD) and catalase catalyze the conversion of the superoxide anion radical (O_2^-) to H_2O and O_2 in a two-step process wherein hydrogen peroxide serves as intermediate. In two studies (*92,93*), the survival rate increased from 30 to 45% and from 12 to 40% when doxorubicin-dosed mice were treated with SOD. In 3-week studies in rats, SOD decreased and catalase increased survival in animals given 4 mg/kg/wk of doxorubicin. An additional increase in survival and in the amelioration of myocardial alterations were observed in animals given both enzymes. In acute single dose (10 mg/kg doxorubicin) studies, the two enzymes provided better protection than either alone (*92*), and the increased survival was likely a function of decreased gastrointestinal toxicity (*92*).

Sulfhydryl antioxidants such as N-acetyl-L-cysteine (NAC) also scavenge ROS, inactivate lipid peroxides and help to maintain biological membranes (*94,95*). Acute studies in mice have revealed that NAC prevents doxorubicin-induced decreases in cardiac sulfhydryl concentrations, increases survival, and ameliorates the severity of myocardial lesions (*96,97*). Glutathione exhibits similar effects (*98*). Additionally, in a comparative study using mice, NAC, isocitrate, and niacin individually protect against doxorubicin-induced weight loss and myocardial alterations (*99*). However, in dogs, NAC fails to reduce doxorubicin-induced alterations in systolic and diastolic function (*100,101*). Comparison of NAC with the bis(dioxopiperazine) ICRF-187 reveals the latter, but not the sulfhydryl compound, protects against chronic doxorubicin-induced cardiotoxicity (*102*). In clinical trials, NAC proved ineffective in reversing long-standing doxorubicin-induced cardiomyopathy (*103*) or chronic cardiotoxicity (*104*) and in tests using human biopsy preparations, NAC was shown not to prevent nuclear degeneration (*105*) or sarcotubular alterations (*106*). NAC does prevent anthracycline-induced inhibition of guanylate cyclase (*107*), but much more work would be required to substantiate the clinical relevance or efficacy of NAC.

The bioflavonoid antioxidant venorutin ameliorates doxorubicin-induced deterioration and loss of body weight in rats (*108*) and prolongs the life of mice given the anthracycline (*109*), but its effects on myocardial toxicity must await further testing. Methylene blue, which like doxorubicin is reduced by NADPH-dependent enzymes,

Superoxide Dismutase (SOD)

$$O_2 \xrightarrow{e^-} O_2^{\overline{\cdot}} \xrightarrow[2H_2O]{SOD} O_2 + H_2O_2$$

Catalase (C)

$$2H_2O_2 \xrightarrow{C} 2H_2O_2 + O_2$$

N-Acetyl-L-cysteine

Ruticide (Venorutin)

Methylene Blue

also increases the survival time of mice following administration of a single high dose of doxorubicin (*110*), but the dye seems not to protect against heart damage.

Additional antioxidants of interest include the captodative olefins (dehydroalanines), ascorbic acid, butylated hydroxytoluene (BHT), and QMDP-66, a quinolyl derivative of N-acetylmuramyl dipeptide. The combined effect of an electron-withdrawing (*captor*) and an electron-releasing (*donor*) substituent (i.e., the captodative effect) on a radical center leads to enhanced stabilization (*111*), and, thus, captodative olefins such as N-acyldehydroalanines scavenge ROS and inhibit lipid peroxidation (*112*). Such compounds also attenuate doxorubicin-induced depression of myocardial mitochondrial enzyme activity and mitochondrial membrane fluidity in vivo (*113*). The 2-methoxyphenylacetyl derivative of dehydroalanine protects against death due to single high or multiple low doses of doxorubicin (*114*), but further work is necessary to determine whether this substance protects against anthracycline-induced cardiomyopathy. The ROS scavenger ascorbic acid also delays death induced by a single high dose of doxorubicin in mice and guinea pigs (*115*). Cardiotoxicity, as evaluated by electron microscopy, is also reduced in guinea pigs treated with low doses of the anthracycline every 5 days and ascorbic acid daily over 20 days (*115*).

Other ROS scavengers also exhibit some protective activity. In isolated rat hearts, BHT reduces doxorubicin-induced acidoses and changes in phosphate metabolism (*116*). The antioxidant increases by two- to fourfold the survival of mice given a single high dose of the anthracycline (*117*). In chronic, low-dose studies (2 weeks) BHT reduces doxorubicin-induced myocardial lipid peroxidation (*118*), and in acute studies BHT partially inhibits anthracycline uptake by the myocardium as well as the concomitant production of malondialdehyde (*117*). QMDP-66, a quinolyl derivative of $\underline{N}$-acetylmuramyl dipeptide, contains a quinone redox functionality. Both the QMDP-66 as well as coenzyme Q_{10} individually attenuate the chronic (23 days) low

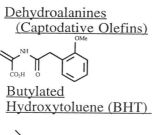

Dehydroalanines (Captodative Olefins)

Butylated Hydroxytoluene (BHT)

Ascorbic Acid

QMDP-66; The quinolyl derivative of N-acetylmuramyl dipeptide

dose doxorubicin-induced changes in ECG and decrease the incidence of myocyte vacuolization (*119*). However, neither antioxidant influences the anthracycline-induced decreases in body and ventricular weight (*119*).

Coenzyme Q_{10} is among those antioxidants (*120*) receiving relatively more attention as an ameliorator of anthracycline-induced toxicities. In addition to its antioxidant chemistry, coenzyme Q_{10} is involved in oxidative metabolism, which is antagonized by doxorubicin in vitro (*121-123*), and has membrane-stabilizing properties (*124*). Further work is required to define coenzyme Q_{10} efficacy in the clinic (*1*), but acute and chronic studies in animals indicate considerable potential. Thus, hearts isolated from rats maintain mechanical function longer when the doxorubicin-treated animals from which the hearts were obtained also received the coenzyme (*125*). Furthermore, coenzyme Q_{10} ameliorates acute toxicity of high dose doxorubicin in mice (*126*), protects against anthracycline-induced mitochondrial lipid peroxidation (*127*), ameliorates anthracycline-induced decreases in respiration in rats (*127*), and attenuates the induced ECG changes and myocardial lesions in chronic animal models (*128-133*).

Animal studies employing vitamin E have produced mixed results, but in vitro the vitamin clearly inhibits lipid peroxidation in doxorubicin-treated rat heart and liver microsomes (*134*) and in anthracycline-treated human platelets (*135*). In animals, vitamin E attenuates high dose doxorubicin-induced increases in cardiac malondialdehyde concentrations (*136*), lethality (*137*), myocardial changes (*14*), and lipid peroxidation (*14*), but in one study no effect was observed after 60 days (*138*). Timing of vitamin E versus anthracycline administration seems to be important (*93*). In another study, mice began to die when treatment with the vitamin was discontinued (*70*). Some have observed the vitamin to be effective when injected, but not when given in the diet (*139*). Others report that acute cardiotoxicity in rats and rabbits is attenuated by pretreatment with the vitamin (*140,141*), while still others observe decreased survival with very high vitamin E doses possibly because of increased doxorubicin aglycone levels in tissue (*142*). Vitamin E protection against chronic doxorubicin-induced toxicity in animals also seems not to be very significant (*22,72*).

<u>Coenzyme Q$_{10}$</u>
(Ubiquinone)

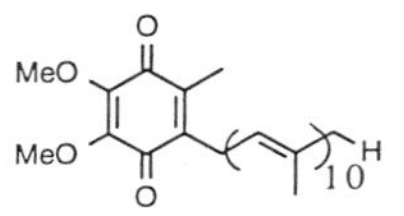

<u>Vitamin E</u>

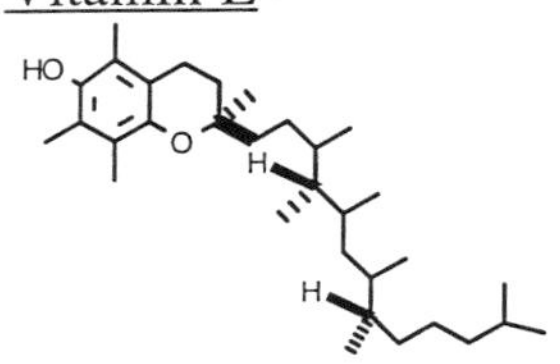

<u>Chelating</u> <u>Antioxidants</u>

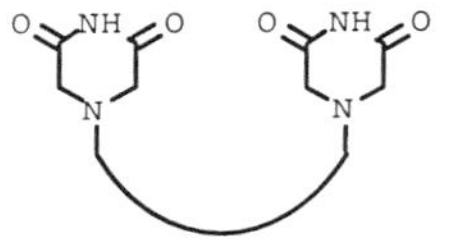

<u>Bis(dioxopiperazine)s</u>

In large doses the vitamin does not ameliorate chronic doxorubicin-induced cardiomyopathy in rabbits (*143*) or dogs (*144,145*), and clinical studies have not been encouraging (*1,144-146*).

Chelating antioxidants such as the bis(2,6-dioxopiperazine)s have received considerable attention as ameliorators of anthracycline-induced cardiotoxicity (*1,2*). Although there are three possible regioisomeric dioxopiperazines (2,3-, 2,5-, and 2,6-dioxopiperazines), only the bis(2,6-dioxopiperazine)s, which are related to EDTA, have these important biological properties (*2*). The 2,3-regioisomers mainly have significance as modifiers of penicillin and cephalosporin activities, whereas the 2,5-dioxocyclodipeptide system is found in numerous natural and synthetic products possessing a vast array of pharmacological properties (*2*).

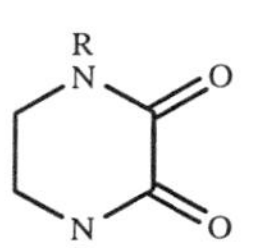

2,3-dioxopiperazines 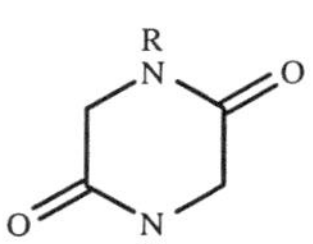   2,5-dioxopiperazines 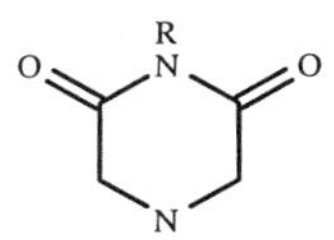   2,6-dioxopiperazines

Whereas protection against anthracycline-induced toxicity is the major topic of this section, it should be noted that these drugs also exhibit antineoplastic, antimetastatic, mutagenic, immunosuppressive, antipsoriatic and embryolethal activities, protect against acetaminophen-induced toxicity and alloxan diabetes; and are synergistic with ionizing radiation (*2*). Both EDTA and the bis(dioxopiperazine) ICRF-159 (R = H, R_2 = Me), when added to perfusate, inhibit anthracycline-induced elevation of coronary perfusion pressure in isolated, blood perfused dog hearts (*147*). In Syrian golden hamsters, ICRF-159 attenuates acute anthracycline-induced toxicity (*148*). ICRF-187, the optically pure, more water soluble dextrorotatory isomer of ICRF-159, is similarly protective (*149*). This protective activity is not attributable to inhibition of anthracycline metabolizing enzymes or to a reduction in DNA-daunorubicin complexation (*150,151*), and although the drug attenuates morphological changes in the heart (*152,153*), myocardial damage is not severe enough to account for high dose anthracycline-induced lethality (*63,149*). Rather, a reduction in gastrointestinal toxicity may be responsible for the protection afforded by ICRF-159 under these conditions (*154*). Additionally, the treatment timing is important for best protection (*154,155*).

EDTA

Biologically Active
Bis(2,6-dioxopiperazine)s

Protection by such bis(dioxopiperazine)s against chronic daunorubicin-induced cardiotoxicity in rabbits (*156,157*) and mice (*152*) is also well characterized. The protection against lethality is greater for daunorubicin than doxorubicin, but attenuation of myocardial alterations is evident when both anthracyclines are used (*158,159*). Similarly, both ICRF-159 and ICRF-187 inhibit daunorubicin, but not doxorubicin-induced HeLa cell colony-forming activity (*160*). Nonetheless, ICRF-187 does provide significant protection against chronic doxorubicin cardiotoxicity in beagle dogs (*161,162*), mice (*163*), miniature swine (*164*), rats (*165-167*), and rabbits (*168*), and also against the cardiomyopathy and nephropathy induced by epirubicin (*169*), the 4'-hydroxy epimer of doxorubicin. In animals, neither ICRF-159 nor ICRF-187 has been shown to alter the antitumor activity of the anthracyclines (*170,171*), and in clinical studies ICRF-187 is clearly cardioprotective (*172,173*). Although efficacy against doxorubicin-induced cardiotoxicity has been established in the clinic, FDA approval of the drug has been denied (June, 1992) because one reported study showed a significant reduction in lung cancer response rates (*173*); future approval is uncertain (*173*).

Bis(dioxopiperazine) protection is not enantioselective. Both optical isomers of ICRF-159 are equally protective in hamsters (*174*). Hydrolysis to mono- and bis(acid-amide) systems prior to reaching the site of action affords highly polar and inactive products (*174*). In neutral solution ICRF-187 exhibits a peak at 205-211 nm, but in basic solution this absorbance is replaced by a peak at 227 nm, which is attributed to the anionic form of the imide (175). ICRF-187 exhibits pK$_a$ of 10 and 9.3 at 25°C and 37°C, respectively. Deprotonation of both rings occurs simultaneously, with the same pK$_a$ values. The hydrolysis of the bis(dioxopiperazine), followed spectrophotometrically, is pseudo-first-order over a wide pH range. The mechanism involves a hydroxide-catalyzed pathway and a pH-dependent pathway similar to the hydrolysis of other imides (Scheme I). The anionic form of the compound is resistant to hydroxide attack; each dioxopiperazine ring undergoes hydrolysis independently of the other. As shown in these studies with ICRF-187 (175), k$_{OH}$(M^{-1}min^{-1}) = 230 ± 10 and 820 ± 50 at 25°C and 37°C, respectively, and is comparable to k$_{OH}$ = 204 M^{-1} min^{-1} for succinimide. At 37°C, the k$_{OH}$ term is better defined than the water term (k$_w$ = 2.2 ± 1.1 x 10^{-4} min^{-1}) because of the smaller contribution (~ 30%) of the latter to the overall rate. The value of k$_{obs}$ at pH 7.4 and 37°C (7.1 x 10^{-4} min^{-1}) is more than 10-fold slower at pH 2.93 (k$_{obs}$ = 6 x 10^{-5} min^{-1}). The t$_{1/2}$ for the first ring opening of ICRF-187 at

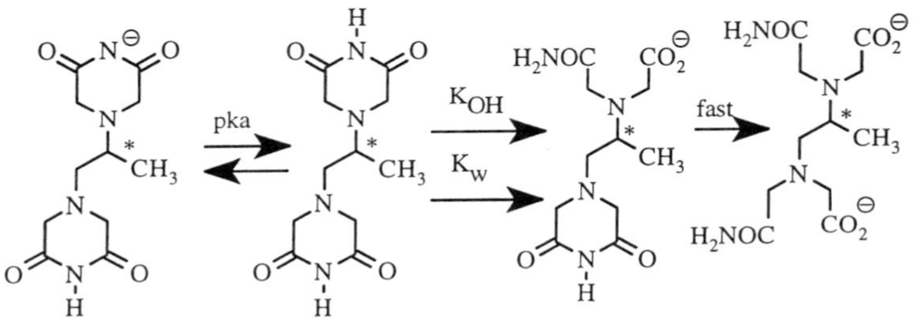

Scheme I

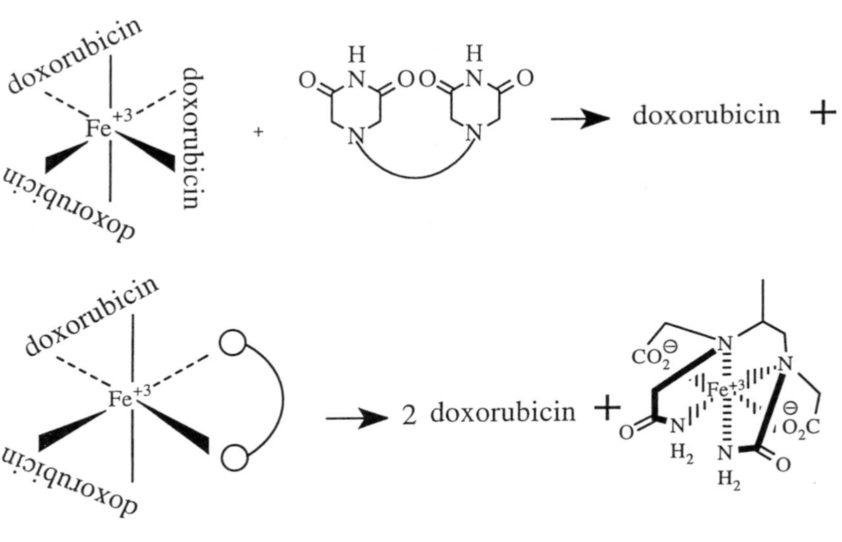

Scheme II

pH 7.4 is 8.2 h, and this is approximately one-half of the $t_{1/2}$ for the decrease in total absorbance change at pH 7.4 and 37°C; i.e., 16.3 h (*175*). Doxorubicin forms complexes with transition metals, and both Fe^{+3} (doxorubicin)$_3$ and Cu^{+2} (doxorubicin)$_2$ react with ICRF-187 to promote hydrolysis of the bis(dioxopiperazine) with concomitant abstraction of the metal ion from the anthracycline complex (*2,176*) (Scheme II). Metal ion complex-promoted hydrolysis is preceded by mixed ligand complex formation. Specifically, the Fe^{+3} (doxorubicin)$_3$ complex exhibits a fast initial drop in absorbance at 600 nm, possibly a function of equilibrium displacement of the most weakly bound anthracycline (*2,176*). This is followed by a slower spectral change relating to removal of the most tightly bound anthracycline (*176*). For the Fe^{+3}(doxorubicin)$_3$ complex k_{obs} shows saturation behavior expected for complex formation preceding a rate-determining step. The k_{obs} for the ion complex occurs at 170 μM bis(dioxopiperazine). The complex-promoted hydrolysis possibly takes place as follows but intermediates and final products have not been rigidly characterized (*2,176*).

Conformationally constrained bis(dioxopiperazine)s of the cyclopropanediyl type (cis or trans) did not show protective activity (*174*) (Scheme III), but reevaluation as a function of stability and bioavailability needs to be carried out. These compounds are of particular interest because of their interesting anti- and prometastatic properties. The cis isomer inhibits and the trans isomer stimulates development of metastases in two different animal models (*177,178*). Interestingly, these rigid analogues mimic crystal structure analyses of racemic ICRF-159 and its dextrorotatory isomer, ICRF-187. The former reveals a cis face-to-face relationship

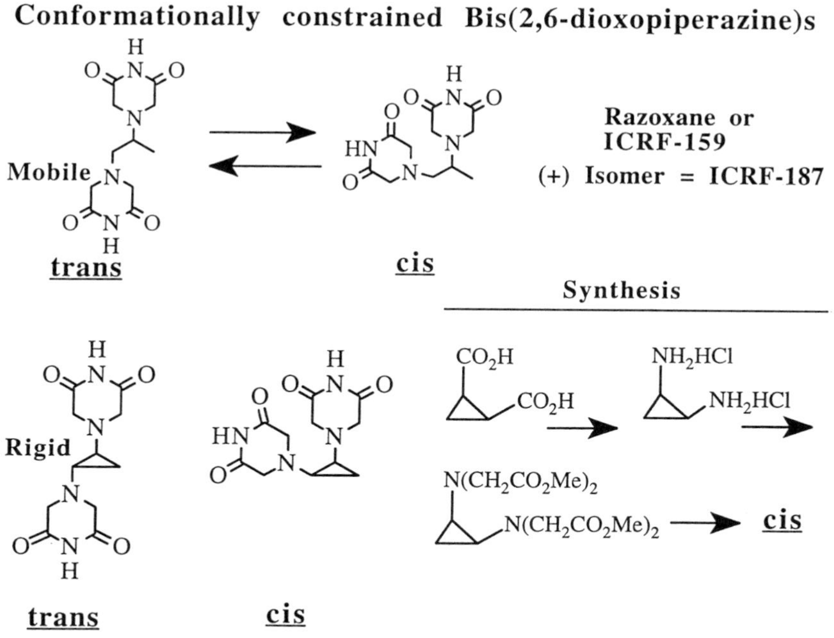

Conformationally constrained Bis(2,6-dioxopiperazine)s

Razoxane or ICRF-159

(+) Isomer = ICRF-187

trans   cis

Synthesis

Rigid

trans   cis

Scheme III

of dioxopiperazine rings, whereas the optical isomer has the extended <u>trans</u> conformation with a parallel arrangement of ring planes in the solid state (*179*). The geometrical relationships of the rigid and mobile forms and the synthesis for the <u>cis</u> isomer are illustrated.   Tables I and II summarize their anti- and prometastatic properties (*2,177,178*).

Table I. Anti- and Prometastatic Activities of Conformationally Constrained Cyclopropanediyl Bis(2,6-dioxopiperazine)s (Adenocarcinoma Model)[a]

Drug	No. of Metastases to Lung
Mobile Drug	174.9 ± 31.5
<u>cis</u> (Rigid)	171.0 ± 26.7
<u>trans</u> (Rigid)	264.8 ± 23.9
CMC Control	205 ± 24.4

[a]Stock suspensions (100 mL) of drug (2.8 mg/mL) containing 3 drops of conc. HCL and 5% CMC. IP injection of drug (15 mg/kg); treatment - q 48 h for 4 wks. Last treatment 48 h prior to tumor excision. Broncogenic adenocarcinoma cells (LG 1002) were inoculated <u>intradermally</u> in the back (10 animals per group). Tumors appeared 3-4 days earlier in the <u>trans</u>-treated animals and grew to a larger size (>100 mm^2) by day 28.

Table II. Anti- and Prometastatic Activities (B16-F10 Model)[a]

Drug	Median Colonies (Lung) μM		Median Colonies (Culture μM	
	0	2	0	2
Mobile drug	117	57	36	9
<u>cis</u> (rigid)	125	53	30	36
<u>trans</u> (rigid)	167	229	54	104

[a]Cells pretreated for 24 h with drug in culture medium. Mice injected iv. (tail vein) with 5 x 10^4 B16-F10 tumor cells in 0.2 mL.   After 20 days black modules enumerated using a dissecting microscope. Colony formation in vitro; 10^2 cells were placed in a 60 mm culture dish for 7-10 days. No effect with B16-F10 Melanoma.

Antimetastatic activity seems to reside in the <u>cis</u> conformation, a conformation likely important to biometal chelation mechanisms (*177,178*). Bis(dioxopiperazine)s of the tetraazaperhydrophenanthrene type (*180*) are related to the <u>cis</u> cyclopropanediyl analogues as illustrated below (Scheme IV).

Elaboration of the Bis(2,6-dioxopiperazine)-Containing Tetraazaperhydrophenanthrenes

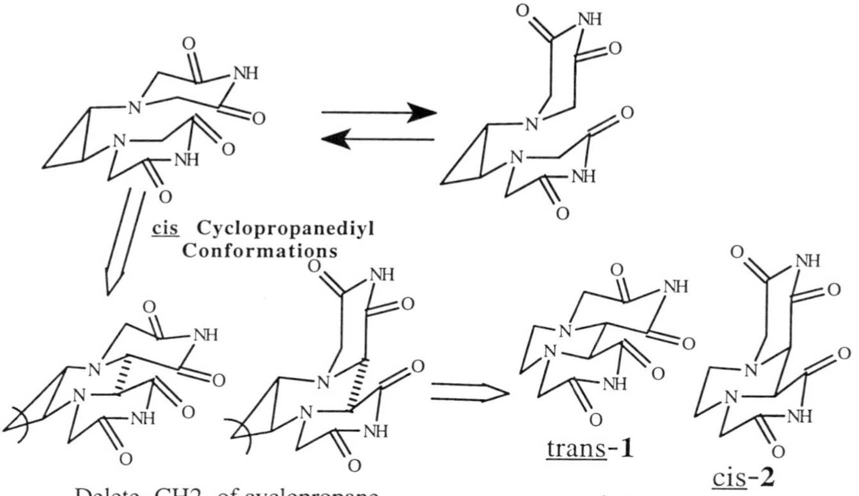

cis Cyclopropanediyl
Conformations

trans-1

cis-2

. Delete -CH2- of cyclopropane ring.
. New bond between the #3 positions on the dioxopiperazine rings

but,
the hetero rings are cisoid

Scheme IV

Removal of the cyclopropyl methylene function and bond connection between the two C(3) carbons of the dioxopiperazine rings provides the desired cis and trans tricycles. Their synthesis from pyrazine-2,3-dicarboxamide is also shown (Scheme V), but modified syntheses generate other related compounds, whose antimetastatic activities are summarized in Table III.

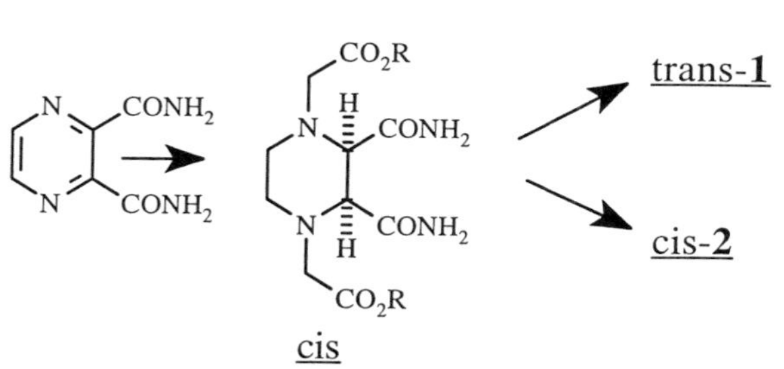

trans-1

cis-2

cis

Scheme V

Table III. Antimetastatic Effects of cis- and trans-Tetraazaperhydrophenanthrenes and Certain Open Chain Systems Following 24-Hour Pretreatment of B16-F10 Melanoma Cells

trans-1   cis-2   cis-3   cis-4   cis-5

Compound	Dose =	Mean No. of Lung Colonies		
		0 μM	2 μM	20 μM
trans-1[a]		137.3 ± 64.2	29.1 ± 24.8	23.3 ± 19.7
cis-3[b]		137.3 ± 64.2	33.2 ± 37.2	43.4 ± 39.0
cis-4			159.7 ± 124.0	160.7 ± 126.8
cis-1[c]		73.5 ± 34.8	39.1 ± 19.8	50.1 ± 23.7
cis-5		73.5 ± 34.8	86.2 ± 31.8	101.3 ± 51.1
cis-2			52.1 ± 13.3	96.1 ± 72.4

[a]Significantly different than controls as determined by Neuman-Keuls test.
[b]Results follow injection of cells into the tail vein of C57B1/6J.
[c]Lung colony formation does not reflect decreased colony formation in vitro.

Trans-1 and cis-3, but not cis-2, were active in this assay (*180*). To further define their geometrical requirements, the tetraazaperhydrophenanthrene diastereomers were synthesized (*181*). In these compounds there cannot be a cisoid relationship of dioxopiperazine rings. The relationship of perhydroanthracenes to the perhydrophenanthrenes as well as their synthesis (*181*) from 2,5-dimethylpyrazine are summarized below. In this case (Scheme VI), formal bond disconnection of one carbonyl group from the central piperazine ring of the perhydrophenanthrenes and rebonding on the opposite carbon of the central ring provide the diastereomeric *cis*- and *trans*-tetraazaperhydroanthracene-type bis(dioxopiperazine)s (*181*).

Elaboration of the Bis(2,6-dioxopiperazine)s Containing Tetraazaperhydroanthracenes

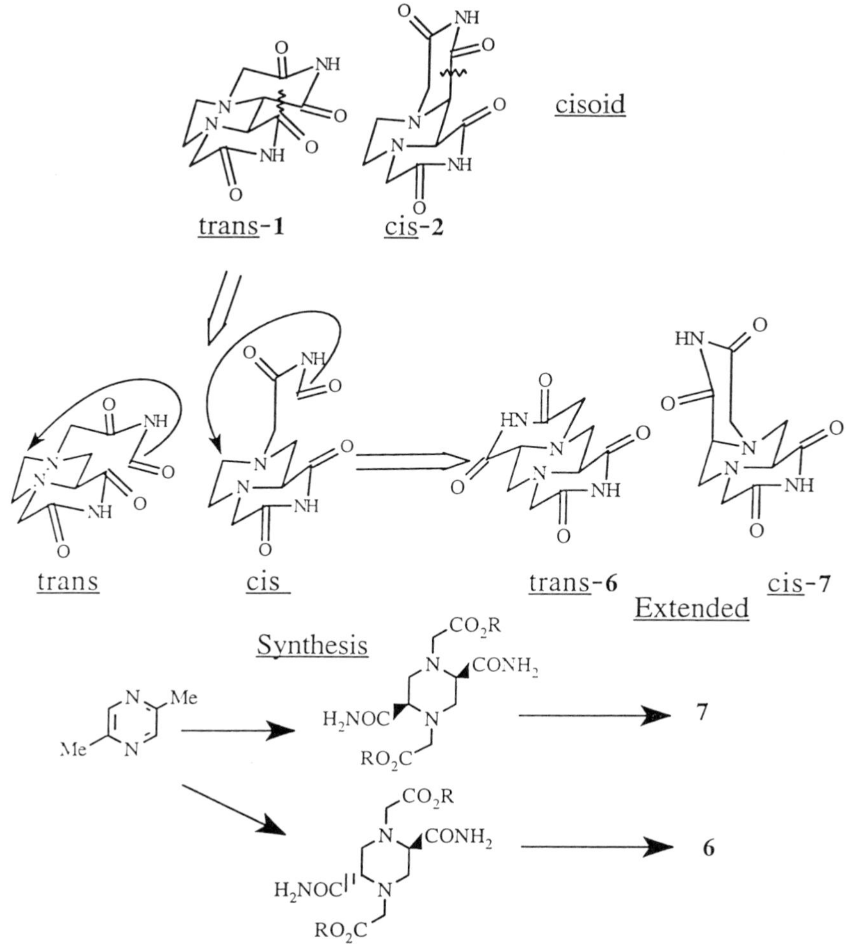

trans-1   cis-2   cisoid

trans   cis   trans-6   cis-7

Extended

Synthesis

7

6

Scheme VI

When compared in the Lewis lung carcinoma metastatic model the cis-tetraazaperhydrophenanthrene was only weakly active and the trans stereoisomer was inactive, but the morpholinomethyl prodrugs **8** and **9** (Tables IV and V) reduced metastases and significantly increased survival in the post, but not preamputation schedule (*182*).   In the post amputation schedule (Table VI) only the bis(morpholinomethyl) derivative of 1,4-cis-tetraazaperhydroanthracene is active (*181*).

Table IV. Lewis Lung Carcinoma Metastasis Study Post Amputation Schedule[a]

Compound	Survival Data			Autopsy Data			
	MST[b]	%ILS[c]	N/T[d]	Av. body wt. g	Av. no of metastases		M/T
					< 2 mm	< 2 mm	
Control	33.5	-	1/8	17.6	0	14	7/8
1	39.0	16	2/9	17.7	2	13	7/9
2	45.0	34	3/8	17.8	1	14	5/8
8	>50	>49	5/9	19.4	0	9	4/9
9	>50	>49	7/9	19.9	0	3	1/9

[a]BDF, female (19-21 g) mice. Implantation; day 0. Amputation; day 9. 160 mg/kg from day 9; q 2 d x 4. Autopsy data - day 50.
[b]MST = medium survival time - days.
[c]% ILS $\geq$ 25 = activity
[d]N/T = number of 50 day survivors/total mice.
[e]M/T = number of mice with metastases/total.

1. R=H   2. R=H

8. R= CH$_2$N⟨O⟩   9. R= CH$_2$N⟨O⟩

Table V. Lewis Lung Carcinoma Metastasis Study

Schedule	Compound	% ILS[b]	N/T[c]	% of mice with no metastases	Av. number of metastases per mouse
	Control	-	0/10	0	17.5
Preamputation	8	15	3/10	30	15.0
	9	16	4/10	40	12.8
Post Amputation[d]	8	4	2/10	20	13.8
	9	>49	6/10	40	7.7

[a]10 BDF, mice (19-21 gm); 160 mg/kg ip from - 1 h on day 0; q 2d x 5; Implantation; day 0; Amputation; day 10.
[b]% ILS $\geq$ 25 indicates activity.
[c]N/T = no. of 50-day survivors/total no.
[d]160 mg/kp ip from day 11; q 2d x 4.

1. R=H   2. R=H

8. R= CH$_2$N(morpholine)   9. R= CH$_2$N(morpholine)

Table VI. Lewis Lung Carcinoma Metastasis Study Post Amputation Schedule[a]

	Survival Data			Autopsy Data			
Compound	MST[b]	%ILS[c]	N/T[d]	Body wt. (g)	Lung wt. (mg)	Av. no. of metastases	M/T[e]
						< 2 mm / > 2 mm	
Control	27	-	0/10	16.3	741	- / 49	10/10
6	28	4	0/5	16/9	853	0 / 48	4/4
7	28	4	1/5	19.3	640	0 / 22	2/3
10	26	>0	1/5	18.9	700	0 / 38	3/4
11	>40	>48	3/5	19.2	358	0 / 6	1/4

[a]BDF, female mice (19-21 g); Implantation; day 0; Amputation; day 8; 160 mg/kg from day 8; q 2 d x 4; Autopsy data; day 40.
[b]MST = medium survival time (days).
[c]% ILS $\geq$ 25 indicates activity.
[d]N/T = number of 40-day survivors/total mice.
[e]M/T = number of mice with metastases/total.

6. R=H
10.R= CH$_2$N(morpholine)

7. R=H
11.R=CH$_2$N(morpholine)

The bis(morpholinomethyl) derivatives of bis(dioxopiperazine)s are of interest because in some cases (but not all) morpholinomethyl substitution increases water solubility and in many instances also increases antitumor activity (1,2) (Scheme VII). Such derivatives undergo hydrolysis in water, may serve as prodrugs, or may have intrinsic antitumor activity because of the presence of an alkylating carbon (2). The morpholinomethyl derivative of ICRF-154 [the bis(imide) of EDTA] reduces daunorubicin-induced lethality in mice 90% and 50% for up to 8 weeks with pretreatment doses of 100 and 50 mg/kg, respectively (*1*).   Morpholinomethyl substitution of ICRF-159 generates a water soluble derivative that is cardioprotective in beagle dogs administered doxorubicin at 3-week intervals over a 21-27 week period (*183*).

Chemical Properties of Bis(morpholinomethyl) Derivatives of Bis(dioxopiperazine)s

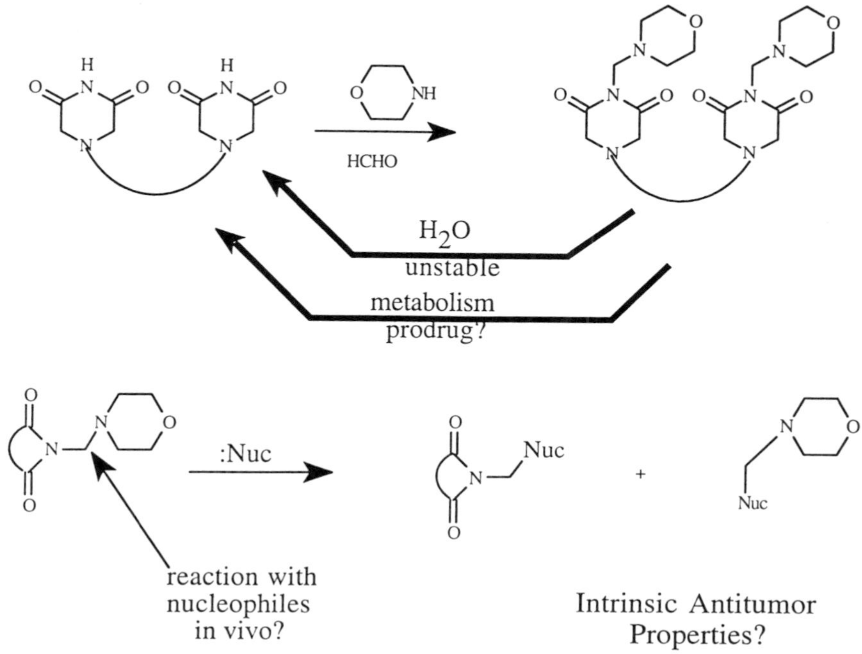

Scheme VII

Recently, the synthesis spectral and physicochemical properties of diastereomeric 4,4'-(4,5-dihydroxy-1,2-cyclohexandiyl)bis(2,6-dioxopiperazine)s have been described (*184*), and these materials are undergoing in-depth biological investigation. Five (**11-15**) of the six (**11-16**) possible diastereomers were prepared in order to determine how hydroxyl group substitution influences water solubility and what conformational or stereochemical components may be related to (1) amelioration of anthracycline-induced toxicity, (2) antimetastatic activity, (3) transport properties, and (4) drug stability.

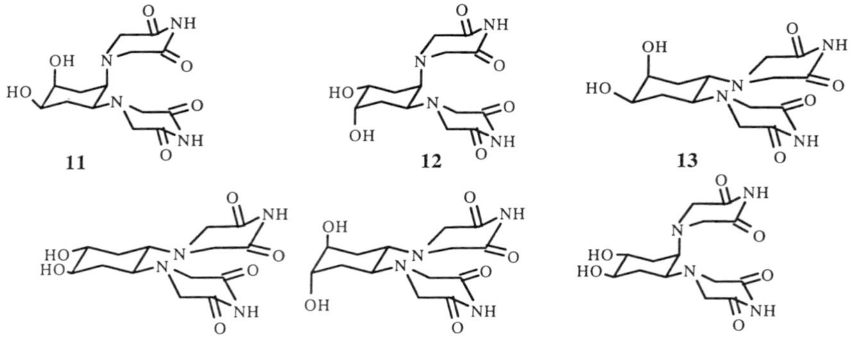

11   **12**   **13**

14   **15**   **16**

Syntheses of five of the six targets proceed from Cbz-protected diamines of known geometric configuration by sequential chemical transformations (*185*) (Scheme VIII).

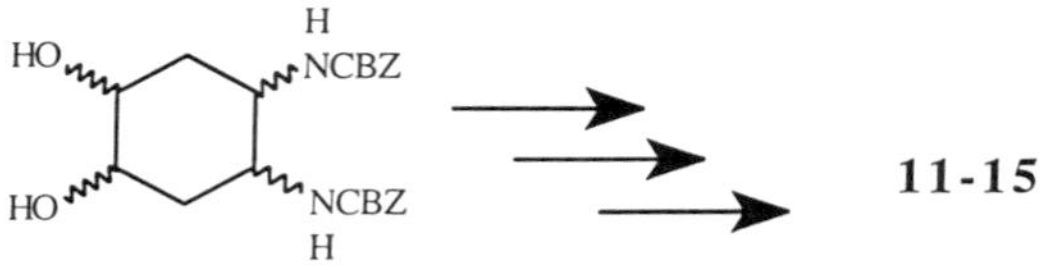

11-15

Scheme VIII

However, an unwanted tricyclic compound was formed in place of sixth diol-protected bis(dioxopiperazine). The rationale for this observation is that equational bis(n-alkylation) is favored over axial bis(N-alkylation). Thus, tris(alkylation) of conformationally rigid precursor for trans-anti-cis isomer, followed by intramolecular cyclization forms the tricyclic compound (Scheme IX).

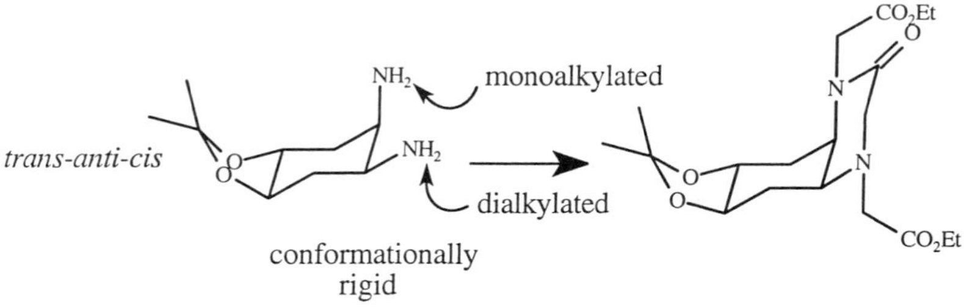

trans-anti-cis   monoalkylated

dialkylated

conformationally rigid

Scheme IX

The conformations (as determined by ^{1}H NMR and NOE studies in DMSO-d_6) solubilities in H_2O, and melting points are listed (Table VII) for the five diol diastereomeric bis(dioxopiperazine)s.

The chair-chair interconversion of <u>cis</u>-syn-<u>cis</u>-**1**, but not <u>cis</u>-anti-<u>cis</u>-**2**, is slow on the NMR time scale.   Isomers <u>cis</u>-anti-<u>trans</u>-**3** and <u>trans</u>-syn-<u>trans</u>-**4** behave as conformationally constrained species at 25°C, whereas <u>trans</u>-anti-<u>trans</u>-**5** likely is in equilibrium with its twist-boat conformer **5a**.   Among these five bis(2,6-dioxopiperazine)s, only <u>cis</u>-anti-<u>cis</u> diastereomer **2** exhibits moderately enhanced water solubility, 29.3 mg/mL at 25°C. The remaining four diastereomers are poorly water soluble. Their solubility is in reverse order to their respective melting points. In addition to the relative competition between intra- and intermolecular hydrogen bonding other forces including differences in crystal packing (*186*) likely affect water solubility. The flip conformers of both isomers **1** and **2** are their respective enantio-

Table VII. Conformation, Water Solubility, and Melting Points

Compound	Solution Conformation Character	Solubility in H$_2$O (mg/mL) at 25°C	Melting Point (°C)
11	In equilibrium with chair conformer	2.1	279-280
12	In equilibrium with chair conformer (fast on NMR time scale)	29.3	221-223.5
13	Conformationally rigid	1.1	>300
14	Conformationally rigid	1.3	>300
15	In equilibrium with twist board conformation. (Twist boat in crystalline state?)	1.1	>300

morphs, and thus these species are effectively meso at room temperature. The 1,4-trans relationship of dioxopiperazine and hydroxyl groups found in cis-anti-cis-**2** provides for enhanced water solubility and the lowest melting point, likely owing to loose crystal packing, a function of conformational flexibility (*187*). Biological studies will be reported at a later date.

Literature Cited

1.   Herman, E.H.; Ferrans, V.J.; Sanchez, J.A. In *Cancer Treatment and the Heart*, Muggia, F.; Green, M.D.; Speyer, J., Eds.; Johns Hopkins University Press: Baltimore, MD, **1992**, pp. 114-169.
2.   Witiak, D.T.; Wei, Y. In *Progress in Drug Research*, Jucker, E., Ed.; Vol. 35, Birkhauser Verlag: Basel, Boston, Berlin, **1991**, pp. 249-363.
3.   Silverman, R.B. DNA, III. In *The Organic Chemistry of Drug Design and Drug Action*, Academic Press: San Diego, New York, Boston, London, Sydney, Tokyo, Toronto, **1992**, pp. 255-257.
4.   Herman, E.; Young, R.; Krop, S. *Agents and Actions*, **1978**, *8*, 551-557.
5.   Bristow, M.R.; Minobe, W.A.; Billingham, M.E.; Marmor, J.B.; Johnson, G.A.; Ishimoto, B.M.; Sageman, W.S.; Daniels, J.R. *Lab. Invest.*, **1981**, *45*, 157-168.
6.   Bristow, M.R.; Thompson, P.D.; Martin, R.P.; Mason, J.W.; Billingham, M.E.; Harrison, D.C. *Am. J. Med.*, **1978**, *65*, 823-832.
7.   Unverferth, B.J.; Magorien, R.D.; Balcerzak, S.P.; Leier, C.V.; Unverferth, D.V. *Cancer*, **1983**, *52*, 215-221.
8.   Formelli, F.; Zedeck, M.S.; Sternberg, S.S.; Philips, F.S. *Cancer Res.*, **1978**, *38*, 3286-3292.
9.   Tritton, T.R.; Murphree, S.A.; Sartorelli, A.C. *Biochem. Biophys. Res. Commun.*, **1978**, *84*, 802-808.
10. Lewis, W.; Kleinerman, J.; Puszkin, S. *Circ. Res.*, **1982**, *50*, 547-553.
11. Myers, C.E. *Anthracycline Antibiotics in Cancer Chemotherapy*, Muggia, F.M.; Young, C.W.; Carter, S.K., Eds., Martinus Nijhoff, Boston, **1982**, pp. 297-305.
12. Myers, C.E.; Gianni, L.; Simone, C.B.; Klecker, R.; Greene, R. *Biochemistry*, **1982**, *21*, 1707-1713.
13. Bachur, N.R. *Anthracycline Antibiotics in Cancer Chemotherapy*, Muggia, F.M.; Young, C.W.; Carter, S.K., Eds., Martinus Nijhoff, Boston, **1982**, pp. 97-102.
14. Myers, C.E.; McGuire, W.P.; Liss, R.H.; Ifrim, I.; Grotzinger, K.; Young, R.C. *Science*, **1977**, *197*, 165-167.
15. Olson, H.M.; Young, D.M.; Prieur, D.J.; LeRoy, A.F.; Reagan, R.L. *Am. J. Pathol.*, **1974**, *77*, 439-454.
16. Maisch, B.; Gregor, O.; Zeuss, M.; Koshsiek, K. *Basic Res. Cardiol.*, **1985**, *80*, 626-635.
17. Combs, A.B.; Acosta, D.; Ramos, K. *Biochem. Pharmacol.*, **1985**, *34*, 1115-1116.

18. Bristow, M.R. *Drug-Induced Heart Disease.* Bristow, M.R., Ed. Elsvier/North-Holland Biomedical Press, New York, **1980**, pp 191-215.
19. Daniels, J.R.; Billingham, M.E.; Gelbart, A.; Bristow, M.R. *Circulation,* **1976**, *54*, Supp. II-20.
20. Piccinini, F.; Monti, E.; Favalli, L.; Villani, F. *Ann. N.Y. Acad. Sci.*, **1988**, *522*, 533-535.
21. Milei, J.; Boveris, A.; Molina, H.; Llesuy, S.; Storino, R.; Milei, S.E. *Acta Cardioe.*, **1985**, *40*, 383-396.
22. Milei, J.; Boveris, A.; Llesuy, S.; Molina, H.A.; Storino, R.; Ortega, D.; Milei, S.E. *Am. Heart J.*, **1986**, *111*, 95-102.
23. Milei, J.; Vazquez, A.; Boveris, A.; Llesuy, S.; Molina, H.A.; Storino, R.; Maranatz, R. *J. Int. Med. Res.*, **1988**, *16*, 19-30.
24. Klugmann, S.; Bartoli, K.F.; Decorti, G.; Gori, D.; Silvestri, F.; Camerini, F. *Pharmacol. Res. Commun.*, **1981**, *13*, 769-776.
25. Giri, S.N.; Marafino, B.J., Jr. *Drug Chem. Toxicol.*, **1984**, *7*, 407-422.
26. Young, D.M.; Mettler, F.P.; Fioravanti, J.L. *Proc. Am. Assoc. Cancer Res.*, **1976**, *17*, 90.
27. Rabkin, S.W.; Otten, M.; Polimeni, P.I. *Can. J. Physiol. Pharmacol.*, **1983**, *61*, 1050-1056.
28. Mochizuki, T.; Okazaki, T.; Ishikura, H.; Isumi, Y.; Tashima, M.; Sawada, H.; Uchino, H. *J. Jpn. Soc. Cancer Ther.*, **1987**, *22*, 539-549.
29. Monti, E.; Paracchini, L.; Piccinini, F.; Rozza, A.; Villani, F. *Pharmacol. Res. Commun.*, **1988**, *20*, 369-376.
30. Villiani, F.; Monti, E.; Piccinini, R.; Favalli, L.; Dionig, A.R.; Lanza, E.; Poggi, P. *Anticancer Res.*, **1988**, *8*, 659-664.
31. Klugmann, F.B.; Decorti, G.; Candussio, L.; Grill, V.; Mallardi, F.; Baldini, L. *Br. J. Cancer*, **1986**, *54*, 743-748.
32. Franco-Browder, S.; Guerrero, M.; Gorodezky, M.; Bravo, L.M.; Aceves, S. *Arch. Inst. Cardioe. Mex.*, **1960**, *30*, 720-728.
33. Ferrans, V.J.; Hibbs, R.G.; Walsh, J.J.; Burch, G.E. *Ann. N.Y. Acad. Sci.*, **1969**, *156*, 309-332.
34. Ferrans, V.J.; Hibbs, R.G.; Weily, H.S.; Weilbaecher, D.G.; Walsh, J.J.; Burch, G.E. *J. Mol. Cell. Cardiol.*, **1970**, *1*, 11-22.
35. Seraydarian, M.W.; Artaza, L. *Cancer Res.*, **1979**, *39*, 2940-2944.
36. Newman, R.A.; Hacker, M.P.; Krakoff, I.H. *Cancer Res.*, **1981**, *41*, 3483-3488.
37. Hacker, M.P.; Newman, R.A. *Eur. J. Cancer Clin. Oncol.*, **1983**, *19*, 1121-1126.
38. Harmsen, E.; de Tombe, P.P.; de Jong, J.W.; Achterberg, P.W. *Am. J. Physiol.*, **1984**, *246*, H37-H43.
39. Czarnecki, W.; Czarnecki, A. *Pharmacol. Res.*, **1989**, *21*, 587-594.
40. Czarnecki, A.; Hinek, A. *Eur. J. Cancer Clin. Oncol.*, **1986**, *22*, 1357-1363.
41. Rigobello, M.P.; Deana, R.; Galzigna, L. In *Advances in Pathology*, E. Levy, Ed.; Oxford Pergamon Press: New York, **1982**, pp 215-217.
42. Lazzarino, G.; Viola, A.R.; Mulieri, L.; Rotilio, G.; Mavelli, I. *Cancer Res.*, **1987**, *47*, 6511-6516.

43.   Bernardini, N.; Danesi, R.; Bernardini, M.C.; Del Tacca, M. *Experientia*, **1988**, *44*, 1000-1002.

44.   Neri, B.; Neri, G.C.; Bandinelli, M. *Oncology*, **1988**, *45*, 242-246.

45.   Vick, J.A.; DeFelice, S.L.; Hassett, C.C. *The Physiologist*, **1975**, *18*, 431.

46.   McFalls, E.O.; Paulson, D.J.; Gilbert, E.F.; Shug, A.L. *Life Sci.*, **1986**, *38*, 497-505.

47.   Alberts, D.S.; Peng, Y-M.; Moon, T.E.; Bressler, R. *Biomedicine*, **1978**, *29*, 265-268.

48.   Payne, C.M. *J. Submicrosc. Cytol.*, **1982**, *14*, 337-345.

49.   Strohm, II, G.H.; Payne, C.M.; Alberts, D.S.; Peng, Y.-M.; Moon, T.E.; Bahl, J.J.; Bressler, R. *Arch. Pathol. Lab. Med.*, **1982**, *106*, 181-185.

50.   Paterna, S.; Furitano, G.; Scaffidi, L.; Barbarino, C.; Campisi, D.; Carreca, I. *Int. J. Tissue React.*, **1984**, *VI*, 91-95.

51.   Vick, J.; DeFelice, S. *The Physiologist*, **1982**, *25*, 191.

52.   DeLeonardis, V.; Neri, B.; Bacalli, S.; Cinelli, P. *Int. J. Clin. Pharmacol. Res.*, **1985**, *5*, 137-142.

53.   Ohuchi, K.; Levine, L. *Prostagland. Med.*, **1978**, *1*, 433-439.

54.   Caldwell, J.; Hutt, A.J.; Fournel-Gigleux, S. *Biochem. Pharmacol.*, **1988**, *37*, 105-114.

55.   Robinson, T.W.; Giri, S.N. *Pharmacol. Res. Commun.*, **1984**, *16*, 409-418.

56.   Gosalvez, M.; Van Rossum G.D.V.; Blanco, M.F. *Cancer Res.*, **1979**, *39*, 257-261.

57.   Arena, E.; D'Alessandro, N.; Dusonchet, L.; Gebbia, N.; Gerbasi, F.; Rausa, L. *J. Antibiotics*, **1973**, *26*, 339-342.

58.   Somberg, J.; Cagin, N.; Levitt, B.; Bounous, H.; Ready, P.; Leonard, D.; Anagnostopoulos, C. *J. Pharmacol. Exp. Ther.*, **1978**, *204*, 226-229.

59.   Necco, A.; Dasdia, D.; Di Francesio, D.; Ferroni, A. *Pharmacol. Res. Commun.*, **1976**, *8*, 105-109.

60.   Guthrie, D.; Gibson, A.L. *Br. Med. J.*, **1977**, *2*, 1447-1449.

61.   Butturini, U.; Deicas, L.; Minco, F.; Baroni, M.C.; Buia, E.; Crotti, G.; Bernandini, B.; Manco, C.; Delsignore, R. *Clin. Ter.*, **1984**, *108*, 389-395.

62.   Whittaker, J.A.; Al-Ismail S.A.D. *Br. Med. J.*, **1984**, *288*, 283-284.

63.   Villiani, F.; Comazzi, R.; DiFronzo, G.; Bertuzzi, A.; Guindani, A. *Tumori*, **1982**, *68*, 349-353.

64.   Weishaar, R.E.; Cain, M.H.; Bristol, J.A. *J. Med. Chem.*, **1985**, *28*, 537-545.

65.   Bossa, R.; Galatulas, I.; Savi, G.; Supino, R.; Zunino, F. *Tumori*, **1982**, *68*, 499-504.

66.   Bossa, R.; Galatulas, I. *Anticancer Res.*, **1986**, *6*, 841-844.

67.   Bossa, R.; Castelli, M.; Galatulas, I.; Ninci, M. *Anticancer Res.*, **1988**, *8*, 1229-1232.

68.   Bossa, R.; Aresca, P.; Galatulas, I.; Ninci, M. *Anticancer Res.*, **1989**, *9*, 605-608.

69.   Facchinetti, T.; Delaini, F.; Salmona, M.; Donati, M.B.; Feuerstein, S.; Wendel, A. *Toxicol. Lett.*, **1983**, *15*, 301-307.

70. Hermansen, K.; Wassermann, K. *Acta Pharmacol. Toxicol.*, **1986**, *58*, 31-37.
71. Van Vleet, J.F., Greenwood, L.; Ferrans, V.J.; Reber, A.H. *Am. J. Vet. Res.*, **1978**, *39*, 997-1010.
72. Van Vleet, J.F.; Ferrans, V.J. *Cancer Treat. Rep.*, **1980**, *64*, 315-317.
73. Dimitrov, N.V.; Hay, M.B.; Siew, S.; Hudler, D.A.; Charamella, L.J.; Ullrey, D.E. *Am. J. Pathol.*, **1987**, *126*, 376-383.
74. Dimitrov, N.V.; Zhang, Z.F.; Sun, J.; Si, L.; Xu, G.L.; Wang, S.C.; Texera, C.; Siew, S. *Proc. Am. Soc. Clin. Oncol.*, **1985**, *4*, 27.
75. Shinozawa, S.; Fukuda, T.; Araki, Y.; Oda, T. *Toxicol. Appl. Pharmacol.*, **1985**, *79*, 353-357.
76. Dasmahapatra, K.S.; Vezeridis, M.; Rao, U.; Perez-Brett, R.; Karakousis, C.P. *J. Surg. Res.*, **1984**, *36*, 217-222.
77. Hamaguchi, T.; Azuma, J.; Awata, N.; Ohta, H.; Takihara, K.; Harada, H.; Kishimoto, S.; Sperelakis, N. *Res. Commun. Chem. Pathol. Pharmacol.*, **1988**, *59*, 21-30.
78. Hamaguchi, T.; Azuma, J.; Harada, H.; Takahashi, K.; Kishimoto, S.; Schaffer, S.W. *Pharmacol. Res.*, **1989**, *21*, 729-734.
79. Arena, E.; Dusonchet, L.; Gabbia, N.; Gerbasi, F.; Picone, M.A.; Traina, A. *Proc. 6th Congress Chemoth.*, University Tokyo Press, **1970**, *2*, 124-129.
80. Arena, E.; D'Allessandro, N.; Dusonchet, L.; Gebbia, N.; Gerbasi, F.; Sanguedolce, R.; Rausa, L. In *International Symposium on Adriamycin*, Carter, S.K.; DiMarco, A.; Ghione, M.; Krakoff, I.; Mathe, G., Eds., Springer, New York, **1973**, pp 96-116.
81. Pour, A.; Cady, W.; Modrak, J. *Toxicol. Lett.*, **1981**, *7*, 379-382.
82. Savaraj, N.; Allen, L.M.; Sutton, C.; Troner, M. *Res. Commun. Chem. Pathol. Pharmacol*, **1980**, *29*, 549-559.
83. Banks, A.R.; Jones, T.; Koch, T.H.; Friedman, R.D.; Bachur, N.R. *Cancer Chemother. Pharmacol.*, **1983**, *11*, 91-93.
84. Averbuch, S.D.; Gaudiano, G.; Koch, T.H.; Bachur, N.R. *Cancer Res.*, **1985**, *45*, 6200-6204.
85. Barone, A.D.; Atkinson, R.F.; Wharry, D.L.; Koch, T.H. *J. Am. Chem. Soc.*, **1981**, *103*, 1606-1607.
86. Averbuch, S.D.; Gaudiano, G.; Koch, T.H.; Bachur, N.R. *J. Clin. Oncol.*, **1986**, *4*, 88-94.
87. Averbuch, S.D.; Boldt, M.; Gaudiano, G.; Stern, J.B.; Koch, T.H.; Bachur, N.R. *J. Clin. Invest.*, **1988**, *81*, 142-148.
88. Nemec, J. *Neoplasma*, **1979**, *26*, 525-528.
89. Klugmann, F.B.; Decorti, G.; Mallardi, F.; Klugmann, S.; Baldini, L. *Eur. J. Cancer Clin. Oncol.*, **1984**, *20*, 405-410.
90. Naganuma, A.; Satoh, M.; Imura, N. *Jpn. J. Cancer Res. (Gann)*, **1988**, *79*, 406-411.
91. Satoh, M.; Naganuma, A.; Imura, N. *Toxicology*, **1988**, *53*, 231-237.
92. McGinness, J.E.; Proctor, P.H.; Demopoulos, H.; Hakanson, J.A.; Van, N.T. In *Pathology of Oxygen*, Autor, A., Ed.; Academic Press: New York, **1982**, pp 191-200.

93. McGuinness, J.E.; Benjamin, R.S.; Wang, Y.-M. *Proc. Am. Assoc. Cancer Res.*, **1980**, *21*, 288.
94. Galvin, M.J.; Lefer, A.M. *Am. J. Physiol.* **1978**, *235*, H657-H663.
95. Gardner, H.W.; Weisleder, D.; Kleiman, R. *Lipids*, **1976**, *11*, 127-134.
96. Doroshow, J.H.; Locker, G.Y.; Ifrim, I.; Myers, C.E. *J. Clin. Invest.*, **1981**, *68*, 1053-1064.
97. Olson, R.D.; MacDonald, J.S.; Van Boxtel, C.J.; Boerth, R.C.; Harbison, R.D.; Slonim, A.E.; Freeman, R.W.; Oates, J.A. *J. Pharmacol. Exp. Ther.*, **1980**, *215*, 450-454.
98. Yoda, Y.; Nakazawa, M.; Abe, T.; Kawakami, Z. *Cancer Res.*, **1986**, *46*, 2551-2556.
99. Schmitt-Graff, A.; Scheulen, M.E. *Pathol. Res. Pract.*, **1986**, *181*, 168-174.
100. Unverferth, D.V.; Mehegan, J.P.; Nelson, R.W.; Scott, C.C.; Leier, C.V.; Hamlin, R.L. *Semin. Oncol.*, **1983**, *10(Suppl 1)*, 2-6.
101. Unverferth, D.V.; Leier, C.V.; Balcerzak, S.P.; Hamlin, R.L. *Am. J. Cardiol.*, **1985**, *56*, 157-161.
102. Herman, E.H.; Ferrans, V.J.; Myers, C.E.; Van Vleet, J.F. *Cancer Res.*, **1985**, *45*, 276-281.
103. Dresdale, A.R.; Barr, L.H.; Bonow, R.O.; Mathisen, D.J.; Myers, C.E.; Schwartz, D.E.; d'Angelo, T.; Rosenberg, S.A. *Am. J. Clin. Oncol.*, **1982**, *5*, 657-663.
104. Myers, C.E.; Bonow, R.; Palmeri, S.; Jenkins, J.; Corden, B.; Locker, G.; Doroshow, J.; Epstein, S. *Semin. Oncol.*, **1983**, *10(Suppl 1)*, 53-55.
105. Unverferth, D.V.; Magorien, R.D.; Unverferth, B.P.; Talley, R.L.; Balcerzak, S.P.; Baba, N. *Cancer Treat. Rep.*, **1981**, *65*, 1093-1097.
106. Unverferth, D.V.; Jagadeesh, J.M.; Unverferth, B.J.; Magorien, R.D.; Leier, C.V.; Balcerzak, S.P. *J. Natl. Cancer Inst.*, **1983**, *71*, 917-920.
107. Unverferth, D.V.; Fertel, R.H.; Balcerzak, S.P.; Magorien, R.D.; O'Doriso, M.S. *Semin. Oncol.*, **1983**, *10(Suppl 1)*, 49-52.
108. Gulati, O.P.; Nordmann, H.; Aellig, A.; Maignan, M.F.; McGinness, J. *Arch. Int. Pharmacodyn.*, **1985**, *273*, 323-334.
109. McGinness, J.E.; Grossie, B., Jr.; Proctor, P.H.; Benjamin, R.S.; Gulati, B.O.; Hokanson, J.A. *Physiol. Chem. Phys. Med. NMR*, **1986**, *18*, 17-24.
110. Hrushesky, W.J.M.; Olshefski, R.; Wood, P.; Meshnick, S.; Eaton, J.W. *Lancet*, **1985**, *1*, 565-567.
111. Viehe, H.G.; Janousek, Z.; Merenyi, R. *Acc. Chem. Res.*, **1985**, *18*, 148-154.
112. Buc-Calderon, P.; Roberfroid, M. *Free Radic. Res. Commun.*, **1988**, *5*, 159-168.
113. Buc-Calderon, P.; Praet, M.; Ruysschaert, J.M.; Roberfroid, M. *Cancer Treat. Rev.*, **1987**, *14*, 379-382.
114. Buc-Calderon, P.; Praet, M.; Ruysschaert, J.M.; Roberfroid, M. *Eur. J. Cancer Clin. Oncol.*, **1989**, *25*, 679-685.
115. Fujita, K.; Shinpo, K.; Yamada, K.; Sato, T.; Niimi, H.; Shamoto, M.; Nagatsu, T.; Takeuchi, T.; Umezawa, H. *Cancer Res.*, **1982**, *42*, 309-316.
116. Ng, T.C.; Daugherty, J.P.; Evanochko, W.T.; Digerness, S.B.; Durant, J.R.; Glickson, J.D. *Biochem. Biophys. Res. Commun.*, **1983**, *110*, 339-347.

117. Daugherty, J.P.; Wheat, M.; Hixon, S.C.; Durant, J.R. *Proc. Am. Assoc. Cancer Res.*, **1981**, *22*, 266.

118. Meerson, F.Z.; Nurmukhambetov, A.N.; Dzhanbaeva, G.E.; Gutkin, D.V. *Pathol. Fizid. Eksp. Ter.*, **1987**, *4*, 66-68.

119. Shimamoto, N.; Tanabe, M.; Shino, A.; Hirata, M.; Kawaji, H.; Azuma, I.; Fukuda, T.; Kobayashi, S.; Yamamura, Y. *Int. J. Immunopharmacol.*, **1983**, *5*, 245-251.

120. Takeshige, K.; Takaganagi, R.; Minakami, S. In *Biomedical and Clinical Aspects of Coenzyme* Q_{10}. Yamamura, Y.; Folkers, K.; Ito, Y., Eds.; Elsevier/North-Holland Biomedical Press: Amsterdam, **1980**, *2*, pp. 15-25.

121. Iwamoto, Y.; Hansen, I.L.; Porter, T.H.; Folkers, K. *Biochem. Biophys. Res. Commun.*, **1974**, *58*, 633-638.

122. Kishi, T.; Folkers, K. *Cancer Chemother. Rep.*, **1976**, *60*, 223-224.

123. Ogura, R.; Toyama, H.; Shimada, T.; Murakami, M. *J. Appl. Biochem.*, **1979**, *1*, 325-335.

124. Shinozawa, S.; Araki, Y.; Oda, T. *Acta Med. Okayama*, **1980**, *34*, 255-261.

125. Ohhara, H.; Kanaide, H.; Nakamura, M. *J. Mol. Cell. Cardiol.*, **1981**, *13*, 741-752.

126. Combs, A.B.; Choe, J.Y.; Truong, D.H.; Folkers, K. *Res. Commun. Chem. Pathol. Pharmacol.*, **1977**, *18*, 565-568.

127. Ogura, R.; Katsuki, T.; Daoud, A.H.; Griffin, A.C. *J. Nutr. Sci. Vitaminol*, **1982**, *28*, 329-334.

128. Zbinden G.; Bachmann, E.; Bolliger, H. In *Biomedical and Clinical Aspects of Coenzyme* Q_{10}. Vol 4, Folkers, K.; Yamamura, Y., Eds.; Elsevier/North Holland Biomedical Press: New York, **1977**, Vol. 4, pp. 219-227.

129. Folkers, K.; Choe, J.Y.; Combs, A.B. *Proc. Natl. Acad. Sci. U.S.A.*, **1978**, *75*, 5178-5180.

130. Bertazolli, C.; Sala, L.; Solcia, E.; Ghione, M. *IRCS Med. Sci.*, **1975**, *3*, 468.

131. Dohmae, N.;, Sawada, H.; Tashima, M.; Uchino, H.; Matsuyama, E.; Konishi, T. *J. Jpn. Soc. Cancer Ther.*, **1979**, *14*, 1009-1028.

132. Domae, N.; Sawada, H.; Matsuyama, E.; Konishi, T.; Uchino, H. *Cancer Treat. Rep.*, **1981**, *65*, 79-91.

133. Usui, T.; Ishikura, H.; Izumi, Y.; Konishi, H.; Dohmae, N.; Sawada, H.; Uchino, H.; Matsuda, H.; Konishi, T. *Toxicol. Lett.*, **1982**, *12*, 75-82.

134. Bachur, N.R.; Gordon, S.L.; Gee M.V. *Mol. Pharmacol.*, **1977**, *13*, 901-910.

135. Stuart, M.J.; deAlarcon, P.A.; Barvinchak, M.K. *Am. J. Hematol.*, **1978**, *5*, 297-303.

136. Lenzhofer, R.; Magometschnigg, D.; Dudczak, R.; Cerni, C.; Bolebruch, C.; Moser, K. *Experientia*, **1983**, *39*, 62-64.

137. Myers, C.E.; McGuire, W.; Young, R. *Cancer Treat. Rep.*, **1976**, *60*, 961-962.

138. Mimnaugh, E.G.; Siddik, Z.H.; Drew, R.; Sikic, B.I.; Gram, T.E. *Toxicol. Appl. Pharmacol.*, **1979**, *49*, 119-126.

139. Tanigawa, N.; Katoh, H.; Kan, N.; Mizuno, Y.; Tanimura, H.; Satomura, K.; Hikasa, Y. *Jpn. J. Cancer Res.* (Gann), **1986**, *77*, 1249-1255.
140. Sonneveld, P. *Cancer Treat. Rep.*, **1978**, *62*, 1033-1036.
141. Wang, Y-M.; Madanat, F.F.; Kimball, J.C.; Gleiser, C.A.; Ali, M.K.; Kaufman, M.W.; Van Eys, J. *Cancer Res.*, **1980**, *40*, 1022-1027.
142. Shinozawa, S.; Gomita, Y.; Araki, Y. *Acta Med. Okayama*, **1988**, *42*, 253-258.
143. Breed, J.G.S.; Zimmerman, A.N.E.; Dormans, J.A.M.A.; Pinedo, H.M. *Cancer Res.*, **1980**, *40*, 2033-2038.
144. Van Vleet, J.F.; Ferrans, V.J.; Weirich, W.E. *Am. J. Pathol.*, **1980**, *99*, 13-24.
145. Weitzman, S.A.; Lorell, B.; Carey, R.W.; Kaufman, S.; Stossel, T.P. *Curr. Ther. Res.*, **1980**, *28*, 682-686.
146. Legha, S.S.; Wang, Y-M.; Mackay, B.; Ewer, M.; Hortobagyi, G.N.; Benjamin, R.S.; Ali, M.K. *Ann. N.Y. Acad. Sci.*, **1982**, *393*, 411-418.
147. Herman, E.H.; Mhatre, R.M.; Lee, I.P.; Waravdekar, V.S. *Proc. Soc. Exp. Biol. Med.*, **1972**, *140*, 234-239.
148. Herman, E.H.; Mhatre, R.M.; Chadwick, D.P. *Toxicol. Appl. Pharmacol.*, **1974**, *27*, 517-526.
149. Herman, E.; Ardalan, B.; Bier, C.; Waravdekar, V.; Krop, S. *Cancer Treat. Rep.*, **1979**, *63*, 89-92.
150. Wang, G.M.; Finch, M. *Proc. Am. Assoc. Cancer Res.*, **1979**, *20*, 23.
151. Wang, G.M.; Finch, M.D. *Drug Chem. Toxicol.*, **1980**, *3*, 213-325.
152. Fischer, V.W.; LaRose, L.S.; Wang, G.M. *Drug Chem. Toxicol.*, **1982**, *5*, 155-164.
153. Fischer, V.W.; Wang, G.M.; Hobart, N.H. *Virchows Arch. B Cell Pathol.*, **1986**, *51*, 353-361.
154. Wang, G.; Finch, M.D.; Trevan, D.; Hellmann, K. *Br. J. Cancer*, **1981**, *43*, 871-877.
155. Herman, E.H.; El-Hage, A.N.; Ferrans, V.J.; Witiak, D.T. *Res. Commun. Chem. Pathol. Pharmacol.*, **1983**, *40*, 217-231.
156. Herman, E.H.; Ferrans, V.J.; Jordan, W.; Ardalan, B. *Res. Commun. Chem. Pathol. Pharmacol.*, **1981**, *31*, 85-97.
157. Herman, E.H.; Ferrans, V.J. *Cancer Chemother. Pharmacol.*, **1986**, *16*, 102-106.
158. Giuliani, F.; Casazza, A.M.; DiMarco, A,; Savi, G. *Cancer Treat. Rep.*, **1981**, *65*, 267-276.
159. Decorti, G.; Klugmann, F.B.; Mallardi, F.; Klugmann, S.; Benussi, B.; Grill, V.; Baldini, L. *Cancer Lett.*, **1983**, *19*, 77-83.
160. Supino, R. *Tumori*, **1984**, *70*, 121-126.
161. Herman, E.H.; Ferrans, V.J. *Cancer Res.*, **1981**, *41*, 3436-3440.
162. Herman, E.H.; Ferrans, V.J.; Young, R.S.K.; Hamlin, R.L. *Cancer Res.*, **1988**, *48*, 6918-6925.
163. Perkins, W.E.; Schroeder, R.L.; Carrano, R.A.; Imondi, A.R. *Br. J. Cancer*, **1982**, *46*, 662-667.
164. Herman, E.H.; Ferrans, V.J. *Lab. Invest.*, **1983**, *49*, 69-77.

165. Hu, S.T.; Brandle, E.; Zbinden, G. *Pharmacology*, **1983**, *26*, 210-220.
166. Herman, E.H.; El-Hage, A.N.; Ferrans, V.J.; Ardalan, B. *Toxicol. Appl. Pharmacol.*, **1985**, *78*, 202-214.
167. Herman, E.H.; El-Hage, A.; Ferrans, V.J. *Toxicol. Appl. Pharmacol.*, **1988**, *92*, 42-53.
168. Herman, E.H.; Ferrans, V.J. *Cancer Chemother. Pharmacol.*, **1986**, *16*, 102-106.
169. Dardir, M.; Herman, E.H.; Ferrans, V.J. *Cancer Chemother. Pharmacol.*, **1989**, *23*, 269-275.
170. Woodman, R.J.; Cysyk, R.L.; Kline, I., Gang, M.; Vendetti, J.M. *Cancer Chemother. Rep.*, **1975**, *59*, 689-695.
171. Verhoef, V.; Bell, V.; Filppi, J. *Proc. Am. Assoc. Cancer Res.*, **1988**, *29*, 273.
172. Speyer, J.L.; Green, M.D.; Kramer, E.; Rey, M.; Sanger, J.; Ward, C.; Dubin, N.; Ferrans, V.; Stecy, P.; Zeleniuch-Jacquotte, A.; Wernz, J.; Feit, f.; Slater, W.; Blum, R.; Muggia, F. *N. Engl. J. Med.*, **1988**, *319*, 745-752.
173. Weiss, R.B. *Semin. Oncol.*, **1992**, *19*, 670-686.
174. Herman, E.H.; El-Hage, A.N.; Creighton, A.M.; Witiak, D.T.; Ferrans, V.J. *Res. Commun. Chem. Pathol. Pharmacol.*, **1985**, *48*, 39-55.
175. Hasinoff, B.B. *Drug Metab. Dispos.*, **1990**, *18*, 344-349.
176. Hasinoff, B.B. *Agents and Actions*, **1990**, *29*, 374-381.
177. Witiak, D.T.; Lee, H.J.; Goldman, H.D.; Zwilling, B.S. *J. Med. Chem.*, **1978**, *21*, 1194-1197.
178. Zwilling, B.S.; Campolito, L.B.; Reiches, N.A.; George, T.; Witiak, D.T. *Br. J. Cancer*, **1981**, *44*, 578-583.
179. Hemple, A.; Camerman, N.; Camerman, A. *J. Am. Chem. Soc.*, **1982**, *104*, 3453-3456.
180. Witiak, D.T.; Trivedi, B.K.; Campolito, L.B.; Zwilling, B.S.; Reiches, N.A. *J. Med. Chem.*, **1981**, *24*, 1329-1332.
181. Witiak, D.T.; Nair, R.V.; Schmid, F.A. *J. Med. Chem.*, **1985**, *28*, 1228-1234.
182. Witiak, D.T.; Trivedi, B.K. Schmid, F.A. *J. Med. Chem.*, **1985**, *28*, 1111-1113.
183. Herman, E.H.; Ferrans, V.J.; Bhat, H.B.; Witiak, D.T. In *Cancer Chemother. Pharmacol.*, **1987**, *19*, 277-281.
184. Witiak, D.T.; Wei, Y. *J. Org. Chem.*, **1991**, *56*, 5408-5417.
185. Witiak, D.T.; Rotella, D.P.; Filippi, J.A.; Gallucci, J. *J. Med. Chem.*, **1987**, *30*, 1327-1336.
186. Grant, D.J.W.; Higuchi, T. *Techniques of Chemistry*, Weissberger, A.; Founding Ed.; Saunders, W.H., Jr., Series Ed.; John Wiley and Sons, Inc.: New York, **1990**, Vol. 21, pp. 22-29.
187. Hempel, A.; Camerman, N.; Camerman, A. *J. Am. Chem. Soc.*, **1982**, *104*, 3453-3456.

RECEIVED July 28, 1994

Chapter 19

Use of Drug Carriers To Ameliorate the Therapeutic Index of Anthracycline Antibiotics

Roman Perez-Soler[1], Steven Sugarman[2], Yiyu Zou[1], and Waldemar Priebe[3]

Departments of [1]Thoracic/Head and Neck Medical Oncology, [2]Gastrointestinal Oncology, and [3]Clinical Investigation, The University of Texas M. D. Anderson Cancer Center, 1515 Holcombe Boulevard, Box 80, Houston, TX 77030

Anthracycline antibiotics are some of the most effective antitumor agents. However, their use is limited by lack of tumor specificity and natural or acquired resistance. Different tumor targeting strategies have been explored to enhance their therapeutic index by actively or passively target them to the tumor site. Immunoconjugates of doxorubicin and different anthracyclines incorporated into liposomes are being evaluated in clinical trials. Based on the preclinical studies, some of these agents have a definite potential of being superior antitumor agents for specific indications.

Anthracyclines are some of the most effective antitumor agents (*1*). Doxorubicin is being used as front-line therapy for breast cancer, sarcoma, and lymphoma. Daunorubicin, and more recently its analogue idarubicin, are standard front-line therapy for adult acute leukemia (*2*).

The use of anthracyclines, like many anticancer agents, is limited by side effects due to the lack of tumor specificity. In the case of anthracyclines, these are bone marrow suppression and mucositis after a single course and chronic cardiotoxicity after multiple courses. In addition, their use is further limited by lack of activity against some of the most common malignancies such as colon carcinoma and non-small cell lung cancer and by the development of acquired resistance after several courses of therapy in patients whose tumors initially respond to them.

During the last few years, important advances have been made in understanding the cellular mechanisms of resistance to anthracyclines. The overexpression of a cell membrane drug efflux pump (P-glycoprotein) has been associated with intrinsic and acquired resistance to anthracyclines and other structurally unrelated anticancer agents in numerous *in vitro* and *in vivo* experimental systems (*3*). This phenomenon has been called multidrug resistance (MDR). Cellular resistance, via the P-glycoprotein pump, to specific plant alkaloids including vincristine, taxol, and colchicine, as well as anthracyclines, is associated with the MDR-1 phenotype (*4*). However, overexpression of P-glycoprotein is not the only mechanism of MDR. Quantitative and qualitative changes in topoisomerase II have been described (*5*) as well as membrane-based mechanisms mediated by overexpression of other transport-related proteins such as the recently discovered multidrug resistance protein (MRP) (*6*).

Classic analogue development through systematic structure-activity studies was done extensively in the past; the objectives were to select more active and less toxic analogues. In the case of doxorubicin, efforts were aimed at reducing or eliminating its cardiotoxic potential. Most of these efforts were, nonetheless, carried out before the phenomenon of MDR was recognized. Idarubicin is the only tangible result of such efforts. More recently, such efforts have been selectively geared towards the identification of structural features in the anthracycline molecule that mediate the overcoming of MDR (*7-10*).

Manipulation of the pharmacological properties of the clinically available anthracyclines has also been an area of intensive study in an attempt to reduce their cardiotoxicity. The use of a continuous intravenous (iv) infusion schedule was associated with decreased cardiotoxicity but did not compromise the antitumor activity of doxorubicin, thus suggesting that this toxicity may be more closely related to high plasma peak levels than the total dose administered (*11*).

A more sophisticated way to modify the pharmacokinetics and organ distribution of a drug is by using a drug carrier with preselected properties. During the last decade, drug carrier technology has attracted a growing interest as a way to confer to drugs desired pharmacological properties including but not limited to: plasma circulation time, organ distribution, excretion, and tumor targeting. Many of the described drug carrier technologies have been explored for anthracycline antibiotics. These efforts have resulted in preparations which offer important advantages in terms of decreased cardiotoxicity, enhanced tumor targeting, and overcoming MDR.

In this chapter, we review the efforts conducted in the area of drug carrier technology to enhance the therapeutic index of anthracycline antibiotics and discuss rational novel strategies and future directions of this field.

Drug Carriers for Anthracyclines: General Principles

The use of drug carriers has the potential to improve the conventional iv infusions of anthracyclines by, (a) altering their biodistribution, (b) enhancing tumor targeting, and (c) increasing the plasma circulation half-life.

The characteristics of the ideal drug carrier for anthracyclines are:
1. Biodegradability to avoid accumulation in the body of non-biodegradable materials.
2. Nontoxicity at the doses needed to achieve therapeutic levels of anthracyclines.

3. Nonimmunogenicity to avoid allergic reactions thus allowing for repeated drug
    administration.
4. A formulation fulfilling standard criteria of a pharmaceutical product, thus allowing for its widespread distribution, easy handling, and use by any hospital pharmacy.
5. The ability to actively or passively target the drug to the tumor site ("active" tumor targeting implies specific recognition of the tumor by the carrier; "passive" tumor targeting implies preferential drug distribution to the tumor cells in relation to organs at risk for toxicity as a result of changes in drug pharmacokinetics and organ distribution).

To accomplish its tumor targeting function, the drug carrier needs to protect the drug while in the circulation, must not interfere with its antitumor properties, and must be able to cross tumor capillaries.

Types of Drug Carriers Used for Anthracycline Antibiotics

Table I shows the different methods that have been explored to target anthracyclines to tumors, most of them through the use of different drug carriers. A detailed description of each one of these carriers as well as their current developmental status is discussed below.

Active Tumor Targeting of Anthracyclines. Conjugates of anthracyclines with ligands that recognize tumor cell receptors have been explored as potential ways of increasing tumor targeting, with varied but in some cases very promising results. The main potential limitation of these drug-conjugates lies in the fact that there is no receptor or antigen that is absolutely specific for tumor cells. As a result, these conjugates not only tend to target tumors, but also tend to be highly selective for other normal tissues, thus resulting in unexpected toxicities. Several of these developments are currently under clinical investigation (Table II).

Anthracycline-Hormonal Conjugates. Melanocyte stimulating hormone (MSH) binds to specific receptors on the surface of melanoma cells. Conjugates of MSH can specifically recognize and be internalized by melanoma cells. Therefore, they offer a possibility for specific, site-directed chemotherapy of melanomas. MSH-daunomycin conjugates have been synthesized and their cytotoxic effects against mouse melanoma cells tested *in vitro* (12). These conjugates were found to be more toxic to mouse melanoma cells than free daunomycin, and in the same concentration range the conjugates were nontoxic to mouse fibroblasts. Unconjugated daunomycin was equally toxic in both cell lines. Selective uptake of MSH-daunomycin conjugates by melanoma cells alone was demonstrated by fluorescence measurements. Endocytosis was assumed to be the mechanism of intracellular accumulation of daunomycin. Although these studies demonstrated a selective cytotoxicity of an anthracycline by coupling it to a tumor targeting moiety, the conjugates were not pursued further.

Table I. Methods for Targeting Anthracyclines to Tumors

1. **Anatomical**	Loco-regional administration intra-arterial intratumoral intracavitary

2. **Pharmacological**

    a. <u>Passive</u>
- Altering plasma pharmacokinetics and organ distribution
  Particulate drug carriers
      Micropheres
      Nanoparticles
      Lipid-based carriers:
          liposomes
          lipoproteins

- Altering cellular pharmacokinetics
  Drug-polymers

    b. <u>Active</u>
- Targeting tumor cell membrane receptors or antigens
      Conjugates
          hormonal
          transferrin
          monoclonal antibodies

3. **Combined**
    1 and 2a temperature-sensitive liposomes
    1 and 2a magnetic liposomes or microspheres
    2a and 2b immunoliposomes

Table II. Tumor-Targeted Anthracycline Conjugates

Tumor Moiety	Tumor Type	Anthracycline	Developmental Status	Reference
Hormone (MSH)	Melanoma	Daunorubicin	Increased in vitro cytotoxicity	Varga et al. (12)
Monoclonal antibody:				
BR 64	Carcinoma	Doxorubicin	Increased in vivo antitumor	Braslawsky et al.(18)
BR 96	Carcinoma		activity in human xenografts	Trail et al. (13)
G 28.1	B-cell lymphoma			
N P-4	Carcinoma			Shih et al.(20)
92-27	Melanoma			Yang et al.(14)
Other:				
Transferrin	Wide variety of human tumors	Doxorubicin	Increased in vitro cytotoxicity Preliminary clinical experience	Faulk et al.(27)

With the major advances in tumor receptor biology and immunology that have occurred during the last few years, this approach could now be expanded to include conjugation with growth factors involved in the development or maintenance of several human malignancies, for example EGF, TGF-α, and IL2.

Anthracycline-Monoclonal Antibody Conjugates. This approach has recently received a great deal of publicity as a result of reports of cure of animals bearing xenografted human carcinomas by BR96-doxorubicin (BR96-Dox) immunoconjugates developed by Bristol-Myers Squibb scientists (*13*). The validity of using anthracycline-antibody conjugates has been confirmed by several investigators during the last few years in *in vitro* and *in vivo* systems (*14-21*). Several different coupling technologies have been described. The BR96 conjugate is the culmination of these efforts. BR96 is a chimeric monoclonal antibody that contains a human framework region and murine binding sites. The antibody binds to an antigen related to Lewis Y that is expressed on the surface of many human carcinomas; it appears to have a high degree of tumor selectivity and is internalized after binding. The BR96-Dox conjugate is synthesized by selective reduction of the chimeric BR96 antibody interchain disulfide bonds, followed by conjugation of malieimidocaproyl doxorubicin-hydrazone to the thiolation sites. This confers stability in plasma; however, the acid-labile hydrazone bond liberates doxorubicin in the acidic environment of lysosomes and endosomes. Preclinical studies of BR96-Dox demonstrate tumor localization in tissues that express the specific antigen, but low drug deposition in cardiac tissue. BR96-Dox induced complete regressions and cures of xenografted human lung, breast, and colon carcinomas growing subcutaneously (sc) in athymic mice and cured 70% of mice bearing extensive metastases of a human lung carcinoma (*13*). More important, BR96-Dox cured 94% of athymic rats with sc human lung carcinomas, even though these rats, like humans and in contrast to mice, express the BR96 target antigen in normal tissues (*13*).

These results, although very encouraging, need to be viewed with caution since several normal human tissues, mainly the gastrointestinal tract, bind to BR96, which suggests that toxicity to the gastrointestinal tract may be dose-limiting in humans. Gastrointestinal toxicity was actually observed in preclinical dog toxicity studies. In addition, similar efficacy results were obtained years ago with vinblastine-monoclonal antibody conjugates, and in a Phase I study, unacceptable gastrointestinal toxicity was observed (*22,23*). In spite of these observations, BR96-Dox represents a remarkable accomplishment in rational anticancer drug development and has a reasonable chance of constituting an important advance in cancer therapy. Clinical trials have recently been initiated at M. D. Anderson Cancer Center and The University of Alabama.

Transferrin-Anthracycline Conjugates. Most tumor cells express transferrin receptors on their plasma membrane, while most adult normal cells do not. Doxorubicin has been coupled to human transferrin by using a glutaraldehyde cross-linking method (*24-26*). TRF-doxorubicin conjugates were found to inhibit the proliferation of both K562 and HL60 cells in culture more efficiently than free doxorubicin. Interestingly, their cytotoxicity was found to be mediated by a mechanism other than intercalation with nuclear DNA, as in the case of doxorubicin-polymers conjugates.

TRF-doxorubicin conjugates have been administered iv to seven patients with acute leukemia at very low doses (27). The preliminary clinical experience is encouraging and suggests that such conjugates may provide a valid alternative approach to monoclonal antibody drug targeting.

Passive Tumor Targeting of Anthracyclines. Different drug carriers have no ability to specifically recognize tumor cells but may effectively alter the pharmacokinetics and organ distribution of drugs in a way that results in a relatively preferential drug distribution to tumor tissue rather than to organs sensitive to the drug. As a result, they may enhance antitumor activity while decreasing organ targeted toxicities such as cardiotoxicity in the case of anthracyclines. Liposomal incorporation of drug represents the leading method to passively target anthracyclines to tumors. They have advantages over monoclonal antibodies, including being nonimmunogenic and easy to prepare, but lack the ability to target tumors actively. However, different approaches to solve this latter limitation are being explored.

Polymer-Bound Anthracyclines (Table III). The observation by Tritton et al. (28,29) and Tokes et al. (30,31) in the early 1980s that doxorubicin covalently bound to polymers had significant cytotoxicity against sensitive and resistant cells without entering the cell remains one of the most interesting observations from the mechanistic standpoint in the field of anthracyclines and suggests that some drug carriers can actually alter the cellular properties of anthracyclines by targeting different cellular structures, in this case the cell membrane.

These investigators observed that anthracyclines may be cytotoxic by interfering with membrane functions, probably signal transduction pathways, and that by targeting the cell membrane preferentially, cellular resistance to anthracyclines may be overcome. In contrast, more recent studies by Cera et al. (32) have suggested that membrane effects of anthracyclines contribute little to their cytotoxicity. The polymers used by these investigators (agarose beads and polyglutaraldehyde microspheres) were not developed further for preclinical studies. However, more recently, Duncan et al. (33,34) have developed metacrylamide copolymers bound to doxorubicin and have shown enhanced in vivo antitumor activity in animal models and reduced toxicity as a result of a prolonged drug circulation time and decreased heart levels.

Nanoparticles. Biodegradable polyalkylcyanoacrylate nanoparticles (mean size 0.3 μm) were explored as carriers of anthracyclines in order to overcome the problems of instability and drug leakage initially encountered with doxorubicin encapsulated in liposomes. *In vivo* organ distribution and antitumor activity studies were performed by Belgian investigators (35,36) (Table IV). Doxorubicin was found to localize preferentially in the RES system, and heart drug levels were significantly reduced. In a model of liver metastases of M5076 reticulosarcoma, doxorubicin in nanoparticles was more effective than free doxorubicin. The hepatic tissue was found to be an efficient reservoir of drug as a result of initial loading of Kupffer cells followed by a prolonged diffusion of free drug towards the neoplastic tissue.

Table III. Polymer-Bound Anthracyclines

Polymer	Anthracycline	Effect	Reference
Agarose beads	Doxorubicin	In vitro cytotoxicity without cellular internalization. Decreased in vivo toxicity	Tritton et al.(28)
Polyglutaraldehyde microspheres	Doxorubicin	In vitro cytotoxicity without cellular internalization	Tokes et al.(30)
Metacrylamide copolymers	Doxorubicin	Increased in vivo antitumor activity and circulating time. Decreased heart levels and toxicity	Duncan et al.(34)

Table IV. Particulate Drug Carriers For Anthracyclines

Material	Drug	Effect	Reference
Microspheres (size 20-50 m)			
Albumin	Doxorubicin	Increased in vivo antitumor activity by i.a. administration in animals	Goldberg et al.(40)
Albumin	Doxorubicin and Magnetite	Increased tumor targeting	Gupta et al.(44,45)
Poly (L-lactic acid)	Doxorubicin	Decreased systemic absorption after intrapleural administration in humans	Ike et al.(39)
Ion-exchange resins	Doxorubicin	Decreased heart levels	Napoli et al.(41)
Nanoparticles (0.2 - 0.3 μm)			
Polyalkylcyanoacrylate	Doxorubicin	Decreased heart and bone marrow levels Increased antitumor activity against mouse liver tumors	Chiannikulchai et al.(35)
Liposomes (0.05 - 5 μm)			
PC:Chol:SA:Cardiolipin	Doxorubicin	Decreased toxicity In vitro lack of cross-resistance	Rahman et al. (51,52,54,61,81)
PC:Chol	Doxorubicin	Decreased toxicity in humans	Mayer el al.(58,60)
PC:PG:Chol	Doxorubicin	Increased tumor targeting	Gabizon et al.(50,64)
DSPC:Chol	Daunorubicin	Increased tumor targeting	Forssen et al.(73)
DMPC:DMPG	Annamycin	Increased tumor targeting In vivo lack of cross-resistance	Zou et al.(87,88)
PEG-DSPE:DSPC:Chol	Doxorubicin	Increased circulation time Increased tumor targeting	Gabizon et al.(71,74)

Microspheres.   Poli-L-lactic or albumin microspheres (median size 2-50 μm) have been used as carriers of doxorubicin for loco-regional administration as a depot system or as a means to exploit a chemoembolization effect consisting of prolonged drug diffusion into the tumor after intraarterial administration (*37-41*).   A brief clinical study of doxorubicin in microspheres administered intrapleurally indicated minimal systemic absorption of the drug while intrapleural drug levels were sustained for a long period of time (*39*).

Magnetic microspheres offer the potential advantage of being amenable to external control of their anatomical distribution by creation of a well designed magnetic field (*42-45*).   In rat experiments, doxorubicin in magnetic microspheres showed a markedly preferential distribution towards areas subjected to a magnetic field, thus indirectly resulting in lower drug levels to the heart and bone marrow. No clinical studies have been performed so far with these materials.

Liposomes.   Liposomes have been the most widely explored drug carrier for anthracycline antibiotics (*46-60*) (Table IV).   The rationale for using liposomes as carriers of doxorubicin was based on their ability to decrease the drug distribution to the heart and, therefore, to diminish its cardiotoxic potential (*50,53,56*).   The natural affinity of liposomes for the liver and RES tends to target the drug to common sites of tumor spread (*49-54*).   Unfortunately, doxorubicin is not the easiest drug to encapsulate efficiently in liposomes, and as a result, the development of liposomal-doxorubicin was slowed down for several years by cumbersome formulation problems.   With the development of more sophisticated techniques, however, several different liposomal formulations of doxorubicin and daunorubicin have recently been developed and tested in humans.   Some of these formulations appear to meet the criteria of a pharmaceutical product.

Three formulations developed by three different groups were initially developed. The first formulation tested in humans was developed by Rahman et al. (*61-63*) and used cardiolipin as one of the key components of the liposomal structure because of its natural affinity for doxorubicin.   About the same time, a second formulation was developed at Hadassah University by Gabizon et al. (*64,65*).   Subsequently, scientists at The Liposome Company, Inc., developed a formulation composed of phosphatidyl choline and cholesterol using a remote loading technique (*66-68*).   All of these formulations used small unilamellar vesicles (median size 0.1-0.2 μm).   The clinical experience with them has shown a maximum tolerated dose, dose-limiting toxicity, and spectrum of antitumor activity similar to that of free doxorubicin.   The gastrointestinal toxicity seems to be reduced, and no significant cardiotoxicity has been observed, although the experience with prolonged treatments using these formulations is still very limited.   Therefore, these formulations appear to represent a moderate advantage over free doxorubicin.

With increasing sophistication in the development of tumor-targeted liposomes, second generation formulations with enhanced passive tumor targeting properties have been developed (*69-73*) and one, developed by the Hadassah group, has reached clinical trials.   This formulation also uses small unilamellar vesicles but contains distearoylphosphatidyl choline and PEG 2000-phosphatidyl ethanolamine, which confer a higher rigidity to the membranes and protect the vesicles from recognition by the RES.   The end result is a markedly increased circulation time and a higher tumor uptake. Preclinical studies indicated a marked tumor targeting ability associated with a markedly increased antitumor activity in different murine tumor models (69-72). The clinical experience has shown a maximum tolerated dose similar to that of free doxorubicin but a different toxicity spectrum, i.e., reduced myelosuppression but markedly increased mucositis and skin toxicity, which are the limiting toxicities (74). Another formulation, using daunorubicin and containing distearoylphosphatidyl choline and cholesterol, is also being actively tested in patients with Kaposi's sarcoma, with promising results (73).

Another approach that uses the regional administration of liposomal doxorubicin has been studied in Japan (75,76). By injecting liposomes containing doxorubicin into the gastric submucosa or the portal vein, higher levels of liposomal drug will be trapped in the target organ, resulting in a higher likelihood of tumor response.

With the potential exception of the Rahman formulation (77-83), these preparations are not active in vivo against tumors with the MDR-1 phenotype. Rahman's observations are the only ones to date that support the idea that liposomes may significantly alter the cellular pharmacology of anthracyclines and consequently their cellular effects. The MDR-1 phenotype is one of the best studied mechanisms of acquired tumor resistance to anthracyclines and other unrelated compounds. It is mediated by the overexpression of a P-glycoprotein membrane glycoprotein which acts as an efflux pump. During the last several years, there has been an intensive effort to synthesize anthracycline analogues that are not good substrates for P-glycoprotein and, therefore, that are non cross-resistant with doxorubicin. Several compounds with these characteristics have been identified and tested (7-10). Most of them are highly lipophilic and, therefore, difficult to administer iv. Yet to capitalize on this lipophilicity, liposomes have been studied as a means of delivering and modulating the pharmacology of some of these compounds (84-85). From more than 30 lipophilic compounds, we selected annamycin as an anthracycline antibiotic with a marked affinity for lipid membranes that can be easily entrapped within the lipid bilayers of different types of liposomes. Preclinical studies have shown that annamycin is not cross-resistant in vivo with doxorubicin and is more active than doxorubicin against murine sc tumors or tumors that metastasize to the lungs and liver when entrapped in large or small vesicles composed of dimyristoyl phosphatidyl choline and dimyristoyl phosphatidyl glycerol (86-90). Since modifications of the formulation are relatively simple, current studies are aimed at improving the tumor targeting properties of the liposomal carriers of annamycin. A multilamellar vesicle formulation of annamycin will enter clinical trials at M. D. Anderson Cancer Center in 1994.

Liposomes also offer additional opportunities to target tumors like the incorporation of tumor-specific ligands such as monoclonal antibodies (immunoliposomes) to the vesicle surface (91-94), using lipids with transition temperatures just above the body temperature, thus allowing for controlled drug release with hyperthermia (temperature-sensitive liposomes) (95-99), or as carriers of magnetite, thus allowing their preferential distribution to areas under the effect of an external magnetic field (magnetoliposomes). Although these approaches are still at an early stage of development because of their technical complexities, they represent futuristic but realistic possibilities for enhancing tumor targeting of anthracyclines and other antitumor agents by combining active and passive drug tumor targeting approaches.

In other series of studies, lipoproteins themselves are being used as carriers to target anthracyclines to tissues with a high number of specific receptors (*100*).

Summary and Prospects

Because anthracyclines continue to be some of the most effective antitumor agents, approaches to increase their therapeutic index by using different drug carrier technologies and tumor targeting strategies appear well justified. Extensive work performed during the last few years has generated a few promising leads. Should any of these technologies demonstrate an improved efficacy at the clinical level as a result of their passive or active tumor targeting properties, the next step will be to use them as carriers of anthracyclines that are intrinsically effective against multidrug resistant cells. With this strategy, two of the major impediments for effective cancer chemotherapy, namely lack of tumor specificity and MDR, will be addressed with a reasonable chance of success.

Most problems with manufacturing many of the drug carriers used have recently been solved. In the case of monoclonal antibodies, humanization has decreased the immunogenic potential of these molecules and hopefully will allow for repeated administration of these agents. The major potential obstacles to the successful development of these tumor targeted preparations lies in the occurrence of toxicities that are not predicted in the preclinical in vivo studies. The effectiveness of enhancing tumor targeting may not be associated with a proportional and homogeneous ability to reduce drug levels in all other organs. Because no molecule is able to recognize tumor cells with complete specificity, there is a significant risk of toxicity related to targeting of other tissues recognized by the targeting molecule.

If that turns out to be the case for the leading preparations described in this chapter, potential alternatives to mitigate them exist. Effective tumor targeting requiring the combination of two interventions, both with moderate or high targeting ability but a different spectrum of toxicity, or one of them using the principle of anatomical targeting may offer a significant therapeutic advantage. The following are two possible examples of such an approach: (a) doxorubicin entrapped in stealth and temperature-sensitive liposomes is combined with local hyperthermia for the treatment of tumors accessible to local hyperthermia; and (b) doxorubicin incorporated in stealth liposomes (limiting toxicity is mucositis, and lesser extent myelosuppression), combined with doxorubicin immunoconjugates.

The information generated from ongoing clinical trials should provide an indication of the potential clinical applications and directions of this field in the near future.

Acknowledgments

We wish to thank Helen Gard for her excellent secretarial assistance in preparing the manuscript, and for support, in part, by National Institutes of Health (NIH) Grant CA50270.

Literature Cited

(1)   *Principles and Practice of Oncology, Ed. 3*; DeVita, V.; Hellman, S.; and Rosenberg, S., Eds.; J.B. Lippincott Co.: Philadelphia, PA, 1985.

(2)   Berman, E. A review of idarubicin in acute leukemia. *Oncology* **1993**, Vol. 7, pp 91-98, 104.

(3)   Moscow, J.A.; and Cowan, K.H. Multidrug resistance. *J. Natl. Cancer Inst.* **1989**, Vol. *80*, pp 14-20.

(4)   Dalton, W.S.; Miller, T.P. *Multidrug resistance*; Principles and Practice of Oncology Ed. 3; J.B. Lippincott Co.: Philadelphia, PA, **1991**; Vol. *5*, pp 1-13.

(5)   Danks, M.K.; Yalowich, J.C.; and Beck, W.T. Atypical multiple drug resistance in human leukemic cell line selected for resistance to teniposide (VM-26). *Cancer Res.* **1987**, Vol. *47*, pp 1297-1301.

(6)   Cole, S.P.; Bhardwaj, G.; Gerlach, J.H.; Mackie, J.E.; Grant, C.E.; Almquist, K.C.; Stewart, A.J.; Kurz, E.U.; Duncan, A.M.; Deeley, R.G. Overexpression of a transporter gene in a multidrug-resistant human lung cancer cell line. *Science* **1992**, Vol. *258*, pp 1650-1654.

(7)   Acton, E.M.; Tong, G.L.; Mosher, C.W.; Wolgemuth, R.L. Intensely potent morpholinyl anthracyclines. *J. Med. Chem.* **1984**, Vol. *27*, pp 638-645.

(8)   Ganapayhi, R.; Grabowski, D.; Sweatman, T.W.; Seshardi, R.; Israel, M. N-benzyladriamycin-14-valerate versus progressively doxorubicin-resistant murine tumors: cellular pharmacology and decharacterization or cross-resistance in vitro and in vivo. *Brit. J. Cancer* **1989**, Vol. *60*, pp 819-826.

(9)   Barbieri, B.; Giuliani, F.C.; Bordoni, T.; Casazza, A.M.; Geroni, C.; Bellini, O., Suarato, A.; Gioia, B.; Penco, S.; Arcamone, F. Chemical and biological characterization of 4'-iodo-4'-deoxydoxorubicin. *Cancer Res.* **1987**, Vol *47*, pp 4001-4006.

(10)   Priebe, W.; Van, N.T.; Burke, T.G.; Perez-Soler, R. Removal of the basic center from doxorubicin partially overcomes multidrug resistance and decreases cardiotoxicity. *Anticancer Drugs* **1993**, Vol. *4*, pp 37-48.

(11)   Legha, S.; Benjamin, R.S.; Mackay, B.; Ewer, M.; Wallace, S.; Valdivieso, M.; Rasmusen, S.L.; Blummenschein, G.R.; and Freireich, E.J. Reduction of doxorubicin cardiotoxicity by prolonged continuous i.v. infusion. *Ann. Intern. Med.* **1982**, Vol. *96*, pp 133-139.

(12)   Varga, J.M.; Asato, N.; Lande, S.; Lerner; A.B. Melanotropin-daunomycin conjugate shows receptor-mediated cytotoxicity in cultured murine melanoma cells. Nature **1977**, Vol. *267*, pp 56-58.

(13)   Trail, P.A.; Willner, D.; Lasch, S.J.; Henderson, A.J.; Hofsted, S., Casazza, A.M.; Firestonre, R.A., Hellström, K.E. Cure of xenografted human

carcinomas by BR96-doxorubicin immunoconjugates. *Science* **1993**, Vol. *261*, pp 212-214.

(*14*) Yang, H.M.; Reisfeld, R.A.   Doxorubicin conjuated with a monoclonal antibody directed to a human melanoma-associated proteglycan suppresses the growth of established tumor xenografts in nude mice. *Proc. Natl. Acad. Sci. USA* **1988**, Vol. *85*, pp 1189-1193.

(*15*) Shouval, D.; Adler, R.; Wands, J.R.; Hurwitz, E.; Isselbacher, K.J.; and Sela, M.   Doxorubicin conjugates of monoclonal antibodies to hepatoma-associated antigens. *Proc. Natl. Acad. Sci. USA* **1988**, Vol. *85*, pp 8276-8280.

(*16*) Ohyanagi, H.; Ishida, H.; Ishida, T.; Soyama, N.; Yamamoto, M.; Okumura, S.; Kano, Y.; Ueda, Y.; and Saitoh, Y.   A monoclonal antibody KM10 reactive with human gastrointestinal cancer and its application for immunotherapy. *Jpn. J. Cancer Res. (Gann)*, **1988**, Vol. *79*, pp 1349-1358.

(*17*) Sheldon, K.; Marks, A.; Baumal, R. Sensitivity of multidrug resistant KB-C1 cells to antibody-dextran-adriamycin conjugate. *Anticancer Res.* **1989**, Vol. *9*, pp 637-642.

(*18*) Braslawsky, G.R.; Edson, M.A.; Pearce, W.; Kaneko, T.; and Greenfield, R.S. Antitumor activity of adriamycin (hydrazone-linked) immunoconjugates compared with free adriamycin and specificity of tumor cell killing. *Cancer Res.* **1990,** Vol. *50*, pp 6608-6614.

(*19*) Greenfield, R.S.,; Kaneko, T.; Daues, A.; Edson, M.A.,; Fitzgerald, K.A.,; Olech, L.J.; Grattan, J.A.; Spitalny, G.L.; and Braslawsky, G.R. Evaluation *in vitro* of adriamycin immuncon-jugates synthesized using an acid-sensitive hydrazone linker. *Cancer Res.* **1990**, Vol.. *50*, pp 6600-6607.

(*20*) Shih, L.B.; Goldenberg ,D.M.; Xuan, H.; Lu, H.; Sharkey, R.M.; and Hall, T.C.   Anthra-cycline immunoconjugates prepared by a site-specific linkage via an amino-dextran intermediate carrier. *Cancer Res.* **1991**, Vol. *51*, pp 4192-4198.

(*21*) Trail, P.A.; Willner, D.; Lasch, S.J.; Henderson, A.J.; Greenfield, R.S.; King, D.; Zoeckler, M.E.; and Braslawsky, G.R.   Antigen-specific activity of carcinoma-reactive BR64-doxorubicin conjugates evaluated *in vitro* and human tumor xenograft models. *Cancer Res.* **1992**, Vol. *52*, pp 5693-5700.

(*22*) Petersen, B.H.; DeHerdt, S.V.; Schneck, D.W.; Bumol, T.F.   The human imune response eto KS1/4-desacetylvinblastine (LY256787) and KS1/4-desacetylvinblastine hydrazide (LY203728) in single and multiple dose clinical studies. *Cancer Res.* **1991** Vol. *51*; pp 2286-2290.

(*23*) Apelgren, L.D.; Zimmerman, D.L.; Briggs, S.L.; Bumol, T.F.   Antitumor activity of the monoclonal antibody-Vinca alkaloid immunoconjugate LY203725 (KS1/4-4-desacetylvinblastine-3-carboxhydrazide) in a nude mouse model of human ovarian cancer. *Cancer Res.* **1990**, Vol. *50*, pp 3540-3544.

(*24*) Barabas, K.; Sizensky, J.A.; Faulk, W.P.   Transferrin conjugates of adriamycin are cytotoxic without intercalating nuclear DNA. *J. Biol. Chem.* **1992**, Vol. *267*, pp 9437-9442.

(*25*) Berczi, A.; Barabas, K.; Sizensky, J.A. Adriamycin conjugates of human transferrin bind transferrin receptors and kill K562 and HL 60 cells. *Arch. Biochem. Biophys.* **1993**, Vol. *300*, pp 356-363.

(*26*) Sun, I.L.; Sun, E.E.; Crane, F.L.   Inhibition of transplasma membrane electron transport by transferrin-adriamycin conjugates. *Biochimica et Biophysica Acta* **1992**, Vol. *1105*, pp 84-88.

(*27*) Faulk, W.P.; Taylor, C.G.; Yeh, C.-J.G.; and McIntyre, J.A.   Preliminary clinical study of transferrin-Adriamycin conjugate for drug delivery to acute leukemia patients. *Mol. Biother.* **1990**, Vol. *2*, pp 57-60.

(*28*) Tritton, T.R.; Yee, G.   The anticancer agent adriamycin can be actively cytotoxic without entering cells. *Science* **1982**, Vol. *217*, pp 248-250.

(*29*) Hacker, M.P.; Lazo, J.S.; Pritsos, C.A.; and Tritton, T.R.   Immobilized Adriamycin: Toxic potential *in vivo* and *in vitro*. *Sel. Cancer Ther.* **1989**, Vol. *5*, pp 67-72.

(*30*) Tokes, Z.A.; Rogers, K.E.; Rembaum, A.   Synthesis of adriamycin-coupled polyglutaradelhyde microscopheres and evaluation of their cystotatic activity. *Proc. Natl. Acad. Sci .USA*   **1982**, Vol. *79*, pp 2026-2030.

(*31*) Jeannesson, P.; Trentesaux, C.; Gerard, B.; Jardillier, J.C.; Ross, K.L.; and Tökes, Z.A. Induction of erythroid differentiation in human leukemic K-562 cells by membrane-directed action of adriamycin covalently bound to microspheres. *Cancer Res.* **1990**, Vol. *50*, pp 1231-1236.

(*32*) Cera, C.; Palumbo, M.; Stefanelli, S.; Rassu, M.; Palu, G.   Water-soluble polysaccharide--anthracycline conjugates: biological activity. *Anti-Cancer Drug Des.* **1992**, Vol. *7*, pp 143-151.

(*33*) Seymour, L.W.; Ulbrich, K.; Strohalm, J.; Kopecek, J.; and Duncan, R. The pharmacokinetics of polymer-bound adriamycin. *Biochem. Pharmacol.* **1990**, Vol. *39*, pp 1125-1131.

(*34*) Duncan, R.; Seymour, L.W.; O'Hare, K.B.; Flanagan, P.A.; Wedge, S.; Hume, I.C.; Ulbrich, K.; Strohalm, J.; Subr, V.; Spreafico, F.; Grandi, M.; Ripamonti, M.; Farao, M.; and Suarato, A.   Preclinical evaluation of polymer-bound doxorubicin. *J. Controlled Release* **1992** Vol. *19* pp 331-346.

(*35*) Chiannikulchai, N.; Ammoury, N.; Caillou, B.; Devissaguet, J.P.; and Couvreur, P. Hepatic tissue distribution of doxorubicin-loaded nanoparticles after i.v. administration in reticulosarcoma M5076 metastasis-bearing mice. *Cancer Chemother. Pharmacol.* **1990**, Vol. 26, pp 122-126.

(*36*) Verdum, C.; Brasseur, F.; Vranckx, H.; Couvreur, P.; and Roland, M. Tissue distribution of doxorubicin associated with polyisohexylcyanoacrylate nanoparticles. Cancer Chemother. Pharmacol. **1990**, Vol. *26*, pp 13-18.

(*37*) El Hag, I.A.; Teder, H.; Roos, G.; Christensson, P.I.; Stenram, U. Enhanced effect of adriamycin on a rat liver adenocarcinoma after hepatic artery injection with degradable starch microspheres. *Sel. Cancer Ther.* **1990**, Vol. 6, pp 23-39.

(*38*) Gupta, P.K.; Hung, C.T.; Lam, F.C. Application of regression analysis in the evaluation of tumor response folowing the administration of adriamycin either as a solution or via albumin microspheres to the rat. *J. Pharm. Sci.* **1990**, Vol. *79*, pp 634-637.

(*39*) Ike, O.; Shimizu, Y.; Hitomi, S.; Wada, R.; and Ikada, Y. Treatment of malignant pleural effusions with doxorubicin hydrochloride-containing poly(l-Lactic Acid) microspheres. *Chest* **1991**, Vol. *99*, pp 911-915.

(40)   Goldberg, J.A.; Willmott, N.; Kerr, D.J.; Sutherland, C.; and McArdle, C.S. An *in vivo* assessment of Adriamycin-loaded albumin microspheres. *Br. J. Cancer* **1992**, Vol. *65*, pp 393-395.

(41)   Napoli, S.; Burton, M.A.; Martins, I.J.: Dose response and toxicity of doxorubicin microspheres in a rat tumor model. *Anti-Cancer Drugs* **1992**, Vol. 3, pp 47-53.

(42)   Gallo, J.M.; Hung, C.T.; Gupta, P.K.; et al: Psysiological pharmacokinetic model of adriamycin delivered via magnetic albumin microspheres in the rat. *Pharmacokinet. and Biopharm.* **1989**, Vol. *17*, pp 305-326.

(43)   Gallo, J.M.; Gupta, P.K.; Hung, C.T.; et al: Evaluation of drug delivery following the administration of magnetic albumin microspheres containing adriamycin to the rat. *J. Pharm. Sci.* **1989**, Vol.*78*, pp 190-194.

(44)   Gupta, P.K.;Hung, C.T. Comparative disposition of adriamycin delivered via magnetic albumin microspheres in presence and absence of magnetic field in rats. *Life Sci.* **1990**, Vol. *46*, pp 471-479.

(45)   Gupta, P.K.; Hung, C.T.; Rao, N.S. Ultrastructural disposition of adriamycin-associated magnetic albumin microspheres in rats. *J. Pharm. Sci.* **1989**, Vol. *78*, pp 290-294.

(46)   Forssen, E.A.; Tökes, Z.A. *In vitro* and *in vivo* studies with adriamycin liposomes. *Biochem. Biophys. Res. Commun.* **1979**, Vol. *91*, pp 1295-1301.

(47)   Olson, F.; Mayhew, E.; Maslow, D.; Rustum, Y., and Szoka, F. Characterization, toxicity and therapeutic efficacy of adriamycin encapsulated in liposomes. *Eur. J. Cancer Clin. Oncol.* **1982**, Vol. *18*, pp 167-176.

(48)   Forssen, E.A.; Tökes, Z.A. Attenuation of dermal toxicity of doxrubicin by liposome encapsulation. *Cancer Treat Rep.* **1983**, Vol. *67*, pp 481-484.

(49)   Mayhew, E.; Rustum, Y.; and Vail, W.J. Inhibition of liver metastases of M 5076 tumor by liposome-entrapped adriamycin. *Cancer Drug Delivery* **1983**, Vol. *1*, pp 43-57.

(50)   Gabizon, A.; Dagan, A.; Goren, D.; Barenholz, Y.; Fuks, Z. Liposomes as *in vivo* carriers of adriamycin: reduced cardiac uptake and preserved antitumor activity in mice. *Cancer Res.* **1982**, Vol. *42*, pp 4734-4739.

(51)   Raham, A.; White, G.; More, N.; Schein, P.S. Pharmacological, toxicological and therapeutic evaluation in mice of doxorubicin entrapped in cardiolipin liposomes. *Cancer Res.* **1985**, Vol. *45*, pp 796-803.

(52)   Rahman, A.; Carmichael, D.; Harris, M.; Roh, J.K. Comparative pharmacokinetics of free doxorubicin and doxorubicin entrapped in cardiolipin liposomes. *Cancer Res.* **1986**, Vol. *46*, pp 2295-2299.

(53)   Herman, E.H.; Rahman, A.; Ferrans, V.J.; Vick, J.A.; Schein, P.S. Prevention of chronic doxorubicin cardiotoxicity in beagles by liposomal encapsulation. *Cancer Res.* **1983**, Vol. *43*, pp 5427-5432.

(54)   Rahman, A.; Fumagalli, A.; Barbier, B.; Schein, P.S., and Casazza, A.M. Antitumor and toxicity evaluation of free doxorubicin and doxorubicin entrapped in cardiolipin liposomes. *Cancer Chemother. Pharmacol.* **1986**, Vol. *16*, pp 22-27.

(55)   Gabizon, A.; Goren, D.; Fuks, Z.; Meshorer, A., and Barenholz, Y. Superior therapeutic activity of liposome-associated adriamycin in a murine metastatic tumor model. *Br. J. Cancer* **1985**, Vol. *51*, pp 681-689.

(56)   van Hoesel, Q.G.C.M.; Steerenberg, P.A.; Crommelin, D.J.A.; van Dijk, A.;
        van Oort, W.; Klein, S.; Douze, J.M.C.; de Wildt, J.; Hillen, F.C. Reduced
        cardiotoxicity and nephrotoxicity with preservation of antitumor activity of
        doxorubicin entrapped in stable liposomes in the LOU/M Wsl rat. *Cancer
        Res.* **1984**, Vol. *44*, pp 3698-3705.

(57)   Gabizon, A.; Meshorer, A.; Barenholz, Y. Comparative long-term study of
        the toxicities of free and liposome-associated doxorubicin in mice after
        intravenous administration. *J. Natl. Cancer Inst.* **1986**, Vol. *77*, pp 459-469.

(58)   Bally, M.B.; Nayar, R.; Masin, D.; Cullis, P.R., and Mayer, L.D. Studies on
        the myelo-suppressive activity of doxorubicin entrapped in liposomes.
        Cancer Chemother. Pharmacol. **1990**, Vol. *27*, pp 13-19.

(59)   Mayhew, E.; Goldrosen, M.; Vaage, J., and Rustum, Y. Liposomal-
        adriamycin and survival of mice bearing liver metastases of colon carcinomas
        26 or 38. *Proc. Am. Assoc. Cancer Res.* **1986**, Vol. *27*, p 402.

(60)   Mayer, L.D.; Tai, L.C.L.; Ko, D.S.C.; Masin, D.; Ginsberg, R.S.; Cullis, P.R.,
        and Bally, M.B. Influence of vesicle size, lipid composition, and drug-to-
        lipid ratio on the biological activity of liposomal doxorubicin in mice.
        Cancer Res. **1989**, Vol. *49*, pp 5922-5930.

(61)   Rahman, A.; Treat, J.; Roh, J.-K.; Potkul, L.A.; Alvord, W.G.; Forst, D., and
        Woolley, P.V. A phase I clinical trial and pharmacokinetic evaluation of
        liposome-encapsulated doxorubicin. *J. Clin. Oncol.* **1990**, Vol. *8*, pp 1093-
        1100.

(62)   Treat, J.; Greenspan, A.; Forst, D.; Sanchez, J.A.; Ferrans, V.J.; Potkul, L.A.;
        Woolley, P.V.; Rahman, A. Antitumor activity of liposome-encapsulated
        doxorubicin in advanced breast cancer: phase II study. *J. Natl. Cancer Inst.*
        1990, Vol. *82*, pp 1706-1710.

(63)   Delgado, G.; Potkul, R.K.; Treat, J.A.; Lewandowski, G.S.; Barter, J.F.;
        Forst, D.; and Rahman, A. A phase I/II study of intraperitoneally
        administered doxorubicin entrapped in cardiolipin liposomes in patients with
        ovarian cancer. *Am. J. Obstet. Gynecol.* **1989**, Vol. *160*, pp 812-819.

(64)   Gabizon, A.; Peretz, T.; Sulkes, A.; Amselem, S.; Ben-Yosef, R.; Ben-
        Baruch, N.; Catane, R.; Biran, S.; and Barenholz, Y. Systemic administration
        of doxorubicin-containing liposomes in cancer patients: a phase I study. *Eur.
        J. Cancer Clin. Oncol.* **1989**, Vol. *25*, pp 1795-1803.

(65)   Gabizon, A.; Chisin, R.; Amselem, S.; Druckmann, S.; Cohen, R.; Goren, D.;
        Fromer, I.; Peretz, T.; Sulkes, A.; and Barenholz, Y. Pharmacokinetic and
        imaging studies in patients receiving a formulation of liposome-associated
        adriamycin. *Eur. J. Clin. Oncol.* **1989**, Vol. *25*, pp 1795-1803.

(66)   Brown, T.; Kuhn, J.; Marshall, M.; Rodriguez, G.; Kneuper, R.; Cagnola, J.;
        Von Hoff, D. A phase I clinical and pharmacokinetic trial of liposomal
        doxorubicin (NSC 620212). *Proc. Am. Soc. Clin. Oncol.* **1991**, Vol. *10*, p
        93.

(67)   Embree, L.; Lohri, A.; Gelmon, K.; Mayer, L.; Cullis, P.; Pilkiewicz, F.;
        Hudon, N.; Heggie, J.; and Goldie, J. Liposome encapsulated doxorubicin
        (TLC D-99) pharmacokinetics in patients with non-small cell lung cancer.
        Proc. Am. Soc. Clin. Oncol. **1991**, Vol. *10*, p 102.

(68) Conley, B.; Egorin, M.J.; Zuhowski, E.G.; Carter, D.C.; and Van Echo, D.A. Phase I & comparative pharmacokinetic study of liposome-encapsulated doxorubicin. *Proc. Am. Soc. Clin. Oncol.* **1991**, Vol. *10*, p 92.

(69) Gabizon, A.; Paphadjopoulos, D.: Liposome formulations with prolonged circulation time in blood and enhanced uptake by tumors. *Proc. Natl. Acad. Sci. USA*. **1988**, Vol. *85*, pp 6949-6953.

(70) Gabizon, A.; Price, D.C.; Huberty, J.; Bresalier, R.S.; and Papahadjopoulos, D. Effect of liposome composition and other factors on the targeting of liposomes to experimental tumors: biodistribution and imaging studies. *Cancer Res.* **1990**, Vol. *50*, pp 6371-6378.

(71) Gabison, A.A. Selective tumor localization and improved therapeutic index of anthracyclines encapsulated in long-circulating liposomes. *Cancer Res.***1992**, Vol. *52*, pp 891-896.

(72) Huang, S.K.; Mayhew, E.; Gilani, S.; Lasic, D.D.; Martin, F.J.; and Papahadjopoulos, D. Pharmacokinetics and therapeutics of sterically stabilized liposomes in mice bearing C-26 colon carcinoma. *Cancer Res.* **1992**, Vol. *52*, pp 6774-6781.

(73) Forssen, E.A.; Coulter, D.M.; Proffitt, R.T. Selective *in vivo* localization of daunorubicin small unilamellar vesicles in solid tumors. *Cancer Res.* **1992**, Vol. *52*, pp 3255-3261.

(74) Gabizon, A., Catane, R., Uziely, B., Kaufman, B., Safra, T., Cohen, R., Martin, F., Huang, A., Barenholz, Y. Prolonged circulation time and enhanced accumulation in malignant exudates of doxorubicin encapsulated in polyethylene-glycol coated liposomes. *Cancer Res.* **1994**, Vol *54*, pp 987-992.

(75) Akamo, Y.; Yotsuyanagi, T.; Mizuno, I.; Ichino, T.; Tanimoto, N.; Kurahashi, S.; Saito, T.; Yamamoto, T.; Yashui, T.; Itabashi, Y.. Delivery of lymph node-targeted *adriamycin* by gastric submucosal liposomal injection in rabbits. *Jpn. J. Cancer Res.* **1993**, Vol. *84*, pp 208-213.

(76) Ichino, T.; Yotsuyanagi, T.; Mizuno, I.; Akamo, Y.; Yamamoto, T.; Saito, T.; Kurahashi, S.; Tanimoto, N.; and Yura, J. Antitumor effect of liposome-entrapped adriamycin administered via the portal vein. *Jpn. J. Cancer Res.* **1990**, Vol. *81*, pp 1052-1056.

(77) Thierry, A.R.; Jorgensen, T.J.; Forst, D.; Belli, J.A.; Dritschilo, A.; and Rahman, A. Multidrug resistance in Chinese hamster cells: effect of liposome-encapsulated Doxorubicin. *Cancer Communications.* **1989**, Vol. *1*, pp 311-31.

(78) Thierry, A.R.; Vigé, D.; Coughlin, S.S.; Belli, J.A.; Dritschilo, A.; and Rahman A. Modulation of doxorubicin resistance in multidrug-resistant cells by liposomes. FASEB J. , **1993**, Vol. *7*, pp 572-579.

(79) Oudard, S.; Thierry, A.; Jorgensen, T.J.; and Rahman, A. Sensitization of multidrug-resistant colon cancer cells to doxorubicin encapsulated in liposomes. *Cancer Chemother. Pharmacol.* **1991**, Vol. *28*, pp 259-265.

(80) Sadasivan, R.; Morgan, R.; Fabian, C.; and Stephens, R. Reversal of multidrug resistance in HL-60 cells by verapamil and liposome-encapsulated doxorubicin. *Cancer Lett.* **1992**, Vol. *57*, pp 165-171.

(81)   Thierry, A.R.; Dritschilo, A.; and Rahman, A. Effect of liposomes on P-glycoprotein function in multidrug resistant cells. *Biomed. and Biophysical. Res. Comm.* **1992**, Vol. *187*, pp 1098-1105.

(82)   Warren, L.; Jardillier, J.-C.; Malarska, A.; and Akeli, M.-G. Increased accumulation of drugs in multidrug-resistant cells induced by liposomes. *Cancer Res.* **1992**, Vol. *52*, pp 3241-3245.

(83)   Rahman, A.; Husain, S.R.; Siddiqui, J.; Verma, M.; Agresti, M.; Center, M.; Safa, A.R.; Glazer, I. Liposome-mediated modulation of multidrug resistance in human HL-60 leukemia cells. *J. Natl. Cancer Inst.* **1992**, Vol. *84*, pp 1909-191.

(84)   Bekers, O.; Beijnen, J.H.; Storm, G.; Bult, A.; and Underberg, W.J.M. Chemical stability of N-trifluoroacetyldoxorubicin-14-valerate (AD-32) in aqueous media and after liposome encapsulation. Int. J. Pharm. **1989**, Vol. *56*, pp 103-109.

(85)   Perez-Soler, R.; and Priebe, W. Liposomal formulation and antitumor activity of 14-0-palmitoyl-hydroxyrubicin. *Cancer Chemother. Pharmacol.* **1992**, Vol. *30*, pp 267-271.

(86)   Perez-Soler, R.; Priebe, W. Anthracycline antibiotics with high liposome entrapment: structural features and biological activity. *Cancer Res.* **1990**, Vol. *50*, pp 4260-4266.

(87)   Zou, Y.; Priebe, W.; Ling, Y.H.; Perez-Soler, R. Organ distribution and tumor uptake of Annamycin entrapped in multilamellar vesicles. *Cancer Chemother. Pharmacol.* **1993**, Vol. *32*, pp 190-196.

(88)   Ling, Y.H.; Priebe, W.; Yang, L.Y.; Burke, T.G.; Pommier, Y.; and Perez-Soler, R. In vitro cytotoxicity, cellular pharmacology, and DNA lesions induced by Annamycin, an anthracycline derivative with high affinity for lipid membranes. *Cancer Res.* **1993**, Vol. *53*, pp 1583-1589.

(89)   Ling, Y.H.; Priebe, W.; Perez-Soler, R. Apoptosis induced by anthracycline antibiotics in P388 parental and multi-drug resistant cells. *Cancer Res.* **1993**, Vol. *53*, pp 1845-1852.

(90)   Zou, Y.; Ling, Y.H.; Van, N.T.; Priebe, W.; Perez-Soler, R. Antitumor activity of the lipophilic and partially non-cross resistant anthracycline annamycin entrapped in liposomes. *Cancer Res.* **1994**, Vol *54*, pp 1479-1484.

(91)   Onuma, M.; Odawara, T.; Watarai, S.; Aida, Y.; Ochiai, K.; Syuto, B.; Matsumoto, K.; Yasuda, T.; Fujimoto, Y.; Izawa, H. Antitumor effect of adriamycin entrapped in liposomes conjugated with monoclonal antibody against tumor-associated antigen of bovine leukemia cells. *Jpn. J. Cancer Res.* **1986**, Vol. *77*, pp 1161-1167.

(92)   Kawano, K.   Tissue distribution of adriamycin-encapsulated liposomes conjugated with monoclonal antibody. *Drug Delivery System* **1992**, Vol. *7*, pp 23-29.

(93)   Ahmad, I.; Allen, T.M. Antibody-mediated specific binding and cytotoxicity of liposome-entrapped doxorubicin to lung cancer cells *in vitro. Cancer Res.* **1992**, Vol. *52*, pp 4817-4820.

(94)   Mori, A.; Kennel, S.J.; Huang, L. Immunotargeting of liposomes containing lipophilic antitumor prodrugs. *Pharmacol. Res.* **1993**, Vol. *10*, pp 507-514.

(*95*) Tomita, T.; Watanabe, M.; Takahashi, T.; Kumai, K.; Tadakuma, T.; and Yasuda, T. Temperature-sensitive release of adriamycin, an amphiphilic antitumor agent, from dipalmitoylphosphatidylcholine-cholesterol liposomes. *Biochim. et Biophys. Acta* **1989**, Vol. *978*, pp 185-190.

(*96*) Ueno, M.; Zou, Y.; Yamagishi, M.; and Horikoshi, I. Basic study on hepatic artery chemoembolization and tumor selective drug targeting by temperature-sensitive liposome with local hyperthermia. *Jpn. J. Cancer Chemother.* **1989**, Vol. *16*, pp 3735-3738.

(*97*) Merlin, J.-L. *In vitro* evaluation of the association of thermosensitive liposome-encapsulated doxorubicin with hyperthermia. *Eur. J. Cancer* **1991**, Vol. *27*, pp 1031-1034.

(*98*) Merlin, J.-L. Encapsulation of doxorubicin in thermosensitive small unilamellar vesicle liposomes. *Eur. J. Cancer* **1991**, Vol. *27*, pp 1026-1030.

(*99*) Zou, Y.; Yamagishi, M.; Horikoshi, I.; Ueno, M.; Gu, X.; and Perez-Soler, R. Enhanced therapeutic effect against liver w256 carcinosarcoma with temperative-sensitive liposomal adriamycin administered into the hepatic artery. *Cancer Res.* **1993**, Vol. *53*, pp 3046-3051.

(*100*) Kerr, D..J.; Hynds, S.A.; Shepherd, J. Comparative cellular uptake and cytotoxicity of a complex of daunomycin-low density lipoprotein in human squamous lung tumor cell monolayers. *Biochem. Pharmacol.* **1988**, Vol. *37*, pp 3981-3986.

RECEIVED August 11, 1994

Author Index

Affiliation Index

Subject Index

2´-Bromo-3´-deaminodaunorubicins
  structure–activity relationships, 30–31
  synthesis, 29–30
Butylated hydroxytoluene, amelioration
  of anthracycline-induced cardiotoxicity,
  276–277

C

Calcium channel blockers, amelioration
  of anthracycline-induced cardiotoxicity,
  269–270
Captodative radical reducing agents, 120
Cardioprotective agents
  ICRF-187, 262–263
  use, 227
Cardiotoxicity
  anthracycline structural modification to
    overcome cardiotoxicity, 228–229
  cardiac tissue composition vs.
    anthracycline drug localization, 226
  cardioprotective agents to overcome
    cardiotoxicity, 227
  drug interactions with redox enzymes
    and generation of reactive species,
    226–227
  liposomal formulations to overcome
    cardiotoxicity, 229–230
  manifestation and dose dependency,
    225–226
  role of membrane interactions of
    anthracyclines, 222–234
  synthetic options for reversal for
    anthracyclines, 78–97
Carminomycin, structure, 248–249
Catalase, amelioration of anthracycline-
  induced cardiotoxicity, 275–276
Cell cycle control, role in cellular
  sensitivity to topoisomerase
  inhibitors, 191*f*,193
Cellular pharmacology, ICRF-187,
  263,264*f*
Cellular sensitivity, determinants for
  topoisomerase inhibitors,
  189,191*f*,192–194

Chelating antioxidants, amelioration of
  anthracycline-induced cardiotoxicity,
  277–278
Chemical ameliorators of anthracycline-
  induced cardiotoxicity,
  categories, 269
Chemotherapy, resistance development,
  14–15
2´-Chloro-3´-deaminodaunorubicins
  structure–activity relationships, 31
  synthesis, 30–31
Chloropheniramine, amelioration of
  anthracycline-induced cardiotoxicity,
  269–270
Classical multidrug resistance, *See*
  Multidrug resistance
Coenzyme Q_{10}, amelioration of
  anthracycline-induced cardiotoxicity,
  277–278
Cromolyn, amelioration of anthracycline-
  induced cardiotoxicity, 269–270
N,N-Cyclic anthracycline derivatives,
  antihelicase action, 209,214
Cytotoxicity
  daunorubicin–DNA adduct, 177–178,179*f*
  evaluation procedure, 143

D

Damvar, amelioration of anthracycline-
  induced cardiotoxicity, 274–275
Daunomycin, *See* Daunorubicin
Daunorubicin
  anticancer activity, 259
  antihelicase action, 209,210*f*,211*t*
  antitumor activity, 223–225,300
  biological activity, 115
  cardiotoxicity, 259–261
  comparison of activity to that of
    5-iminodaunorubicin, 4,5*f*
  development of multidrug resistance, 230
  DNA as target, 157
  DNA interactions
    macroscopic binding, 157–160
    microscopic binding, 160–163,164*t*
    structure and activity, 168–181

Non-cross-resistant anthracyclines—
Continued
3′-mercaptodoxorubicin synthesis, 25–26
structure–activity relationship of
  2′-halogenated anthracyclines, 32–35
water-soluble analogues of
  2′-halogenated anthracyclines, 35–37

O

OMDP-66, amelioration of anthracycline-induced cardiotoxicity, 276–277
Organofluorine compounds, interest by medicinal chemists, 47
Ouabain, amelioration of anthracycline-induced cardiotoxicity, 271–273
Oxomorpholinyl radicals, use for elucidation and control of redox chemistry of anthracyclines, 115–130
Oxygen metabolism role in cardiotoxicity of anthracycline antibiotics, reactive, *See* Reactive oxygen metabolism role in cardiotoxicity of anthracycline antibiotics

P

P-glycoprotein
  alteration in multidrug resistant cell membranes, 230–231
  functions, 250
  mechanism of drug removal, 248
  role
    in multidrug resistance, 301
    in reducing drug levels in resistant cells to sublethal levels, 222
    in uptake and release of anthracycline derivatives by drug-resistant K562 cells, 248–257
P-glycoprotein^{MDR}, role in cellular sensitivity to topoisomerase inhibitors, 191f,192
P-glycoprotein-mediated multidrug resistance, structural considerations of anthracyclines, 233–234

Passive tumor targeting, drug carriers for anthracycline antibiotics, 306
Permeability glycoprotein, *See* P-glycoprotein
Phenazine dioxide
  structure–activity relationships, 9–10,11–12f
  synthesis, 9
Phenoxybenzamine, amelioration of anthracycline-induced cardiotoxicity, 269–270
Phentolamine, amelioration of anthracycline-induced cardiotoxicity, 269–270
Poly(dG-dC)–DNA cross-linked adduct, immunogenicity, 179–180
Polyether-linked precursors, synthesis, 134–137
Polyethylene glycol 400, amelioration of anthracycline-induced cardiotoxicity, 274–275
Polymer-bound anthracyclines, drug carriers for anthracycline antibiotics, 306,307t
Prenylamine, amelioration of anthracycline-induced cardiotoxicity, 269–270
Prodrugs
  anthracyclines, use in antibody-directed enzyme prodrug therapy, 89,92–97
  semisynthetic rhodomycins and anthracycline, 59–75
Propranolol, amelioration of anthracycline-induced cardiotoxicity, 269–270
Protein kinase C pathways between cell surface and nucleus, signal transduction systems in doxorubicin mechanism of action, 243–247

Q

Quinone(s), use of bi(3,5,5-trimethyl-2-oxomorpholin-3-yl) as low-toxicity reducing agent, 115–116
Quinone isosteres
  structure(s), 4,6f